ACCIDENTAL RELEASES OF AIR TOXICS

ACCIDENTAL RELEASES OF AIR TOXICS

Prevention, Control and Mitigation

by

Daniel S. Davis, Glenn B. DeWolf, K.A. Ferland
D.L. Harper, R.C. Keeney, Jeffrey D. Quass

Radian Corporation
Austin, Texas

NOYES DATA CORPORATION
Park Ridge, New Jersey, U.S.A.

Copyright © 1989 by Noyes Data Corporation
Library of Congress Catalog Card Number: 89-8858
ISBN: 0-8155-1210-4
ISSN: 0090-516X

Published in the United States of America by
Noyes Data Corporation
Mill Road, Park Ridge, New Jersey 07656

Transferred to Digital Printing, 2011
Printed and bound in the United Kingdom

Library of Congress Cataloging-in-Publication Data

Accidental releases of air toxics.

    (Pollution technology review, ISSN, 0090-516X ;
no. 170)
    Bibliography: p.
    Includes index.
    1. Chemical industry--Accidents--Environmental
aspects. 2. Chemical industry--Safety measures.
3. Air--Pollution. I. Davis, Daniel S. (Daniel
Scott), 1958-   . II. Series.
TD888.C5A23   1989   660'.2804   89-8858
ISBN 0-8155-1210-4

Foreword

Accidental air releases of toxic chemicals must be prevented by all reasonable means. If and when they do occur, appropriate measures must be taken to reduce their consequences. This book presents an overview of the methods available for identifying, evaluating, preventing, controlling and mitigating hazards in facilities that use, manufacture, or store acutely toxic chemicals that could be released into the air.

A toxic chemical release occurs when equipment breakdown or loss of process control in a chemical process operation causes a loss of containment of the chemical(s). The probability that an accidental release will occur depends on the extent to which deviations (in magnitude and duration) in the process can be tolerated before a loss of chemical containment occurs. Development of satisfactory control systems and equipment capable of withstanding deviations from the design intent requires adherence to the principles of sound process and physical plant design and to appropriate procedures and management practices. These areas are discussed in terms of their relation to specific categories of hazards, their effectiveness in reducing the probability of a release, and the relative economics of their application.

Reducing the probability of accidental toxic chemical releases reduces the possibility of harm to human health and to the environment. When such a release does occur, however, its consequences must be reduced. This can be accomplished by means of a variety of "mitigation" measures that contain, capture, destroy, divert, or disperse the released toxic chemical. Mitigation measures are described and discussed in terms of their applicability, performance, state of development, and general application costs.

The book presents a brief history of accidental releases such as Bhopal and Chernobyl. Hazardous chemicals and their key properties of interest are defined. Hazards in process operations that relate to process design, physical plant design, and to management and maintenance procedures and practices are examined. Formal methods of hazard identification are described and evaluated, and major features of the most common formal methods are compared. The principles of prevention, protection, and mitigation control are discussed, and example control technologies are listed. A guide to facility inspections is also presented.

The information in the book is from the following documents:

Prevention Reference Manual: User's Guide Overview for Controlling Accidental Releases of Air Toxics, prepared by Daniel S. Davis, Glenn B. DeWolf, and Jeffrey D. Quass of Radian Corporation for the U.S. Environmental Protection Agency, July 1987.

Prevention Reference Manual: Control Technologies—Volume 1. Prevention and Protection Technologies for Controlling Accidental Releases of Air Toxics, prepared by D.S. Davis, G.B. DeWolf, and J.D. Quass of Radian Corporation for the U.S. Environmental Protection Agency, August 1987.

Prevention Reference Manual: Control Technologies—Volume 2. Post-Release Mitigation Measures for Controlling Accidental Releases of Air Toxics, prepared by D.S. Davis, G.B. DeWolf, K.A. Ferland, D.H. Harper, R.C. Keeney, and J.D. Quass of Radian Corporation for the U.S. Environmental Protection Agency, January 1989.

The table of contents is organized in such a way as to serve as a subject index and provides easy access to the information contained in the book.

Advanced composition and production methods developed by Noyes Data Corporation are employed to bring this durably bound book to you in a minimum of time. Special techniques are used to close the gap between "manuscript" and "completed book." In order to keep the price of the book to a reasonable level, it has been partially reproduced by photo-offset directly from the original reports and the cost saving passed on to the reader. Due to this method of publishing, certain portions of the book may be less legible than desired.

NOTICE

Contents and Subject Index

**PART II
PREVENTION AND PROTECTION TECHNOLOGIES**

PART III
MITIGATION MEASURES

Part I

Overview—User's Guide

The information in Part I is from *Prevention Reference Manual: User's Guide Overview for Controlling Accidental Releases of Air Toxics,* prepared by Daniel S. Davis, Glenn B. DeWolf, and Jeffrey D. Quass of Radian Corporation for the U.S. Environmental Protection Agency, July 1987.

Acknowledgments

This manual was prepared under the overall guidance and direction of T. Kelly Janes, Project Officer, with the active participation of Robert P. Hangebrauck, William J. Rhodes, and Jane M. Crum, all of U.S. EPA. In addition, other EPA personnel served as reviewers. Sponsorship and technical support was also provided by Robert Antonpolis of the South Coast Air Quality Management District of Southern California, and Michael Stenberg of the U.S. EPA, Region 9. Radian Corporation principal contributors involved in preparing the manual were Graham E. Harris (Program Manager), Glenn B. DeWolf (Project Director), Daniel S. Davis, Nancy S. Gates, Jeffrey D.Quass, Miriam Stohs, and Sharon L. Wevill. Contributions were also provided by other staff members. Secretarial support was provided by Roberta J. Brouwer and others. Special thanks are given to the many other people, both in government and industry, who served on the Technical Advisory Group and as peer reviewers.

1. Introduction

1.1 GENERAL BACKGROUND

Increasing concern about the potentially disastrous consequences of accidental releases of toxic chemicals has resulted from the Bhopal, India methyl isocyanate release on December 3, 1984, which killed approximately 2,000 people and injured thousands more. Concern about the safety of process facilities that handle hazardous materials increased further after the accident at the Chernobyl nuclear power plant in the Soviet Union in April of 1986.

While headlines of these incidents have created the current awareness of toxic release problems, there have been other, perhaps less dramatic, incidents in the past. These previous accidents contributed to the development of the field of loss prevention as a recognized specialty area within the general realm of engineering science. Interest in reducing the probability and consequences of accidental toxic chemical releases that might harm workers within a process facility and people in the surrounding community prompted the preparation of this manual and series of companion manuals. The other manuals in the series cover:

- Prevention control technologies (Part II of this book.)
- Mitigation control technologies (Part III of this book.)

These manuals compile the technical information that is necessary for developing approaches to preventing and controlling accidental releases. They cover various aspects of release control, including identification of causes; methods of hazard identification and evaluation; and prevention, protection, and mitigation measures. Prevention involves design and operating measures

3

applied to a process to ensure that primary containment is not breached. Protection focuses on the capture or destruction of a toxic chemical involved in an incipient release after primary containment has been breached, but before an uncontrolled release of the toxic chemical to the environment has occurred. Mitigation measures reduce the consequences of a release once it has occurred. The manuals are based on current and historical technical literature and they address fundamental considerations of the design, construction, and operation of chemical process facilities where accidental releases of toxic chemicals could occur.

Four types of releases are encountered in facilities that use, manufacture or store toxic chemicals:

- Releases from limited process upsets,
- Process vents,
- Fugitive emissions, and
- Accidental, sudden, large releases.

Accidental releases are the primary subject of this manual and of other manuals in the series.

The User's Guide is a general introduction to the subject of toxic chemical releases and to the broad concepts addressed in more detail in the other manuals. The manual gives a brief history of toxic chemical releases and overview of the accidental release problem. Primary industrial chemicals of concern are identified and the fundamental causes of toxic releases are summarized. Methods commonly used in hazard identification and evaluation are briefly discussed and an overview of the general principles of hazard control are presented. An example of the kind of guide to facility hazard inspections that can be developed from the information in this and other manuals is also presented.

The control technologies manual will focus on the fundamentals of process design, equipment design, and procedures and on how changes in these areas can help prevent and reduce the probability and magnitude of accidental releases. The mitigation manual will discuss ways of reducing the consequences of accidental releases.

The ultimate objective of controlling accidental chemical releases is to reduce adverse consequences to human health and to the general environment. The place of various controls in achieving this objective is illustrated conceptually in Figure 1-1. Each category of controls contributes to the reduction of the consequences of an accidental release. A full accidental release control program will contain some control methods from each of the categories listed in the figure.

1.2   HISTORICAL BACKGROUND AND ACCIDENTAL EVENTS OVERVIEW

The Bhopal incident, one of the most dramatic chemical accidents in history, eclipsed the 1976 major toxic discharge in Seveso, Italy, and numerous other significant, but less disastrous chemical releases over the years.

The historical development of concern about accidental releases tracks the general advance of loss prevention in the process industries. Much of that development has focused on fire and explosion protection. Since physical property losses are largest in these incidents, the magnitude of losses from fire and explosions has dominated industrial insurance issues for years. Accidental toxic releases have not been ignored, however. Lees (1) presents a summary table of major fire, explosion, and toxic release incidents in the chemical industry from the early years of this century through 1979. A listing of toxic releases taken from that table is presented in Table 1-1.

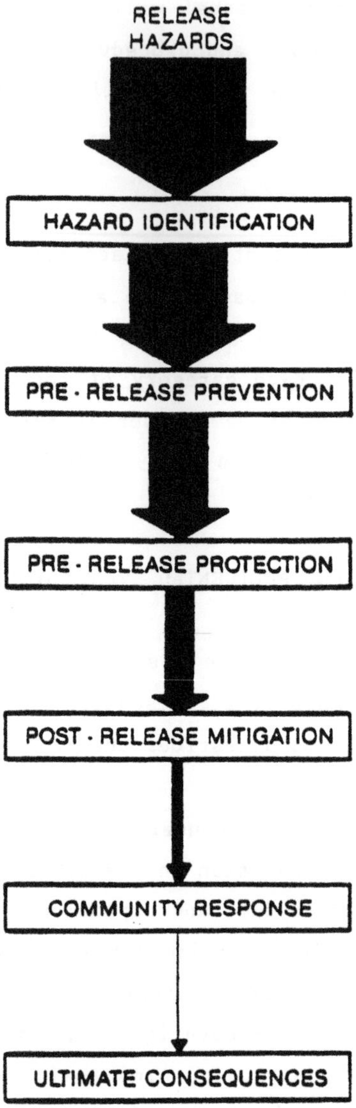

Figure 1-1.   The role of various accidental release control measures in reducing the consequences of an accidental release.

TABLE 1-1. MAJOR TOXIC RELEASE INCIDENTS BETWEEN 1950 AND 1980

Date	Location	Chemical	Deaths	Injuries
1950	Poza Rica, Mexico	Hydrogen sulfide	22	320
1952	Wilsum, Germany	Chlorine	7	Unknown
1961	Billingham, Great Britain	Chlorine	0	Unknown
1961	La Barre, Lousiana	Chlorine	1	114
1961	Morganza, Lousiana	Chlorine	0	17
1962	Cornwall, Ontario	Chlorine	0	89
1963	Brandtsville, Pennsylvania	Chlorine	0	Unknown
1963	Philadelphia, Pennsylvania	Chlorine	0	430+
1966	La Spezia, Italy	Chlorine	0	Unknown
1967	Bankstown, Australia	Chlorine	0	5
1967	Newton, Alabama	Chlorine	0	Unknown
1968	East Germany	Vinyl chloride	24	Unknown
1968	Lievin, France	Ammonia	5	Unknown
1969	Crete, Nebraska	Ammonia	ca. 8	20
1969	Glendora, Mississippi	Vinyl chloride	0	Unknown

Source: Reference 1   (continued)

TABLE 1-1 (Continued)

Date	Location	Chemical	Deaths	Injuries
1970	Blair, Nebraska	Ammonia	0	Unknown
1971	Floral, Arkansas	Ammonia	0	Unknown
1973	Loos, British Columbia	Chlorine	0	Unknown
1973	McPherson, Kansas	Ammonia	0	Unknown
1973	Potchefstroom, South Africa	Ammonia	18	Unknown
1974	Nebraska	Chlorine	Unknown	Unknown
1976	Baton Rouge, Louisiana	Chlorine	Unknown	Unknown
1976	Houston, Texas	Ammonia	6	Unknown
1976	Seveso, Italy	Dioxin (TCDD)	0	Unknown
1977	Columbia	Ammonia	30	Unknown
1978	Baltimore, Maryland	Sulfur trioxide	Unknown	Unknown
1978	Chicago, Illinois	Hydrogen sulfide	Unknown	Unknown
1978	Youngstown, Florida	Chlorine	8	Unknown

Source: Reference 1.

Various other surveys also address accidental releases. A recent publication by the U.S. Environmental Protection Agency (EPA), the Acute Hazardous Events Data Base, examines the causes of toxic chemical accidental release events (2). Along with other statistics, this report presents the locations of toxic release events within a process facility, as shown in Figure 1-2. The causes of losses in the chemical and allied industries as shown in an insurance survey are presented in Figure 1-3 (3). Table 1-2 summarizes similar information from two other sources (4).

Examination of this information shows that it is difficult to absolutely quantify the distribution of accidental release causes; every survey results in a somewhat different distribution of causes. A likely reason for these differences is that every survey uses a different and fairly limited data set. Another possible reason is the difficulty in consistently defining the actual causes of a release. As an example, if a valve were to fail because of corrosion and result in an accidental release, the cause might be classified as valve failure, maintenance failure or design failure, depending on the classification criteria of the specific survey. Even within this complexity, however, some trends can be observed. A comparison of the information shows that faulty pipes and fittings are common causes of accidental releases.

The occurrence of such incidents tends to obscure the long and dedicated activity of numerous individuals and organizations who have contributed to the field of loss prevention. Some organizations that have been very active in this area, especially in the last two decades, include the Institute of Chemical Engineers (Britain), the American Institute of Chemical Engineers, the American Petroleum Institute, other organizations and many major corporations, especially in the chemical industry. Two leaders in this field have been Dow, which developed the well-known Dow Index for ranking process facilities for fire and explosion potential, and Imperial Chemical Industries, for modifying this index to the Mond Index, which includes toxicity in the ranking. Activity in both the public and private sectors in this area is intense,

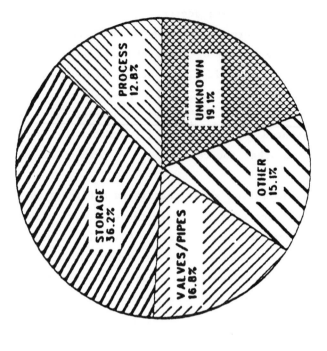

DEATH/INJURY EVENTS

PROCESS 12.8%
UNKNOWN 19.1%
STORAGE 36.2%
OTHER 15.1%
VALVES/PIPES 16.8%

NUMBER OF EVENTS = 304

ALL EVENTS

PROCESS 14.1%
UNKNOWN 17.9%
STORAGE 20.7%
OTHER 27.8%
VALVES/PIPES 19.4%

NUMBER OF EVENTS = 5179

Figure 1-2.   In-plant acute hazardous events by location.

Source: Reference 2.

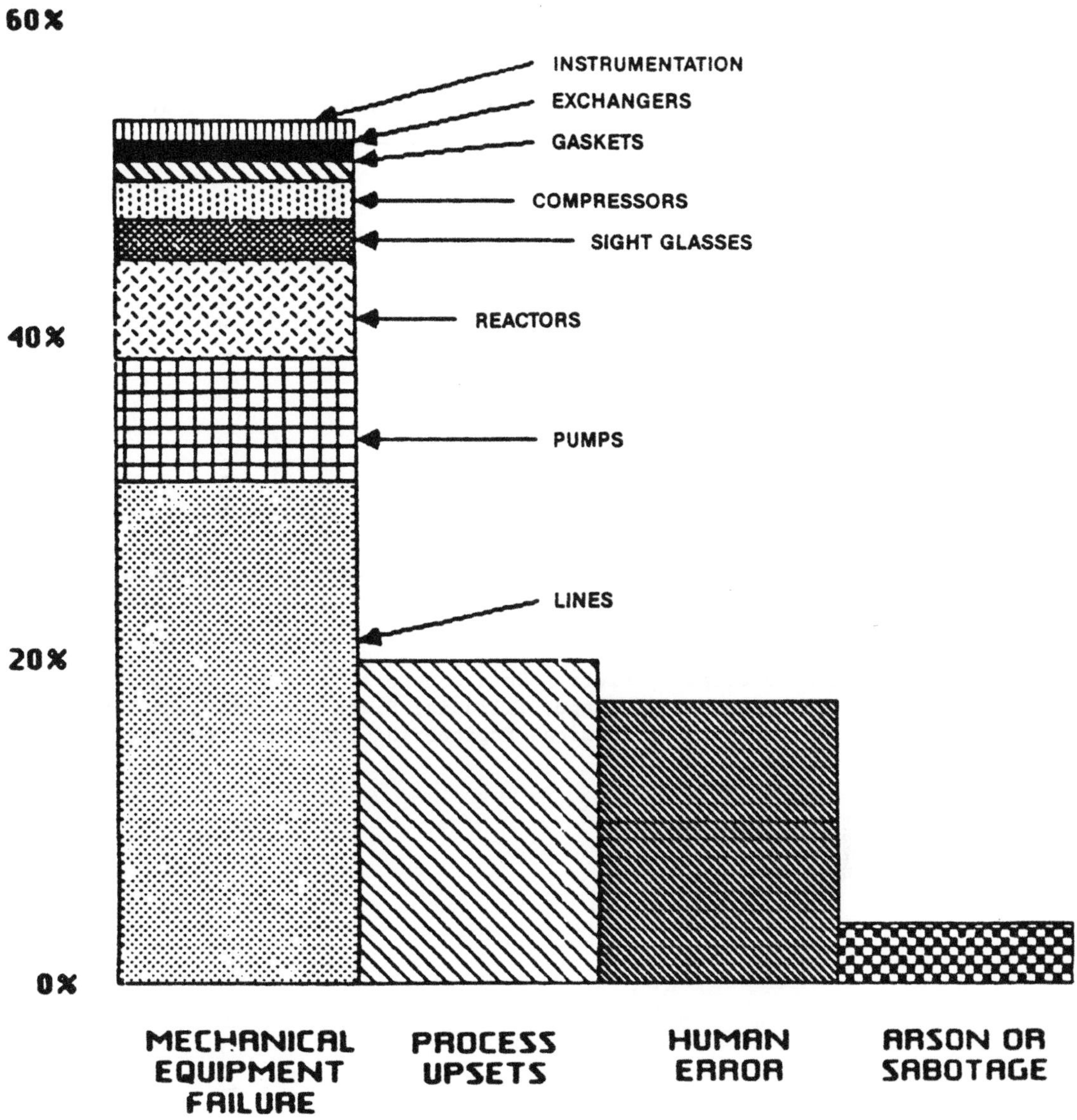

Figure 1-3.   Causes of loss in the chemical and allied
industries according to insurance survey.

Source: Adapted from Reference 3.

TABLE 1-2. DISTRIBUTION OF EVENT LOCATIONS AS REPORTED BY KLETZ

Location	Percentage Attributable to Each Location[a]	
	(Data Source A)	(Data Source B)
Pipes and fittings	34	61
Vessels	--	9
- internal reactions	22	--
- other	21	--
Relief valves, vents, drains	11	22
Pumps	6	4
Equipment under maintenance	6	4
	100	100

Source: Reference 4.

[a]Two references were cited by Kletz. Data source A is: One Hundred Largest Losses, Marsh and McLennan, Chicago, Illinois, Sixth Edition, 1985. Data source B is: Davenport, J., Chemical Engineering Progress, September, 1977, p. 54.

as each strives to apply the latest knowledge and technology to the prevention and mitigation of accidental toxic chemical releases.

A number of public and private organizations have developed codes and standards, recommended practices, design criteria, or guidelines establishing at least minimum standards for equipment and systems potentially involved in accidental releases. Table 1-3 presents a list of some of these organizations. Some of the codes and standards developed by these organizations present recommended design criteria for individual pieces of equipment or for entire plant systems. Others present recommended practices for conducting a safe operation. Some of these organizations have published specific hazard evaluation and reduction information. A complete presentation of such information is beyond the scope of this manual, but the reader may contact individual organizations for additional information. Table 1-4 summarizes general areas addressed by the various organizations.

1.3   PURPOSE OF THIS MANUAL

The User's Guide is an introduction to the overall area of accidental chemical releases prevention, protection, and mitigation for government agency personnel, industry managers, technical people, and other persons concerned with reducing the risk of accidental toxic chemical releases. It is intended to assist and inform the reader about where and how to seek additional information. The manual is also a guide to the more detailed information in the companion set of manuals and to the general technical literature.

Government agencies will probably continue to become more involved in this area as awareness specific to accidental releases increases and more regulations are promulgated. Past involvement by industry has been broad. Future participation in the area of accidental release prevention, protection, and mitigation is expected to increase in response both to new regulations and to increased awareness of toxic air release issues on the part of company management, technical staff and the general public.

TABLE 1-3. MAJOR ORGANIZATIONS PROVIDING CODES AND STANDARDS, RECOMMENDED PRACTICES, DESIGN CRITERIA, OR GUIDELINES FOR EQUIPMENT IN CHEMICAL AND ALLIED INDUSTRY PROCESS PLANTS

Name	Abbreviation
Technical and Trade Groups	
Air Conditioning & Refrigeration Institute	ARI
Air Moving and Conditioning Association	AMCA
American Association of Railroads	AAR
American Gas Assocation	AGA
American Petroleum Institute	API
American Water Works Association	AWWA
Chemical Manufacturer's Association (formerly	CMA
Manufacturing Chemists Association)	(MCA)
Chlorine Institute	CI
Compressed Gas Association	CGA
Cooling Tower Institute	CTI
Manufacturers Standardization Society	MSS
National Electrical Manufacturers Association	NEMA
Pipe Fabrication Institute	PFI
Scientific Apparatus Makers Association	SAMA
Society of Plastics Industry	SPI
Steel Structure Painting Council	SSPC
Tubular Exchanger Manufacturers Association	TEMA
U.S. Government Agencies	
Bureau of Mines	BM
Department of Transportation	DOT
U.S. Coast Guard	USCG
Hazardous Materials Regulation Board	HMRB
Federal Aviation Administration	FAA
Environmental Protection Agency	EPA
National Bureau of Standards	NBS
Occupational Safety and Health Administration	OSHA
Testing Standards and Safety Groups	
American National Standards Institute	ANSI
American Society for Testing and Materials	ASTM
National Fire Protection Association	NFPA
Underwriters Laboratories, Inc.	UL
National Safety Council	NSC

(Continued)

TABLE 1-3 (Continued)

Name	Abbreviation
Insuring Associations	
American Insurance Association	AIA
Factory Insurance Association	FIA
Factory Mutual System	FM
Oil Insurance Association	OIA
Professional Societies	
American Conference of Governmental Industrial Hygienists	ACGIH
American Industrial Hygiene Association	AIHA
American Institute of Chemical Engineers	AIChE
American Society of Mechanical Engineers	ASME
Amer. Soc. of Htg. Refrig. & Air-Cond. Engs.	ASHRAE
Illumination Engineers Society	IES
Institute of Chemical Engineers (Britian)	IChE
Institute of Electrical and Electronic Engineers	IEEE
Instrument Society of America	ISA

Source: Adapted from Reference 6.

TABLE 1-4.   AREAS COVERED BY CODES, STANDARDS, RECOMMENDED PRACTICES, DESIGN
CRITERIA, OR GUIDELINES OF DESIGNATED ORGANIZATIONS (SEE TABLE
1-3 FOR SYMBOLS DEFINITIONS)

Accident Case History	AGA, AIA, AIChE, API, FIA, FM, NFPA, NSC, OIA, OSHA, USCG
Air Compressors	AIA, ANSI, FM, USCG
Air-Fin Coolers	ARI, ASHRAE, OIA, USCG
Boilers	ANSI, NFPA, NSC, UL
Combustion Equipment and Controls	ANSI, FIA, FM, NFPA, NSC, OIA, UL, USCG
Compressors	AIA, ARI, ASHRAE, ASME, FM, OIA, USCG
Cooling Towers	CTI, FM, NFPA, OIA
Drain and Waste Systems	AICHE, AWWA, MCA, USCG
Dust Collection Equipment	FIA, FM, NFPA, USCG
Dust Hazards	ACGIH, AIHA, ANSI, BM, FIA, FM, NFPA, NSC, UL, USCG
Electric Motors	ANSI, IEEE, MCA, NFPA, UL, USCG
Electrical Area Classification	AIA, ANSI, API, FIA, FM, MCA, NFPA, NSC, OJA, OSHA, USCG
Electrical Control and Enclosures	AIA, ANSI, ARI, FIA, FM, IEEE, ISA, MCA, NEMA, NFPA, NSC, OIA, OSHA, UL, USCG
Emergency Electrical Systems	AGA, AIA, FM, IEEE, NEMA, NFPA, USCG
Fans and Blowers	ACGIH, AIHA, AMCA, ARI, ASME, FM, USCG
Fire Protection Equipment	AIA, ANSI, API, AWWA, BM, CGA, FIA, MCA, NEMA, NFPA, NSC, OIA, OSHA, UL, USCG

(Continued)

TABLE 1-4 (Continued)

Fire Pumps	ANSI, FM, HI, IEEE, NFPA, UL, USCG
Fired Heaters	ANSI, ASME, FIA, FM, NFPA, OIA, UL, USCG
Gas Engines	FM, NFPA, OIA, USCG
Gas Turbines	AGA, FIA, FM, NFPA, OIA, USCG
Gear Drives Power Transmission	AGMA, AIA, ANSI, NSC, USCG
Grounding and Static Electrical	AIA, ANSI, API, FIA, FM, IEEE, NEMA, NFPA, NSC, OIA, OSHA, UL, USCG
Inspection and Testing	ABMA, AGMA, AIChE, AMCA, API, ASHRHE, ASTM, AWWA, CGA, CTI, DOT, HEJ, HI, IEEE, MSS, NFPA, NSC, PFI, USCG
Instrumentation	AIA, ANSI, API, ARI, ASTM, AWWA, CGA, FIA, FM, HMRB, IEEE, ISA, NBS, NFPA, OIA, SAMA, UL, USCG
Insulation and Fireproofing	AIA, ANSI, ASHRAE, ASTM, FM, OIA, UL, USCG
Jets and Ejectors	HEI, USCG
Lighting	ANSI, FM, IEEE, IES, NEMA, NFPA, NSC, UL, USCG
Lubrication	AMCA, ANSI, ASME, NFPA
Material Handling	MCA, NFPA, NSC, OSHA
Materials of Construction	AIA, ANSI, ASTM, AWWA, CGA, CI, CTI, FM, HMRB, ISA, MCA, NBS, NFPA, NSC, OIA, TEMA, UL, USCG
Noise and Vibration	AGA, AIChE, AIHA, AMCA, ANSI, API, ARI, ASHRAE, ASTM, EPA, ISA, NFPA, NSC, OSHA, UL

(Continued)

TABLE 1-4 (Continued)

Painting and Coating	AIChE, ANSI, ASTM, AWWA, HMRB, OSHA, NBS, SSPC, UL
Piping Materials and Systems	AGA, AIA, ANSI, API, ARI, ASHRAE, ASTM, AWWA, CGA, CI, FIA, FM, HMRB, IES, MSS, NBS, NFPA, NSC, PFI, SPI, UL, USCG
Plant and Equipment Layout	AAR, AIA, API, AWWA, CGA, FIA, FM, HMRB, MCA, NFPA, NSC, OIA, USCG
Pneumatic Conveying	ANSI, FIOA, NFPA, USCG
Power Wiring	ANSI, API, FIA, FM, IEEE, NEMA, NFPA, OIA, OSHA, UL, USCG
Pressure Relief Equipment Systems	AIA, API, ASME, CGA, CI, FIA, FM,
Pressure Vessels	AIA, API, ASME, CGA, DOT, NFPA, NSC, OSHA, USCG, HMRB, OIA, OSHA, USCG
Product Storage and Handling	AAR, AIChE, AIA, ANSI, API, CGA, CI, FIA, FM, MCA, NFPA, OIA, OSHA, USCG
Pumps	AIChE, ANSI, AWWA, CI, NFPA, OIA, UL, USCG
Refrigeration Equipment	ANSI, API, ASHRAE, FM, NFPA, UL, USCG
Safety Equipment	ACGIH, AIHA, ANSI, BM, CGA, CI, FM, MCA, NSC, OSHA, UL, USCG
Shell and Tube Exchangers	AGA, AIChE, API, ASHRAE, ASME, CGA, PFI, USCG
Shutdown System	AIA, API, FIA, NFPA, OIA, UL, USCG
Solids Conveyors	MCA
Stacks and Flares	FAA, OIA, USCG

(Continued)

TABLE 1-4 (Continued)

Steam Turbines	AIA, FM, IEEE, OIA, USCG
Storage Tanks	AWWA, CI, NBS, NFPA, OIA, OSHA, UL, USCG
Ventilation	ACGIH, AIHA, ANSI, BM, FIA, FM, NFPA, NSC, UL, USCG
Venting Requirements	API, FIA, FM, HMRB, NFPA, USCG

Source: Adapted from Reference 6.

Table 1-5 illustrates the potential needs of government agency personnel and individual companies if called on to enforce or comply with a regulation specifically geared toward accidental release prevention. These needs include the resources that will be required for such a task. The table summarizes approaches to meeting these needs. The User's Guide and companion manuals will provide one information resource for responding to accidental release regulations established by local authorities. For needs or approaches not specifically covered in the manuals, or for areas where more detail is required than the manuals can provide, the references cited in the manuals can help the reader obtain more detailed information.

For both regulators and companies interested in release prevention, an overall concept of a process facility must be developed in terms of the physical and functional areas that would be covered by a complete release prevention evaluation or control plan. Such a concept is shown in Figure 1-4. These aspects of a facility must be addressed at each stage in the total life cycle of the facility.1 The basic stages, design, construction, startup, operation and shutdown, are discussed further in Section 3.4.

Chemicals are transported to the facility and pass through a sequence of transfer, storage, and process operations that produce more chemicals that pass through a similar sequence until they leave the facility. To develop a company hazard control plan or for a regulatory review of such a plan, each step in the sequence must be scrutinized for the following:

- The toxic chemicals used and where,
- Hazardous processes and operations, and
- Control measures.

Later sections of this manual address each of these considerations by summarizing the fundamental principles of chemical hazards, process and operational hazards, hazard identification and evaluation, and hazard control.

TABLE 1-5.   EXAMPLES OF NEEDS AND APPROACHES FOR REGULATORS AND COMPANIES ADDRESSED BY PREVENTION REFERENCE MANUALS

	Government		Industry	
Needs	**Approach**		**Needs**	**Approach**
Basic Information			**Basic Information**	
Types of Facilities	Identify facilities within the regulators jurisdiction that are subject to safety concerns covered by the accidental release guidelines or regulations.		Types of Processes	Evaluate the requirements of the accidental release safety objectives, guidelines, or regulation in light of the specific processes in use at the facility.
	Gather historical background on previous accidental releases at each facility and accidental releases at similar facilities.			Gather historical background on previous accidental releases at similar facilities.
Types of Processes	Obtain information about the processes used at each facility.		Hazard Identification	Become aware of the range of accidental release hazards associated with the processes used at the facility.
Hazard Identification	Become aware of the range of accidental release hazards associated with the processes used at each of the facilities covered by the regulation.			Use hazard identification and evaluation methods to evaluate the adequacy of the accidental release prevention measures already in place and to pinpoint areas of deficiency.
Prioritization	Develop a criteria for ranking the facilities for accidental release risk.		**Control Measure Selection and Evaluation**	
	Compare the available resources to the ranking and select which facilities require further investigation and specifically which processes within a facility should be examined for compliance.		Prioritization	Identify the control measures that could be applied to the areas where the present control measures are deficient.
	Identify the range of potential control techniques that could be applied to each process to be examined.			Evaluate the resources that are required to perform each level of control.
Control Measure Verification				Evaluate the effectiveness of each control method.
Document Review	Identify which accidental release control techniques are in place at each facility by requesting summary documentation of a control plan.			
	If the documentation provided does not indicate compliance then request additional documentation.			

(Continued)

TABLE 1-5 (Continued)

Needs	Government		Industry
	Approach	Needs	Approach
Facility Inspection	If the additional documentation does not indicate compliance then visit the facility and perform a detailed review of control plan documentation.		Apply measures that will bring the facility into compliance and will provide the most efficient use of resources.
	If compliance is not confirmed then inspect those portions of the facility that are of interest.		Prepared both detailed and summary documentation of control practices.
			Periodically reinspect the facility; especially important when process changes are made.
Resources Required/Available		**Resources Required/Available**	
Historical literature	Newspapers; local news stations.	Historical literature	Newspapers; local news stations.
Technical literature	Loss prevention journal articles and text books; Accidental Release Prevention Reference Manual (ARPM) series.	Technical literature	Loss prevention journal articles and text books; Accidental Release Prevention Reference Manual (ARPM) series.
Public records	State or local data base.	Public records	State or local data base.
Organizations	Technical societies and trade organizations; local, state and federal regulatory organizations; consultants.	Organizations	Technical societies and trade organizations; local, state and federal regulatory organizations; consultants.
Regulations	Related regulations that impact accidental release prevention; other regulations that specifically apply to accidental release prevention.	Regulations	Related regulation that impact accidental release prevention; other regulation that specifically apply to accidental release prevention.
Personnel	Experienced personnel within the regulatory agency.	Personnel	Experienced personnel within the plant.
	Personnel from other regulatory agencies with experience in accidental release prevention.		Outside consultants with experience in accidental release prevention.
	Outside consultants with experience in accidental release prevention.		
Costs	The costs the agency will incur by performing document and facility reviews at various levels of detail.	Costs	The cost to perform different varieties of internal hazard evaluation, the cost to implement various control measures, and the cost of assembling a comprehensive control plan.
	The facility's cost to perform different varieties of internal hazard evaluation, the cost to implement various control measures, and the cost of assembling a comprehensive control plan.		

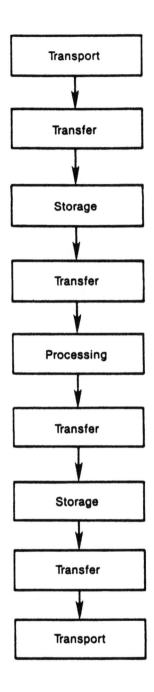

Figure 1-4. Functional areas of a typical chemical process facility.

Once an accidental release control plan is developed, the review of the control plan by a regulator or the company itself may follow the kind of logic flow shown in Figure 1-5. A regulator will probably go into less detail than a company reviewer, but the logic of analysis will probably be the same. This diagram highlights major decision points in the review process and illustrates the possible iterations that may be involved before a plan can be considered acceptable. Procedures may differ from this in detail in individual circumstances.

The remainder of this manual discusses chemical hazards in Section 2, hazards in process operations in Section 3, methods for hazard identification and evaluation in Section 4, an overview of principles of control in Section 5, a guide to facility inspections in Section 6, costs of prevention in Section 7, and references in Section 8.

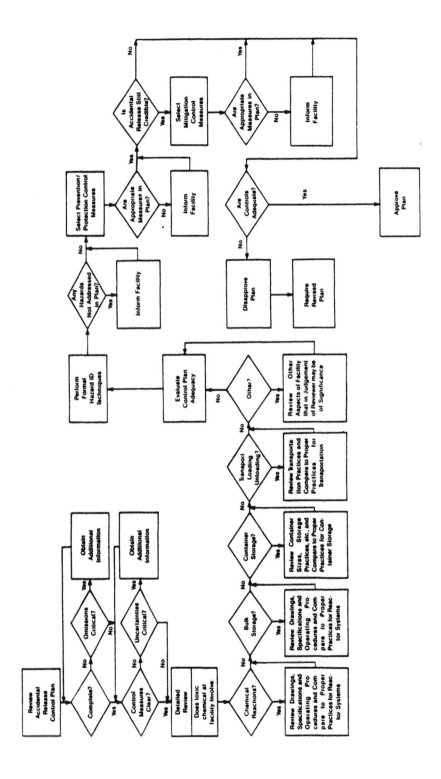

Figure 1-5. Example of logic flow for accidental release control plan review.

2. Chemical Hazards

A fundamental need from either a regulator's or company's point of view is the identification and ranking of chemical hazards within a process facility. The primary basis for selecting chemicals that pose a significant danger if accidentally released is their acute toxicity, but other properties also enter the selection process. Various lists of hazardous chemicals have been prepared by numerous organizations, including the U.S. Environmental Protection Agency (EPA) and the European Economic Community (EEC). Current hazardous chemicals of primary interest to the U.S. EPA are listed in an EPA publication (5). This list is reproduced in Appendix A of this manual.

This section of the manual discusses major considerations for identifying and classifying toxic chemicals and serves as a guide for setting up priority lists of the chemicals themselves. An example ranking system is also illustrated.

2.1 TOXIC CHEMICALS

In the context of accidental releases, hazardous chemicals are materials with acute toxic and other properties that make them an imminent threat to human health and/or the general environment, even after brief exposure. Releases of chemicals that have the potential for long-term health effects and/or environmental damage are also of concern, and while many of the principles discussed in this manual would apply to these, the primary focus is on acute toxic chemicals.

The primary methods of expressing toxicity are:

- Immediately Dangerous to Life and Health (IDLH), defined as the maximum level to which a healthy male worker can be exposed for 30 minutes and escape without suffering irreversible health effects or impairing symptoms. These values have been developed and specified by the National Institute for Occupational Safety and Health (5).

- Low Lethal Concentration (LC_{LO}) is the lowest lethal concentration observed in tests on laboratory animals or in accidental human exposure.

- 50% Lethal Concentration (LC_{50}) is the concentration for which 50% of the test animals died when exposed for a specified period of time.

- Permissible Exposure Limit (PEL), defined as the maximum air concentration to which a healthy male worker can be exposed for 8 hours per day, 40 hours per week without adverse effects as determined by the Occupational Safety and Health Association (OSHA).

- Short-term Exposure Limit (STEL) is the maximum concentration to which workers can be exposed for up to 15 minutes, provided no more than four exposures per day are permitted with at least 60 minutes between exposure periods.

The relative acute toxic hazard of different chemicals may be ranked using any one of these criteria. In a recent document, the U.S. EPA suggests using the IDLH as the primary criterion for estimating consequences of accidental releases, with the LC_{LO} and LC_{50} as second and third choices, or other criteria if these are not available (5).

The ultimate adverse consequences of a release result from the toxicity of the chemical, but other physical and chemical properties are also important when considering the causes and prevention of releases, and their consequences.

2.2   PHYSICAL AND CHEMICAL PROPERTIES

Significant physical and chemical properties include boiling point, vapor pressure, heat of vaporization, density, viscosity, and reactivity. Low boiling points and high vapor pressures increase the quantity of chemical emitted to the air during a release. A low heat of vaporization results in an increased rate of vaporization from spills or liquid releases of volatile toxic chemicals. High vapor or gas densities hinder dispersion and result in low lying clouds which imperil people at ground level. In the case of liquids, viscosity is important because a spilled or leaked low viscosity material will flow and spread more rapidly than a high viscosity material. Other physical properties, also significant in the evaporation and dispersion behavior of liquids, vapors, and gases, include surface tension, diffusivity, and heat capacities. For solids, particle size is important. Reactivity includes properties such as flammability, explosivity, exothermicity, and corrosiveness.

Flammability is the ability of a chemical to burn. Explosivity is the ability of a chemical to react rapidly enough with itself or other materials including oxygen in ambient air, to cause an explosion. A common destructive manifestation of reaction with air is a vapor cloud or dust explosions when a material is within its explosive limits. Exothermicity refers more to specific chemical reactions than to the chemicals themselves, but this ability for the reactions to generate significant amounts of heat means that if they get out of control, they can result in thermal runaway, overpressure in containment equipment, and possibly to explosions. Finally, corrosiveness, another aspect of reactivity, can damage equipment and cause equipment failures and chemical releases. A more detailed discussion of the relationship between specific chemical properties and release hazards can be found in the technical literature (1, 6).

Vapor pressure, vapor density, and the IDLH are the minimum property data needed to establish that a specific chemical is an acute toxic, air release hazard. Values for some common hazardous chemicals are presented in Table 2-1.

TABLE 2-1. SELECTED PROPERTIES OF SOME COMMON HAZARDOUS CHEMICALS

Chemical	Boiling Point °F	Vapor (psi) Pressure 68°F	Vapor Density (Air = 1) 32°F	IDLH (ppm)
Ammonia	−28.03	128.9	0.597	500
Carbon Tetra-chloride	170.1	1.74	5.32	300
Chlorine	−29.29	92.8	2.49	25
Chloropicrin	233.6	0.35	5.70	4
Anhydrous Hydrogen Chloride	−121.1	0.505	1.27	100
Anhydrous Hydrogen Fluoride	67.10	15.0	1.56	20
Phosgene	45.46	23.5	3.40	2
Sulfur Dioxide	13.96	47.9	2.26	100
Hydrogen Cyanide	78.26	12.8	0.95	50

Sources of data: References 7, 8, 9, 10 and 11.

Table 2-2 shows a more complete list of properties as would be required for a complete comparative evaluation of chemicals.

Once the chemicals of concern have been identified, the next step in controlling accidental releases is to examine the operations in which these chemicals are used. Hazards in process operations is the subject of the next section of this manual.

TABLE 2-2.   CHEMICAL PROPERTY DATA PERTINENT TO ACCIDENTAL RELEASE EVALUATIONS

Chemical name, synonyms, and Chemical Abstracts Service Registry Number (CAS #)
Chemical formula and/or drawing of structure
Phase at room temperature
Boiling point (normal and/or at _____ pressure)
Melting point
Liquid density and/or specific gravity at _____ temperature
Vapor pressure at _____ temperature(s) (e.g., Antoine equation)
Gas density and/or specific gravity at _____ temperature
Solubility in water, alcohol, ether, and other reactants/products
Liquid viscosity
Enthalpy at _____ temperature(s)
Specific heat at constant volume
Specific heat at constant pressure
Critical temperature
Critical pressure
Critical volume
Flash point
Limits of flammability and explosivity
  NEPA 704M Safety Hazard Rating (health, flammability, reactivity, special)
Differential thermal analysis
  CHETAH (Chemical Thermodynamics and Hazards) evaluation (or equivalent)

Hazard Properties: Oral Poison   Inhalant Poison
  Contact Poison   Lachrymator   Contact Irritant
  Inhalation Irritant   Pyrophoric   Teratogen
  Carcinogen   Mutagen
  Releases Vapors   Hyroscopic

Reaction Properties: Reacts with Air   Reacts with Water
  Reacts with Acids   Reacts with Bases   Reacts with Alkanes
  Exothermic   Releases Gases   Foams
  Solidifies   Violent Reaction   Forms Toxic Products
  Reacts with Metals   Polymerizes   Autocatalytic
  Decomposes   Light Sensitive   Shock Sensitive

3. Hazards in Process Operations

Once a regulator or company has identified the toxic chemicals manufactured, used, or stored at a facility, the process situations that could lead to a release should be identified so that the adequacy of proposed controls (in the case of the regulator), or select appropriate control measures (in the case of the company) can be evaluated. Before identifying process hazards, discussed in Section 4, and their corresponding control measures for specific facilities, discussed in Section 5, an understanding of the kinds of hazards, and available control measures is required.

This section of the manual presents an overview of hazards and failure modes of process systems for toxic chemicals as they relate to process design considerations, physical plant design considerations, and operational procedures and practices at any stage in the life cycle of a facility. Examples of specific hazards and possible causes are cited for each of these key areas.

3.1 BACKGROUND

The preceding sections have presented an overview of accidental releases. The hazard of a release increases with the toxicity and reactivity of the chemicals involved, the process energy content, the inventory, and the complexity of the process system. The more toxic a material, the more severe may be the consequences of release. Reactive materials are more dangerous than less reactive materials because physical containment may be more difficult (e.g., corrosion problems), and operational problems may be more severe (e.g., hard to control rapid chemical reaction). Operating pressure and temperature determine process energy content. The higher the energy content, the higher the potential driving force for release and the more difficult it is to design the containment equipment. Large inventories are a greater hazard than small inventories because more material can be released. The more complex a process is, the more physical components there are that can fail, making control more difficult. These broad categories of hazards must be addressed in the safe design, construction, and operation of process systems for toxic chemicals.

They can be addressed in the following general categories:

- Process design considerations;
- Physical plant design considerations; and
- Operational procedures and practices.

Each of these categories organizes specific hazards, failure modes, and control measures in terms of prevention, protection, and mitigation. Examples of specific hazards and failure modes are discussed in this section. Prevention, protection, and mitigation control measures are discussed in Section 5.

Basic causes of releases can be summarized as follows:

- Process or operational failures causing pressure or temperature to exceed limits of the process equipment;

- Equipment containment failures at normal process conditions;

- Operational or maintenance errors, omissions, or deliberate criminal acts (e.g., vandalism) leading to either of the above two conditions, or to direct releases (e.g., inadvertently opening a valve and releasing material); and

- Imposition of external damaging factors such as fire, explosion, flooding, or mechanical stress, which directly lead to equipment failure. Natural phenomena such as flooding, earthquakes, or wind storms can be contributing factors here.

Each of these general causes can have many specific initiating and enabling events or contributing causes, forming a chain leading to the final event which physically results in the release. Prevention and protection measures interdict the event chain at a point before the final release event can occur.

Some specific areas of process facilities that should always receive close attention are:

- Large inventories of toxic materials such as storage areas, and inventories of flammable or explosive materials near large toxic inventories,

- Exothermic chemical reactors, and inventories of chemicals prone to exothermic reactions with other materials, even if by contamination rather than by design,

- Any process areas with high energy content; high temperature and pressure operations,

- Any process operations with positive energy input such as distillation,

- Processes with complex sequencing or unit operations interactions such as recycles,

- Processes involving toxic chemicals in combination with highly corrosive, flammable, or explosive materials, and

- Existing process facilities that have recently been modified, are very old, or very new.

There are many other specific areas that can be listed, but the above represent areas of high priority.

Table 3-1 presents examples of a few typical process hazard areas and possible corresponding control technologies. Other hazard areas and corresponding controls could be listed.

3.2 PROCESS DESIGN CONSIDERATIONS

Process design considerations encompass technology, procedures and practices associated with the sequence and conditional state of all of the process steps and operations in a chemical process. These considerations include the nature of the chemical materials used in the process and the fundamental manipulated process variables. Process design considerations address the relationships between physical variables and time; in other words, the physical states of the process as a function of time and the means and characteristics of process control. Process design considerations include:

- Process characteristics and chemistry,
- Overall process control,
- Flow control,
- Pressure control,
- Temperature control,
- Quantity control,
- Mixing effects,
- Composition control,
- Energy systems,
- Detection and alarm systems, and
- Fire and explosion protection.

TABLE 3-1.   SOME TYPICAL PROCESS HAZARD AREAS AND EXAMPLES OF
CORRESPONDING CONTROL TECHNOLOGIES

Hazard Area	Example Control Technologies
Large inventories of toxic materials	• Change process or procedures to reduce need for large inventories. • Use substitute. • Design storage for higher containment reliability.
Exothermic chemical reactors	• High reliability cooling systems, including backup cooling. • Change process chemistry. • High reliability feed process control. (e.g., feed interlocks, ratio control) • Emergency relief systems. • Emergency dump systems. • Emergency quench and inhibition systems.
Contamination of stored chemicals	• Special backflow protection. • Special isolation valves. • Equipment/process segregation.
Distillation processes	• High reliability process control. • Venting and pressure relief to emergency scrubbers or flares. • High reliability heating systems, with emergency shutdown interlocks. • Distillation under vacuum.

Process failures leading to an accidental release may be related to deficiencies in any of these areas. Such failures would cause the conditional state of the system to exceed the design limits of the equipment or the ability of a human operator to respond quickly or accurately enough to maintain control for changes that are occurring in the process . For example, the loss of flow control of a reactant to an exothermic chemical reactor could lead to a loss of temperature control, which, in turn, could cause overpressure (loss of pressure control). This could lead to a vessel rupture. If events occurred fast enough, a human operator might not be able to detect and respond quickly enough to take corrective action.

Some examples of specific hazards associated with each of these considerations include:

- Process characteristics and chemistry--potential explosive mixtures, or highly exothermic reactions;

- Overall process control--a control system which is improperly configured for the dynamics of the process, causing sensitivity and difficult-to-control conditions;

- Flow control--significant deviations such as insufficient cooling water rates, excessive reactant feeds, or blockage;

- Pressure control--overpressure or severe cycling or surges;

- Temperature control--overheating equipment to the point of materials failure or runaway chemical reactions;

- Quantity control--incorrect sequence of reactant charge, incorrect reactant ratio, or overfilling a vessel;

- Mixing--inadequate mixing causing poor heat transfer and overheating;

- Composition--contamination leading to unexpected reactions or corrosion;

- Energy systems--loss of critical heating or cooling;

- Detection and alarm systems--inadequacy or instrument failure;

- Fire and explosion protection--inadequacy or equipment failure.

Details of hazards associated with each of these areas will be discussed in a companion manual on prevention and protection control technologies, in this series.

3.3 PHYSICAL PLANT DESIGN CONSIDERATIONS

As stated earlier, a release may result from the conditional state of the process exceeding the physical limits of the equipment, or at normal process conditions when the physical limits of the equipment deteriorate below those required for containment. In either case, equipment will fail. Physical plant design considerations address both situations, as well as the interactions of individual equipment failures with failure of the total system.

Physical plant design considerations include the following:

- Codes and standards,
- Complexity and operability,
- Reliability,
- Materials of construction,

- Vessels (e.g. heat exchanger columns, tanks, and reactors),
- Piping and valves,
- Process machinery,
- Instrumentation, and
- Siting and layout.

Design must ensure that equipment and components can withstand normal operating conditions for the anticipated life of the facility and can tolerate abnormal conditions within certain bounds. A common cause of failure is the deterioration of equipment over time through various materials failure phenomena such as corrosion. Codes and standards provide some basis for ensuring adequate design, but these are primarily minimum standards. In specific cases, design beyond these minimum standards may be important. The complexity and operability of the equipment may also influence how well a process is controlled, and how easily the equipment and its components are maintained. Reliability, the ability of equipment and components to perform their functions with few or no failures, depends on adequate design, construction, and maintenance. A fundamental principle of reliability is that the system will be only as reliable as its most unreliable component, which is the basis for backup or redundancy in physical systems.

Specific causes of failure have been classified in the technical literature (1). Some common causes of mechanical failure are:

- Excessive stress,
- External loading,
- Overpressure,
- Overheating,
- Mechanical fatigue and shock,
- Thermal fatigue and shock,
- Brittle fracture,
- Creep, and

- Chemical attack (e.g., corrosion, hydrogen blistering, etc.)

Conditions leading to such failure modes range from improper design through improper installation and operation. For example, improper alignment can lead to excessive stress on couplings or shafts of rotating equipment. If properly recognized and evaluated, the prevention of each of these failure modes can be incorporated into initial facility design and construction. The first line of prevention of such failures is in the proper selection and use of construction materials. As can easily be seen, however, some of these conditions are related to operating conditions which can change over time.

Recognition of these failure modes is the basis for many of the specific equipment considerations for the categories of vessels, piping and valves, process machinery, and instrumentation. The manual on control technologies will address these considerations in more detail.

Siting and layout are also considered to be within the realm of physical plant design considerations. Hazards associated with siting may include both natural and man-made factors. Some natural factors include floods, earthquakes, and windstorms. Man-made factors include the siting of facilities near other high hazard facilities, or in areas where an adequately educated and trained labor force may not be available. Layout considerations refer to the relationships among equipment and components within a process unit and among process units in an overall facility. Hazards can arise from layout, for example, when highly toxic and flammable materials or incompatible reactive materials are stored close together.

In summary, physical plant design considerations are concerned with the hardware, as opposed to process design considerations, discussed earlier, which are concerned with the "software" of a process facility.

3.4   PROCEDURES AND PRACTICES

The final general hazard area in process operations involves procedures and practices. Hazards in this area arise from human error in decision making, physical actions controlling a process, and in planning, supervising, and other non-physical activities in the design, construction, and operation of process facilities at any stage of a facility's total life cycle. Figure 3-1 illustrates the various phases of a facility life cycle where errors leading to hazards may be introduced. Figure 3-2 illustrates some kinds of errors that may be introduced. The realization of some hazards in this area are indirect; for example, a lax management policy that does not enforce its own safety standards. Others are direct, such as the operator who takes a wrong action at a control panel.

One categorization of procedures and practices includes:

- Management policy,
- Operator training and practices,
- Maintenance practices, and
- Communications.

Management policy is important because the successful prevention of accidental releases requires the commitment of the human, financial, and material resources of an organization to do what is necessary for release prevention. One of the most important roles of management, after an appropriate safety policy has been established, is to enforce it, keep it up to date, and change it as circumstances change. For example, a preventive maintenance program for high hazard process areas may be well defined and properly written up, but if the program is not audited to see that it really works, then the program may not accomplish its objective. The focus on loss prevention can be so much on specific details that the indirect problem of poor management or supervision in failing to create a proper safety environment is overlooked. Management policy should address special safety procedures for toxic chemicals

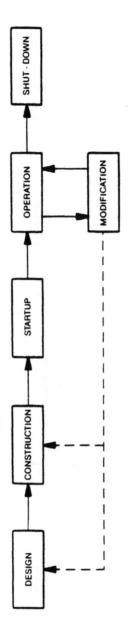

Figure 3-1.   Major phases of facility life cycle.

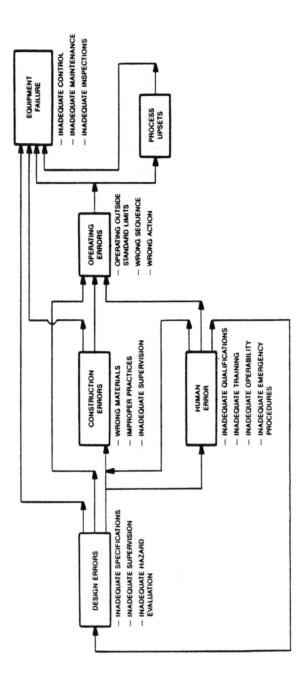

Figure 3-2. Types of errors leading to hazards.

through all parts of the chemical process life cycle that was shown in Figure 3-1.

Operator training and practices have a more direct bearing on release prevention. The skill and knowledge of the operators should be commensurate with the needs of the process, and high hazard processes should, therefore, require a higher standard of operator skill and knowledge than low hazard processes. An often overlooked aspect of operator training is performance auditing. Performance auditing requires a systematic way of obtaining a measurable check on individual operator knowledge is necessary, and especially the operator's knowledge of the possible causes of and means of dealing with unlikely yet potentially serious accidental release events. Emergency response plan drills should be a regular part of operator training.

Maintenance practices are crucial to accidental release prevention. Even if a process facility is originally designed in a way that minimizes the potential for accidental releases, both deliberate and unrecognized changes may occur over time that render a facility unsafe. Proper maintenance is the primary prevention measure that ensures that the original specifications are adhered to and that all special preventive or protective systems are functional.

Finally, the whole human component of accidental release prevention is tied together by effective communications. Information transmitted among the various parts of the organization must be clear, accurate, and timely. Communications procedures for high-hazard facilities should receive high attention and may require different approaches from more routine communications.

A more detailed discussion of each of these areas will be found in the manual on control technologies.

4. Methods for Hazard Identification and Evaluation

Hazard identification is the first step in controlling hazards; the evaluation of hazards follows. Hazard identification is qualitative; hazard evaluation is quantitative. Evaluation seeks to determine the relative importance of two different hazards identified and in some cases the relative probability that a specific hazard will be realized. Various formal and systematic methods for both hazard identification and evaluation are used for facilities manufacturing, using, or storing toxic chemicals.

This section of the manual presents general descriptions of the various methods of formal hazard identification and identifies key features of each of these methods including: the purpose, best times to use, nature of the results, staff size required, and the relative cost. Also discussed are the various methods of hazard evaluation and the purpose they serve in the various steps of the hazard evaluation process.

These methods or variations of them are applicable for use by both regulators and companies. The regulator needs to be familiar with them in order to evaluate their appropriateness, interpret results presented by companies who use the methods, and recommend methods to companies who don't. In some cases a regulator may actually use a technique directly to verify a company's analysis. The company needs to be familiar with the techniques in order to identify, analyze, and control hazards.

4.1 HAZARD IDENTIFICATION

The preceding section summarized where hazards can arise in process operations. The first step in hazard control is hazard identification in a specific process or facility. A publication of the American Institute of

44

Chemical Engineers (AIChE) on guidelines for hazard evaluation lists the following methods of formal hazard identification (12):

- Checklists,
- Safety Review,
- Relative Ranking,
- Preliminary Hazard Analysis,
- What-If Analysis,
- Hazard and Operability Studies,
- Failure Modes, Effects, and Criticality Analysis,
- Fault Tree Analysis,
- Event Tree Analysis,
- Cause Consequence Analysis, and
- Human Error Analysis.

A summary table of key features of these methods is presented in Table 4-1.

The Chemical Manufacturer's Association confirmed the use of these methods in a survey of 39 companies (13). "What-If" analyses and Hazard and Operability (HAZOP) studies were named as the top two methods.

In general, hazard identification procedures can be divided into four main classes:

- Experience,
- Augmented experience,
- Analytical methods, and
- Creative methods.

The experience method compares a new process or equipment situation with knowledge gained from previous experience with similar processes. This method is an inherent part of the principles of design standards and codes, so that hazard identification consists of identifying where deviations from

TABLE 4-1.   SUMMARY OF KEY FEATURES OF HAZARD IDENTIFICATION AND
            EVALUATION METHODS

METHOD	Checklist	Safety Review	Relative Ranking (Mond)
PURPOSE	Identify common hazards Ensure compliance with prescribed procedures	Identify hazards Ensure compliance with prescribed procedures Ensure compliance with design intent. Identifies possible changes Determines applicability of new technology to existing hazards Reviews adequacy of safety maintenance	Provide relative process ranking by risk
WHEN TO USE	Design Construction Startup Operation Shutdown	Startup Operation (i.e., existing facility) Shutdown	Design Operation
NATURE OF RESULTS	Qualitative	Qualitative	Relative qualitative ranking
STAFF SIZE	Small	Small - Moderate	Moderate
RELATIVE COST	Low	Low - Moderate	Moderate

(Continued)

TABLE 4-1.   (Continued)

METHOD	Preliminary Hazard Analysis	What If Analysis	HAZOP
PURPOSE	Identify hazards early in process life cycle prior to final plant design	Identify hazards Identify event sequences Identify possible methods of risk reduction	Identify hazards Identify operability problems Identify event sequences Identify possible methods of risk reduction
WHEN TO USE	Early design	Process development Pre-startup Operation	Late design Operation
NATURE OF RESULTS	Qualitative	Qualitative	Qualitative
STAFF SIZE	Small	Small - Moderate	Moderate - Large
RELATIVE COST	Low	Moderate	Moderate - High

(Continued)

TABLE 4-1. (Continued)

METHOD	Failure Modes, Effects and Criticality Analysis	Fault Tree Analysis
PURPOSE	Identify system/equipment failure modes Identify effect of failure on system/plant Rank criticality of each failure mode	Determine causes and event sequence leading to a defined event Identify combinations of causes including both equipment failures and human errors
WHEN TO USE	Design Construction Operation	Design Operation
NATURE OF RESULTS	Qualitative Quantitative for relative ranking of equipment failures	Qualitative Quantitative
STAFF SIZE	Small	Small - Large
RELATIVE COST	Low - Moderate	Low - High

(Continued)

TABLE 4-1. (Continued)

METHOD	Event Tree Analysis	Cause-Consequence Analysis	Human Error Analysis
PURPOSE	Determine consequences sequence of defined initiating event	Identify both cause sequences and consequence sequences of events	Identify potential human errors Identify effects of human errors Identify cause of human errors
WHEN TO USE	Design Operation	Design Operation	Design Construction Operation
NATURE OF RESULT	Qualitative Quantitative	Qualitative Quantitative	Qualitative
STAFF SIZE	Small - Large	Small - Large	Small
RELATIVE COST	Low - High	Low - High	Low

established safe procedures exist, based on experience. Basically the experience method relies on comparing a new situation to a known past situation.

The augmented experience method relies on various checks on design and
operation beyond mere comparison to standards or previous situations. As
indicated by the previous list, the "What-If" method appears to be one of the
most commonly used. Each step of a process is reviewed to determine what
would happen following equipment failures, process upsets, or operating
errors.

The analytical approach uses either logic diagrams or various types of
checklists. Logic diagrams include fault trees, event trees, and
cause-consequences diagrams. Logic diagrams clearly define cause-effect
relationships and identify combinations of failure that can lead to an undesirable event. These methods can be combined with quantitative data on
probabilities to provide a method of hazard evaluation.

These summary descriptions are based on more detailed descriptions
presented in the AIChE publication cited previously (12). These methods are
also highlighted in some AIChE short courses presented at national meetings.
A general description of each hazard identification method follows. These
same methods apply for hazard evaluation as discussed in Section 4.2 of this
manual.

4.1.1 Checklists

A checklist is a set standard evaluation elements for equipment, materials, or procedures in a chemical process facility. Its purpose is to identify
standard hazards at any time in the life cycle of a chemical process from
research and development through shut-down. In its most basic configuration,
a checklist is merely a memory aid that helps the evaluator remember all the
items he should consider. If the checklist is prepared based on pooled
corporate or industry experience, it becomes a vehicle for transferring a

broader experience base to the evaluator. The level of detail varies with the situation.

4.1.2 Safety Reviews

In the context of accidental releases, a Safety Review is a comprehensive facility inspection to identify facility conditions or procedures that could ultimately lead to a toxic chemical release. This technique is applicable to operating facilities, pilot plants, laboratories, storage facilities, or support functions.

4.1.3 Dow and Mond Hazard Indices

The Dow and Mond Indices are quantitative methods for developing a relative hazard ranking for different chemical processes facilities. Various characteristics of a facility are assigned scores. Features which can lead to an accident are given a negative score and features that can prevent or mitigate the effects of an accident are given positive scores. Scores are combined to yield a ranking index for the process facility being evaluated.

4.1.4 Preliminary Hazard Analysis

The Preliminary Hazard Analysis (PHA) is a hazard identification method for use in the preliminary phase of plant development. The purposes of the PHA are early identification of potential hazards for design and process development personnel. Its special applicability is for the early phases of new processes where there is little past experience. A list of hazards is developed which is used to develop safety guidelines and criteria to be followed as design and development progress.

4.1.5   "What If" Method

A "What If" analysis systematically considers the consequences of unexpected abnormal events that may occur in a process facility. It can include
design, construction, operating, or other deviations from the norm. This
examination can include all parts of a process facility. Its comprehensiveness and success depends on the experience level of the staff conducting the
analysis.

4.1.6   Hazard and Operability (HAZOP) Studies

A HAZOP study can identify process hazards and operability problems. A
HAZOP study involves the multidisciplinary team that works together by search-
ing for deviations from expected design and operating conditions. The team
carefully examines the process facility stream by stream, or component by
component using standard design "guide words." Consequences of deviations are
examined using the guide words. Example guide words are presented in Table
4-2.

4.1.7   Failure Modes, Effects, and Criticality Analysis

Failure Modes, Effects, and Criticality Analysis (FMECA) examines the
ways in which a process system or its equipment could fail, consequences of
failure, and the estimated failure probabilities. FMECA is not efficient for
identifying interactions of combinations of equipment failures that lead to
accidents because it considers each failure individually and traces its
consequences. Criticality rankings can be expressed as probabilities or by
various ranking scores based on evaluators' experience.

TABLE 4-2.   EXAMPLE GUIDE WORDS AND CORRESPONDING DEVIATIONS FOR HAZOP
ANALYSIS

Guide Word	Deviations
None	No forward flow when there should be, i.e., no flow or reverse flow
More of	More of any relevant physical property than there should be, e.g., higher flow (rate or total quantity), higher temperature, higher pressure, higher viscosity, etc.
Less of	Less of any relevant physical property than there should be, e.g., lower flow (rate or total quantity), lower temperature, lower pressure, etc.
Part of	Composition of system different from what it should be, e.g., change in ratio of components, component missing, etc.
More than	More components present in the system than there should be, e.g., extra phase present (vapor, solid), impurities (air, water, acids, corrosion products), etc.
Other than	What else can happen apart from normal operation, e.g., startup, shutdown, uprating, low running, alternative operation mode, failure of plant services, maintenance, catalyst change, etc.
Reverse	Variable or activity is reverse of what it should be, e.g., reverse flow.

4.1.8   Fault Tree Analysis

Fault Tree Analysis (FTA) is a method that constructs a logic tree of events leading to a specifically defined failure event, the "top event." FTA seeks to develop the chain of interrelated events that can lead to a top event. These chains of events include equipment failures and human errors.

The FTA results in identification of combinations of equipment and human failures that are sufficient to result in the top event. The minimum number of independent combinations that can cause the top event are known as minimal cut sets. Probabilities can be assigned to events and the top event probability determined by Boolean algebra.

4.1.9   Event Tree Analysis

An Event Tree Analysis is similar to a FTA except that the logic is to trace the consequences of an initiating event forward to its ultimate consequence. Rather than beginning with the definition of the top event, the event tree finds the top events resulting from initiating and propagating events. A quantitative evaluation can be developed using probabilities in the same manner as with fault trees. The event tree defines multiple consequences of an initial event, whereas a fault tree identifies multiple causes of a final event.

4.1.10   Cause-Consequence Analysis

Cause-consequence analysis combines characteristics of both event tree analysis and fault tree analysis. A cause consequence analysis logically relates both multiple consequences and multiple basic causes or initiating events.

Like a FTA, the cause-consequence analysis leads to minimal cut sets which are all the combinations of basic events that can result in the various top events. Quantitative analysis can be applied to estimate frequencies or probabilities of various top events.

4.2 METHODS FOR HAZARD EVALUATION

Many of the methods for hazard identification discussed in the preceding section also apply to hazard evaluation (also referred to as hazard analysis or "HAZAN"). Hazard evaluation is the next step after hazard identification in the total assessment of risk for accidental releases. The evaluation step attempts to rank the hazards qualitatively, quantitatively, or both, seeks to identify measures that reduce the probability that the hazard will be realized, and examines the potential consequences of the hazard if it is realized. As with the method of identification, evaluation methods are also discussed in depth in the technical literature (12, 13, 15). A recent publication of the American Institute of Chemical Engineers presents a comprehensive summary of procedures with considerable discussion of distinguishing features, such as applicability and expected results (12). A summary table is presented, shown in slightly modified form in Figure 4-1.

It is clear that the various methods all have some features in common and yet are distinguished by differences that reflect the specific suitability of one method over another in certain situations. Only a few of the methods are specifically suited for quantitative hazard evaluations where probabilities of an accident are to be determined. Another significant difference is that some methods are hardware-oriented while others easily accommodate the effects of human interactions with the process being evaluated.

None of the listed methods deal explicitly with consequences outside the process itself, but some methods, such as Fault Tree Analysis, can easily accommodate such an extended analysis. Another approach is to conduct a completely separate consequence analysis, with the accident event being the starting point.

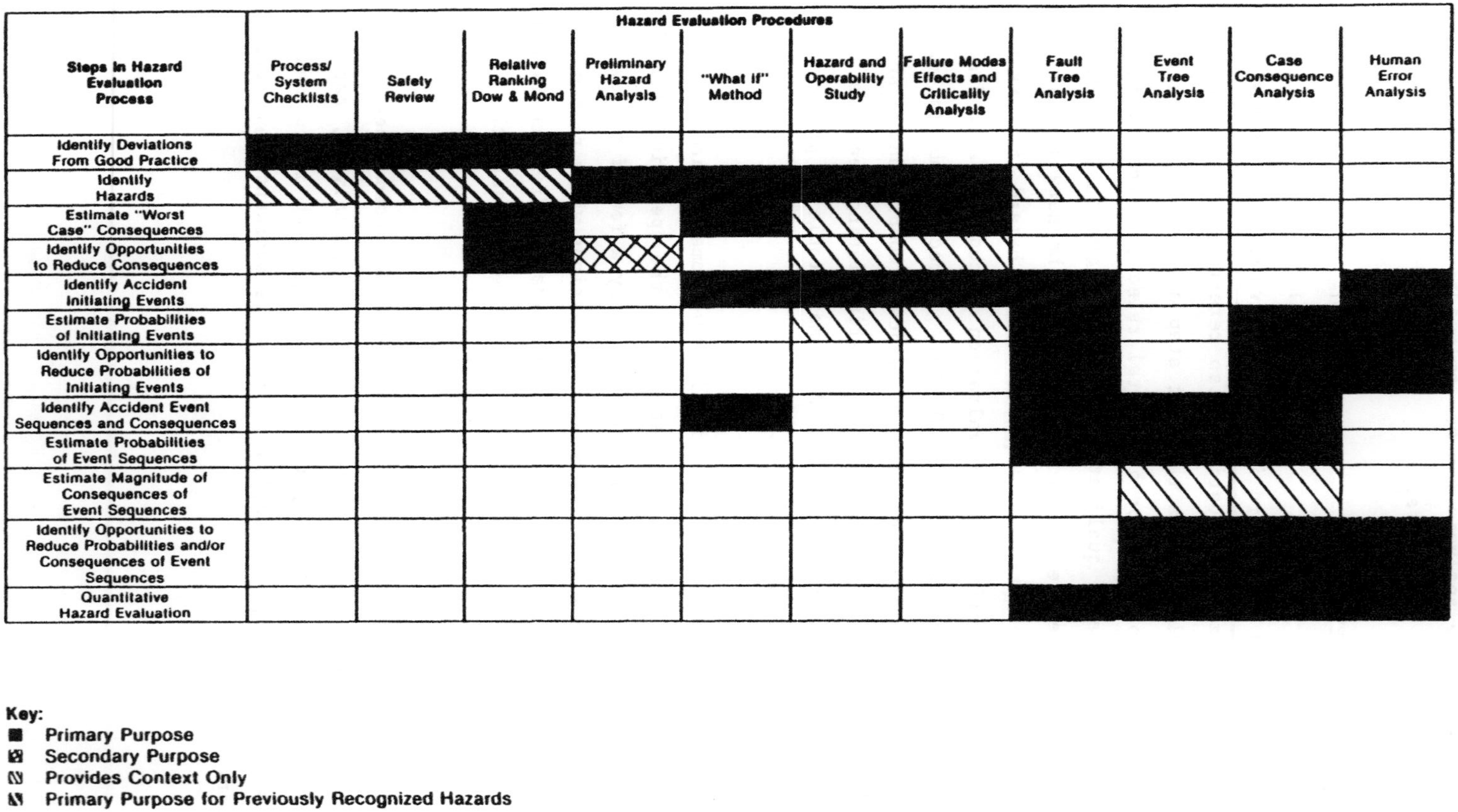

Hazard Evaluation Procedures

Steps In Hazard Evaluation Process	Process/ System Checklists	Safety Review	Relative Ranking Dow & Mond	Preliminary Hazard Analysis	"What If" Method	Hazard and Operability Study	Failure Modes Effects and Criticality Analysis	Fault Tree Analysis	Event Tree Analysis	Case Consequence Analysis	Human Error Analysis
Identify Deviations From Good Practice											
Identify Hazards											
Estimate "Worst Case" Consequences											
Identify Opportunities to Reduce Consequences											
Identify Accident Initiating Events											
Estimate Probabilities of Initiating Events											
Identify Opportunities to Reduce Probabilities of Initiating Events											
Identify Accident Event Sequences and Consequences											
Estimate Probabilities of Event Sequences											
Estimate Magnitude of Consequences of Event Sequences											
Identify Opportunities to Reduce Probabilities and/or Consequences of Event Sequences											
Quantitative Hazard Evaluation											

Key:
- ■ Primary Purpose
- ▨ Secondary Purpose
- ▨ Provides Context Only
- ▨ Primary Purpose for Previously Recognized Hazards

Figure 4-2. Relationship between hazard evaluation procedures and hazard evaluation process.

Source: Adapted from Reference 12.

The consequence analysis can also be qualitative, quantitative or both. Numerous mathematical models have been proposed and some are in common use for predicting the effects of fire, explosion, and accidental releases. Methods for examining the effects of accidental releases are often based on various forms of dispersion models. A summary of procedures available up through the late 1970s is presented in Lees (1). The prediction of affected areas and of concentrations resulting from accidental releases can be combined with health and environmental effects data to estimate the severity of the consequences of any release in the affected area. A recent report reviews major available models (16).

A more detailed discussion of these methods is beyond the scope of the present work. The reader is referred to the general technical literature for more details.

5. Overview of Principles of Control

Both regulators and companies need to be familiar with the fundamental principles of control: the regulator, to intelligently review and evaluate company control plans and practices; companies, to evaluate existing company practices and to implement safer practices.

This section of the manual presents an introductory discussion on control technologies as they relate to reducing the probability and consequences of an accidental chemical release. Three fundamental levels of control are addressed: prevention, protection, and mitigation. Specific process design considerations, physical plant design considerations, and procedures and practices are discussed as they relate to prevention. Various protection and mitigation technologies are also discussed. In addition, examples of possible causes of releases and potential controls are illustrated.

5.1 BACKGROUND

The control of accidental chemical releases involves reducing the probability and consequences of such releases. Such control can be viewed as consisting of three fundamental levels:

- Prevention,
- Protection, and
- Mitigation.

The purpose of prevention is to reduce the probability of accidental releases. Prevention refers to all those measures taken to ensure that the primary containment of the chemical, that is storage, transfer, and process equipment, is not breached. These measures include process design, physical plant design, and operational procedures and practices, and involve considerations ranging from process control and hardware design to operator training and management policy. The ideal result of a successful prevention effort is that accidental releases of a chemical from its primary containment do not occur. In reality, prevention can be successful in reducing the probability of a release to a reasonable minimum.

"Reasonable minimum" is an imprecise criterion; however, the inherent characteristics of any process are that equipment can fail, and people make mistakes, so that from time to time an accidental release event will happen. The probability can be reduced if appropriate prevention measures are taken. These prevention measures enhance the control of a process or the ability of the hardware to tolerate severe process conditions, process upsets, and human error.

When preventive measures fail, a second level of control deals with protection from releases. In the context of this manual, protection means to contain, capture, neutralize, or destroy a toxic chemical subsequent to release from primary containment, but before it escapes into the environment. Protective systems are defined as add-on equipment and processes systems not considered part of the actual chemical process system itself, but which control a potential release. Examples of protection technologies include diking, flares, and scrubbers. Protection systems will also fail from time to time.

A deficiency in a protection system when it is needed may allow a toxic vapor or gas to escape into the environment. Once this occurs, the consequences may be reduced by using effective mitigation measures. Mitigation refers to equipment and procedures that can reduce the concentration of a chemical below levels that would otherwise occur and hopefully below levels harmful to sensitive receptors. Measures include technical approaches such as water sprays, or steam curtains for dilution and dispersion, barriers for diversion and dispersion, and procedures such as closing doors and windows, and evacuation in affected areas. It should be noted that mitigation is not a back-up approach to protection as protection is to prevention. The latter two areas are directed toward stopping a release. Mitigation merely attempts to control a release that has already occurred.

A detailed discussion of prevention and protection is the subject of the manual on control technologies. Mitigation will be the subject of a separate manual. A summary of major considerations in these three areas is presented in the ensuing subsections.

5.2   PREVENTION

Prevention measures can be classified into the general topical areas of:

- Process design considerations,

- Physical plant design considerations, and

- Procedures and practices.

These measures include consideration of both operational and hardware aspects of a chemical process system. Operational aspects include both the inherent characteristics of the process itself and human aspects such as operator training, maintenance procedures, and general management policy and procedures. Prevention measures are applicable to new facilities as well as existing facilities, where they are the basis for process modification.

5.2.1   Process Design Considerations

Process considerations involve the areas listed in Section 3, which can be regrouped as follows:

- Chemical process characteristics;

- Control system characteristics;

- Hazard control for flow, pressure, temperature, quantity
  measurement, composition, mixing, and energy systems; and

- Hazard control for fire and explosion.

The design of a process in which toxic chemicals are used or produced
should be based on sufficient data to ensure a safe operating system. Necessary data to be considered in the design process include, but are not limited
to, the following: chemical, physical, thermodynamic, and toxicological
properties of the individual chemical components used or produced in the
process; the process potential for explosive reaction or detonation under
normal or abnormal conditions; process reactivity with water or other common
contaminants; possibility of self-polymerization or heating; potential side
reactions and conditions under which they are favored; whether reactions are
endothermic, exothermic, or thermodynamically balanced; the explosive range of
volatile or gaseous components and the possibility of explosive mixtures
during storage, processing, or handling; the possibility of dust or mist
explosions; and interactions with materials of construction.

Flow, temperature, pressure, quantity measurement, and composition
control are fundamental process variables. Mixing is a fundamental process
phenomenon. These process elements have associated with them the basic
potential operational hazards of any process. Prevention measures associated
with these elements include the appropriate process control measures discussed
above applied to the specific process effects and hazards of these elements.
The fundamental hazards of all these variables arise from deviations that
exceed the limits of the operational or physical system. For example, excess
flow may be a hazard if it causes a change in a process system faster than the
operator or control system can respond. The excess flow of a reactant in an
exothermic reaction might cause a thermal runaway sooner than emergency
cooling or other safety measures could be activated. Or, flow blockage could

prevent cooling water flow in a critical circuit. For each of the above variables, there are specific measures that can reduce the chances for deviations or minimize their effects.

The control characteristics of the process should be understood and the control system should be appropriate for the system. This means that the process dynamics should be properly considered and the control system must be compatible with the available operating and maintenance staff knowledge and skills. Specific technological prevention measures in all control systems include using control components of greater reliability and accuracy and faster response times. A fundamental principle in control system components is redundancy, or the use of independent backup to critical components. The key word here is independent--to avoid a common mode failure in critical systems. For a specific process, reducing the probability or magnitude of process deviations or upsets that could lead to an accidental release may involve design changes in a control system ranging from individual components through the entire control strategy and hardware.

Fire and explosion protection is basic to release prevention. Preventing the conditions that lead to fire and explosion is inherent in the items already discussed above and also involves physical plant design considerations discussed in the next section. Once a fire has started in or near a process involving toxic chemicals, controlling the fire becomes of paramount concern. Design considerations for fire protection are based on the removal of fuel, oxygen, or heat from the fire. Adequacy and reliability of water supply, possible chemical fire fighting measures, and suitable fire fighting equipment are the primary process design considerations for fire protection beyond initial fire prevention through careful process design.

Explosion protection relies on preventing dangerous process conditions such as the formation of explosive mixtures from occurring, preventing contact of potentially explosive mixtures with ignition sources, and applying specialized explosion suppression systems. Explosion protection also relies on appropriate physical plant design considerations, as discussed in the next subsection.

5.2.2 Physical Plant Design Considerations

Physical plant design considerations address hazards and their control in the following areas:

- Siting and layout,
- Structures and foundations,
- Vessels,
- Piping and valves,
- Process machinery, and
- Instrumentation.

Hazards were discussed in Section 3. Control considerations are discussed here.

At a minimum, codes and standards should be followed in the design of each of these areas. However, additional protection measures above and beyond those specified by the codes and standards should be incorporated into the design based on the specific situation. Problems arise because codes and standards are not developed with specific situations in mind. Thus, code cannot be relied upon unless the basis matches the scenario of concern with great precision.

Siting is the first area considered. A plant's location might affect frequency or severity of an accidental release. Siting considerations include, but are not limited to, the following: drainage systems should prevent the runoff of spilled liquid chemicals onto adjacent properties and prevent the spread of toxic and/or flammable liquid chemicals in a manner that minimizes adverse effects within and outside of the plant boundaries; minimization of the effects of natural calamities such as freezing, fire, wind, floods, earthquakes, and landslides in contributing to an accidental release should be incorporated into equipment design, for example, different foundation designs in earthquake prone areas; the potential impact of accidents such as fires, explosions, or hazardous chemical releases at adjacent industrial facilities, roads, or railways should be recognized as a possibility; reliable water and power supplies should be available with backups where a failure could cause an accidental chemical release; and traffic flow patterns within the plant and around the perimeter should be designed to prevent congestion and allow access by emergency response vehicles and appropriate movement of personnel in an emergency.

The layout of a plant is the next area considered. Layout considerations include, but are not limited to, the following: process units and the equipment and piping within a unit should be arranged to minimize congestion; where possible, hazardous processes should be segregated from other hazardous processes or sensitive areas within the plant or plant property; adequate spacing should be available for access by maintenance and emergency response personnel and equipment; explosion barriers should be applied where appropriate as described, for example, in the Dow guide (17); escape routes for personnel should be easily accessible; and offices, lunchrooms, or other support structures should be located at the perimeter of the facility.

Foundations should ensure the stability of all vessels and non-transportable equipment containing hazardous chemicals. The design should be in accordance with recognized construction and material specification standards in the industry as a minimum requirement. The design should consider all normal and abnormal load and vibration conditions, as well as severe conditions caused by freezing, fire, wind, earthquakes, flood, or landslides. Transportable equipment should be secured to prevent the upset or accidental detachment of process lines conveying hazardous chemicals during use and should not be used to permanently replace a stationary piece of equipment unless a given situation dictates a preference or requirement for such transportable equipment from a safety standpoint.

Structural steel should, at a minimum, be designed and constructed in accordance with appropriate construction and material specification standards in the industry. The design should consider all normal and abnormal dead loads and dynamic loads resulting from wind, collision, earthquake, or other external forces. As a minimum, fireproofing should be used for areas in which hazardous chemicals are manufactured, stored, handled, or generated and such areas should conform with legally applicable codes and standards. Fire protection beyond minimum standards should be considered for hazardous areas in which hazardous chemicals are present.

Vessel design and construction should conform to recognized design and material standards for the specific application in the industry as a minimum. Standards and specifications should be reviewed for adequacy of criteria. Stricter standards may sometimes be appropriate. Design should consider the combination of conditions anticipated for quantity, fill rate, pressure, temperature, reactivity, toxicity, and corrosivity. As a minimum, all vessels should be equipped with the following safety features: overfill and overpressure protection and, where appropriate, vacuum protection; storage cooling systems for low boiling point liquids and liquefied gases; storage vessels should be surrounded by diking, firewalls, or other containment devices unless such features are deemed to create a more severe secondary hazard in specific cases; vessels and vessel fittings should be protected from damage caused by collision or vibration and should be adequately braced to support the weight of piping; columns should be adequately supported to withstand the maximum wind loads expected in the area; and operators should be trained concerning the vessel's limits for pressure, temperature, fill and emptying rates, and incompatible materials.

Additional items such as nitrogen blanketing, improved fire protection, or release reduction equipment (e.g., water or steam curtains) may be appropriate in certain situations.

All pressure vessels and vessel jackets should be fitted with adequate pressure and/or vacuum relief. The relief systems should be designed according to recognized design procedures and standards appropriate in the industry as a minimum. Stricter procedures and standards may sometimes be appropriate. Containment systems should be designed according to recognized design procedures for containment systems. Valves upstream of pressure or vacuum relief devices should be prevented from being closed in such a way that the vessel will be isolated from all pressure relief or vacuum relief. Where possible, a pressure trip system should be used along with a pressure relief system. This will help minimize the frequency of releases of hazardous chemicals through the pressure relief system. All pressure or vacuum relief devices should be inspected and maintained periodically as part of routine maintenance. Testing a values capacity should be done whenever any corrosion, fouling, or scaling has occurred. The adequacy of a pressure or vacuum relief system should be reevaluated when a vessel or process unit is used to handle more material, or a different material, than that for which it was originally designed.

As a minimum, heat exchangers should be constructed in accordance with accepted industry codes and standards. Standards should be reviewed for adequacy of criteria. Stricter standards may sometimes be appropriate. The materials of construction should be selected to minimize corrosion and fouling. All exchangers should be equipped with pressure relief, by-pass piping, and adequate drainage facilities. Exchanger design should allow for thermal expansion and construction without causing excessive stress on connections.

Turbines, drivers, and auxiliary machinery should be designed, constructed, and operated in accordance with applicable industry standards and codes. Standards should be reviewed for adequacy of criteria. Stricter controls may sometimes be appropriate. The equipment should have adequate protective devices to shut down the operation and/or inform the operator before danger occurs. Vibration sensors and/or shutdown interlocks may be required on high speed rotating equipment.

Heaters and furnaces should be located so as to minimize the possibility of bringing an open flame and/or extreme heat too close to a hazardous area. Basic units and controls should be designed in accordance with applicable standards and codes as a minimum. Standard should be reviewed for adequacy of criteria. Stricter standards may sometimes be appropriate. Examples of some of the basic requirements for furnaces include the following: provision for adequate draft; positive fuel ignition; automatic water level controls; pressure relief devices; and fuel controls. Air heaters should have igniters designed to provide positive ignition, proper safety controls on fuel sources, sight glasses for flame observation, monitoring devices for flame-out detection, and high temperature alarms. All heaters and furnaces should be inspected regularly. Where heaters and furnaces handle hazardous process materials, appropriate precautions should be taken to prevent releases in the event of tube failures, such as cracking, rupture, or plugging.

As a minimum, piping, valves, and fittings should be designed according to recognized industry codes and standards pertaining to working pressures, structural stresses, and corrosive materials to which they may be subjected. The thermal stress of repeated heating and cooling cycles or excessive temperatures, either high or low, should be considered. Some additional considerations include, but are not limited to, the following: dead ends or unnecessary and rarely used piping branches should be avoided; the type of pipe appropriate for pumping a hazardous chemical should be selected (e.g., using welded or flanged pipe instead of threaded pipe or using a suitable metal or lined metal piping instead of plastic wall piping); backflow protection should be installed where necessary, but backflow prevention should not be relied on as the only means of avoiding a backflow hazard; materials of construction suitable for the application should be selected and checked before installation to confirm the composition; recordkeeping on critical lines should be provided to prevent incorrect future substitutions; a means of remotely shutting off the flow in lines that carry a large volume of hazardous materials should be provided; adequate structural support should be provided to protect against vibration and other loads and to protect piping from

potential collisions with vehicles in the vicinity; piping should be pitched
to avoid unintentional trapping of liquids; and provisions should be made to
ensure that a liquid-full condition cannot exist in a blocked section of line
unless such a section of line has pressure relief.

Extra precautions should be taken in the design of pumps and compressors
to minimize the potential for an accidental release of a hazardous chemical.
Extra precautions include, but are not limited to, the following: where a
pressurized hazardous material is being pumped or where the consequence of a
seal failure could result in the accidental release of designated chemical,
seals should be suitable to ensure reliable leak prevention (e.g., double mechanical seal with a pressurized barrier fluid that is compatible with the
process fluid and equipment materials of construction); totally enclosed pump
or compressor systems may be appropriate, if safely vented and inerted and
monitored for oxygen where enclosure could result in a secondary hazard such
as an explosive mixture; remotely operated emergency isolation valves and
power shutoff switches may be appropriate on the suction and discharge sides
of a pump or compressor; compressors or positive displacement pumps should be
fitted with adequate overpressure protection; instrumentation to determine
when flow into or out of a pump has ceased may be appropriate; where overheating could result in a fire or explosion, temperature monitoring may be appropriate; a backup power supply should be used for critical pumping systems;
surge protection should be provided for pumps; and pumps, compressors, and
their associated piping should have foundations and supports that protect
against damage caused by vibration and any static and dynamic loads.

Every reasonable effort should be made to maximize the effectiveness of
automatic process control systems for preventing an accidental release. All
systems and instrumentation should be of the "fail-safe" type. Instruments
should be made of materials capable of withstanding the corrosive or erosive
conditions to which they are subjected. Central control rooms should be
protected from fire and explosion hazards. An owner-operator should evaluate
the ability of control systems to operate on manual control and should install

a backup power supply in situations where operating on manual control would be impractical.

A variety of miscellaneous modifications may be appropriate, depending on the needs of the particular process unit. Examples of these modifications include, but are not limited to the following: the addition of control systems where none are presently employed; redundancy of key components; replacing components to improve accuracy, reliability, repeatability, or response time; the addition of a backup control system; simplification of an existing control system to improve operability; replacing a system that indirectly controls the variable of interest with a system that directly measures and controls the variable of interest; the addition of trip systems for emergency situations; and the redesign of a control system to conform to acceptable design standards.

All wiring and electrical equipment should be installed in accordance with the National Electric Code or stricter standards, if applicable. Electrical equipment for use in hazardous locations should comply with acceptability standards of recognized testing organizations. Standards should be reviewed for adequacy of criteria. Stricter standards may sometimes be appropriate. All electrical apparatus should be grounded.

Protection devices should have the capability of warning operating personnel when emissions are not being controlled. Plant alarms such as klaxons and sirens can be used to alert or signal such personnel. If a device is only used on an intermittent basis, then a testing program should be in place to ensure that the system will function when necessary. In addition, alarms should be tested, audited, and inspected to ensure reliability.

As a minimum, plant fire protection systems should be laid out in accordance with recognized codes and standards, such as those prepared by the National Fire Protection Association. A reliable water supply for all portions of the plant should be available. Flammable gas detection systems are recommended for locations where flammable chemicals are used at elevated temperatures and pressures. Central fire alarm systems should be in place. In addition to water, firefighting materials, such as spray foams, dry chemicals, and carbon dioxide, may be appropriate to handle various specialized types of fires.

5.2.3 Procedures and Practices

Operational Controls--

The following types of reactive materials should be stored so that the potential for mixing in the event of an accidental release is minimized by dikes or other physical barriers: materials that react to form a hazardous chemical; hazardous chemicals that react exothermically and thereby contribute to the rate of evaporation of the chemicals; and hazardous chemicals where reactions will contribute to the potential for an accidental release. Chemicals may be mutually reactive or reactive with other materials that may be nearby such as cooling or heating fluids, cleaning agents, and materials of construction.

Extra precautions may be required where there is a potential for mixing two incompatible chemicals within a process. Such precautions could include backflow protection, composition monitoring, and interlocks that prevent valves from being opened in combinations that allow for cross-contamination. Use of common lines for handling such incompatible chemicals should be avoided.

All materials of construction should be capable of withstanding normal operating conditions, normal shutdown conditions and potential deviations from normal operation. Where a specialized material is required, initial construction materials and replacement parts should be tested before use to ensure that the composition is consistent with specifications.

Safe procedures should be established to minimize the risk of an accidental release of a hazardous material during filling or emptying operations for tanks, vessels, tank trucks, or tank cars. Some considerations include, but are not limited to, the following: before material is added to a vessel, tank, tank truck, or tank car, the operator in charge of the addition should be able to verify what material is in the vessel or was last in the vessel; where hoses are used, a system should be in place to ensure that the proper type of hose is used for each application (e.g., different types of fittings for each application); hoses should be regularly inspected and maintained as necessary; efforts should be made to decrease the possibility of materials being sent to the wrong location; a system should be in place to prevent tank trucks or rail cars from moving away with a hose still connected; when a hose is used to transfer materials, it should be possible to stop the flow if the hose should fail; equipment should be grounded and operators trained in the appropriate methods for chemical transfer so as to avoid static charge accumulation.

Procedures and equipment should be in place so that every reasonable effort may be made to prevent an accidental release from the storage, handling, or treatment of wastes containing the hazardous chemicals.

Management Controls--

Programs to train plant personnel to handle normal operating conditions, upset conditions, emergency conditions, and accidental releases should be used. The programs should include written instruction, classroom-type instruction, and field drills. Periodic review and drill exercises should be part of such programs. Printed materials describing standard and emergency procedures should be provided to employees and revised as necessary to be consistent with accepted practices and recent plant modifications.

A plant-wide fire prevention and protection plan should be used. All operating personnel should be instructed concerning fire prevention and fire response. All facility personnel should be instructed in basic first aid and

fire extinguisher use. The formation and training of specialized fire fighting teams and first aid teams should be in accordance with or exceed minimum
specified requirements. All fire protection and prevention plans should be
periodically reviewed and training drills held.

The owner/operator of a facility should formulate a comprehensive contingency plan to handle major plant disasters. All facility personnel should be
trained to participate in plans for controlling facility emergencies related
to accidental releases including emergencies such as large windstorms, earthquakes, floods, power failure, fires, explosions, and accidental releases of
hazardous chemicals.

The contingency plan should describe coordination between the plant and
local police, fire, and other emergency personnel. The plan should be specific in designating responsibilities and in addressing specific high-hazard
situations that are possible for the plant. Communications responsibilities
and procedures for relaying information during emergencies should also be
clearly defined. The plan should include procedures for emergency notification of community and local governments. Where an accidental release could
adversely affect the local community, the plan should include appropriate
community emergency response procedures.

Simulated emergency exercises involving plant personnel should be performed on a regular basis. Disaster exercises that incorporate local emergency response organizations should also be undertaken periodically. Exercises
may include tabletop exercises, emergency operations simulations, drills, and
full field exercises.

An inspection, testing, and monitoring program for process equipment and
instrumentation should be considered for areas of high hazard potential.
Systems and components to which this program can be applied include, but are
not limited to, the following: pressure vessels; relief devices and systems;
critical process instruments; process safety interlocks (trips); isolation,

dump, and drowning valves; process piping systems; electrical grounding and bonding systems; fire protection systems; and emergency alarm and communications systems. Engineering drawings and design specifications should be available for inspection, if requested.

Maintenance staff qualifications, skill level, and numbers should be consistent with the hazard potential at the specific operation.

A process safety review consistent with the magnitude of the modification should be made before implementing any modification. Documentation of modifications should be made and be available for inspection, if required.

5.3 PROTECTION

Protection measures are equipment and systems that prevent or reduce the quantity of chemical that is discharged in an incipient release that has already escaped primary containment. Protection technologies include flares, scrubbers, diking, and enclosures (i.e. containment buildings). Each of these technologies may be appropriate in specific circumstances, but none of them is universally applicable to accidental releases. Much depends on the specific toxic chemical involved, the quantity released, the rate of release, and how it is released.

Flares are commonly used in chemical process plants and in petroleum refineries to dispose of flammable gases and vapors resulting from normal operating upsets. They may be suitable in certain circumstances for the destruction of toxic chemicals that would otherwise be released to the environment. A prerequisite for destruction in a flare system is that the toxic material be flammable, or that it at least thermally decompose to less toxic compounds at flare flame temperatures. The other requirement is that the nature of the emergency discharge be compatible with the overall design and operating requirements of flares, such as the maintenance of specific gas velocities at the flare tip, and flow fluctuations within the design turn-down

capabilities of the particular flare system. There may also be some real
difficulty in the safe design of a flare system for the emergency discharge of
toxic materials combined with normal process discharges of other process
materials. Considerations in this area include the manifold system and how
backflows and inappropriate mixing of incompatible materials might be avoided.
A dedicated flare system is not necessarily a solution to the problem since
keeping a flare on standby for a relatively rare emergency may not be feasible. Flares are an option that can be considered for the right circumstances
for treating emergency discharges that are still confined by a pipeline or
stack. The use of flares in such an application, however, must be carefully
evaluated for possible secondary hazards that could make the use of flares
more dangerous than not using them. For example, improper design could lead
to flash back.

Scrubbers are another alternative for treating confined toxic discharges
before they are released to the environment. Scrubbers have a long history of
success in the process industries. Many of the considerations that apply to
combined versus dedicated flare systems apply also to scrubbers. Scrubbers
may be easier to maintain on a standby basis than flares, however.

The applicability of scrubbers depends on the solubility of a toxic
chemical in a suitable scrubbing medium or the ability of the scrubbing medium
to reactively neutralize the toxic chemical. Chemicals soluble in water or in
various aqueous solutions are not a particular problem. Other toxic materials
that require non-aqueous scrubbing liquids could present more difficulties.
For example, it would be hazardous to use a flammable organic liquid as a
scrubbing medium since one could easily create a flammable mixture. Most
scrubbing systems that would be considered feasible for toxics would probably
be based on aqueous scrubbing chemistry.

The type of scrubber that is suitable for use depends on the nature of
both the discharge circumstances and the scrubbing chemistry. Since a low
system pressure drop appears desirable for emergency scrubbing systems, simple

spray towers may be appropriate in many applications. However, other types of scrubbers, such as packed beds and venturis, may also be appropriate in some situations.

Both scrubbers and flares can be used when the incipient release is still confined by piping or a stack, as might be the case with an emergency discharge from a relief valve. When the release results from equipment failure, such as a vessel or pipeline rupture, containment is required before the chemical can be disposed of by flaring, scrubbing, or in-place neutralization. Such temporary containment can be effected by diking and enclosures. Diking is a physical barrier around the perimeter of process equipment or areas designed to confine the spread of liquid spills and to minimize the liquid surface area. It can be simple earthen berms or it might be concrete walls. The diking might be little more than a high curb, or it could be a high wall rising to the top of a storage tank. The applicability of diking to spills of volatile liquids is readily apparent. By containing the liquid, the dike reduces the surface area available for evaporation, at the same time allowing a liquid to be cooled by evaporation so that the vapor release rate is diminished. In this way, diking can reduce the rate at which a toxic material is released to the air. The material can be allowed to evaporate at a manageable rate, collected into alternate containers, or neutralized in place.

Enclosures or containment buildings directly provide secondary containment for materials that have escaped primary containment. Such buildings can be designed to contain the toxic chemical until it is vented through an appropriate destruction system such as a flare or scrubber, collected into an alternative container, or neutralized in place. The primary difference between diking and a complete enclosure or containment building is the roof. The building confines virtually all of the material, whereas diking permits the continuous release of some of the material.

The applicability of either diking or enclosures must be carefully evaluated to determine if there might be secondary hazards associated with their use. For example, if the toxic material is also flammable, containment could create a fire or explosion hazard which could be as or more serious in its consequences than the original release itself. When properly applied, however, diking and enclosure can be effective protection technologies for accidental releases.

5.4 MITIGATION

Once a toxic chemical has been released to the air or that has the potential for entering the air, the primary concern becomes reducing the consequences to the plant and the surrounding community. Reducing these consequences is referred to as mitigation. Two aspects of mitigation are measures to control the quantity of toxic material that could reach receptors, and protect the receptors by ensuring that they remain in or are evacuated to locations that will prevent or minimize their exposure to the chemical. An emergency response plan addressing the issues is a key part of mitigation. The other part is the technology of controls. Mitigation technologies include such measures as physical barriers, water sprays, steam curtains, and foams. Mitigation technologies divert, limit, or disperse the toxic chemical that has been released to the atmosphere.

An emergency response plan must be based on identifying the consequences of the accidental release based on downwind quantities, concentrations, and duration of exposure at various receptor sites. The plan should include the information required to decide whether evacuation should be undertaken, or whether people in the path of the release might be better protected by remaining indoors in their own homes or other places. The plan should include specific first aid steps to be taken for the exposed population. Finally, the plan should lay out the specific responsibilities and activities of all facility personnel and community response teams in addressing the emergency.

Physical barriers may be specially constructed for the purpose, may be constructed for another purpose but function as barriers, or they may be natural terrain features. An example of a specially constructed structure is a diversion wall. A functional barrier could be a building. A natural terrain feature that might be a barrier is a hill or a line of trees. The primary function of a physical barrier is to protect especially sensitive receptors and provide additional time in which to respond to the accidental release emergency. A physical barrier located in the right place can also contribute to enhancing the dispersion of the released chemical. Physical barriers do not, however, directly capture or neutralize the chemical. They might improve the performance of other mitigation measures such as water sprays, however. Applicability and overall performance will depend on the nature of a specific release and on the meteorological conditions at the time of the release.

Water sprays and steam curtains are methods used to increase the dispersion rate of the released chemical, divert its direction, serve as a barrier between the toxic cloud or plume and potential receptors, and even absorb the chemical. The applicability and effectiveness of these methods depend on the nature of the release, the properties of the specific chemical, and the capability of extending to the highest effective point of a release. Effectiveness will also depend on meteorological conditions at the time these mitigation methods are applied.

A final mitigation technology is the use of foams. Foams are chemical mixtures that can be applied to liquid spills with special foam generating apparatus and, by covering the surface of the spill, reduce the rate of evaporation. Foams can be applied to spills that would otherwise result in the release of large quantities of toxic vapors as long as the foam is physically and chemically compatible with the spilled material. A fundamental requirement is that the foam have a density lighter than the liquid over which it is applied, and that the material over which the foam is applied does not easily diffuse through the foam. Foams may sometimes permit a spill to be transferred to containers for final disposition.

Mitigation technologies and community preparedness are the final line of defense in reducing the risks of accidental releases of toxic chemicals.

5.5   CONTROL TECHNOLOGY SUMMARY

The various technologies discussed in this section of the manual are discussed in greater detail in the manual on control technologies. It is apparent that controls applicable for the prevention of, protection from, and mitigation of accidental chemical releases, cover a wide range of both equipment and procedural measures. These measures may range from changes in the process employing a toxic chemical to the addition of specialized equipment outside of the immediate needs of the process itself.

Table 5-1 summarizes some locations and possible causes of accidental chemical releases, and presents corresponding control measures that might be applied. Selection of appropriate control measures depends on the results of a hazard evaluation that would define the most likely causes for a given process system. For most processes combinations of alternatives could reduce the probability of release and also reduce the consequences of any release that might occur. Additional discussion of this issue is also provided in the manual on control technologies.

TABLE 5-1.   EXAMPLES OF POSSIBLE RELEASES AND CONTROLS

Location of Accidental Release	Cause	Potential Controls
External Causes		
General equipment failure	Fire at adjacent plant	Extra fire protection Protective barriers Coordination of emergency response efforts
	Explosion at adjacent plant	Protective barriers Coordination of emergency response efforts
	Traffic or rail accident from outside the plant	Protective barriers Reroute traffic flow
	Natural disaster, (flood,land-slide, earthquake, windstorm)	Protective barriers Strengthened equipment foundations and structural support Alter surface contours to facilitate drainage around the plant Emergency response plans for each potential event
	Fire in adjacent process unit	Adequate spacing between process units Protective barriers Utility piping arranged in a way that prevents loss in adjacent units when one unit fails
	Explosion in adjacent process unit	Adequate spacing between process units Protective barriers Strengthened equipment foundations and structural support
	Loss of process control as a result of utility failure	Provide local or plant wide backup for crucial utilities Where possible, improve process operability so that the process can be manually operated or shut down when utilities are lost Arrange utility distribution so that utility losses will only effect a small area within the plant
	Trench fire	Install covers that are two thirds closed and one third grate Place flame traps periodically throughout the trenches
General process failure	Control room destroyed by fire/explosion	Protective barriers around control room Relocation of control room Where possible, improve process operability so that the process can be manually operated Pressurize control room with clean air supply Construct control room with extra fire protection
Pipebridge	Collision with plant vehicle	Protective barriers Warning signs of restricted clearance Rerouting in-plant traffic flow Remote shutoff valves on both sides of bridge

(Continued)

TABLE 5-1 (Continued)

Location of Accidental Release	Cause	Potential Controls
Vessel	Overpressure as a result of vessel BLEVE when exposed to fire	Adequate deluge system Water spray monitors available to cool the vessel Adequate pressure relief for handling heating from fire
Operating/Maintenance/Management Error		
General equipment failure	Thermal shock	Flow and temperature control Improve operator training Periodically inspect and test equipment for signs of fatigue
	Replacing a worn part with a part made of a material that is incompatible with the process	Establish testing procedures to certify materials of construction Where materials of construction are critical, have written details as to what materials of construction are appropriate or inappropriate for a given application
General process failure	Loss of process control as when a valve is incorrectly operated	Clearly label lines Physically segregate piping according to use Relieve congestion Eliminate unused and excess piping Improve employee training
Vessel	Tank overfill as a result of operator error	Level gauges and alarms Overflow catch tank Surround tanks with dikes Improve operator training
	Internal explosion as a result of static discharge in the presence of a flammable mixture	Operator training as to the hazards of static discharge Establish specific procedures for transfer operations
	Collapse due to underpressure when the tank is emptied or cooled too quickly	Operator training as to the physical limitations of the vessel and how this effects operating procedures Establish specific procedures for filling, emptying, heating and cooling
	Overpressure as a result of incorrectly adjusted relief valve	Test all relief valves after any adjustments have been made Allow only certified personnel to work on relief devices
Piping	Overpressure caused by reaction between process material and old material held up in a seldom used piping run	Train operators as to potential incompatibilities between process materials Blind off unused sections of pipe

(Continued)

TABLE 5-1 (Continued)

Location of Accidental Release	Cause	Potential Controls
Flange	Improper installation	Replace with a welded joint Maintenance training Establish a check procedure to certify work before operating Alternate variety of gasket
Pipe fitting	Poor weld	Radiographic weld test Personnel training
Threaded joint	Improper installation	Replace with a flanged or welded joint Weld seal threaded joints Leak or pressure test all critical joints Personnel training
Valve	Overpressure caused by water hammer	Operator training
Loading/Unloading operations	Mixing incompatible materials	Establish procedures for verifying heel composition Dedicate tanks for only one use
	Truck leaves without disconnecting hose	Block wheels during operation Install remotely operated emergency shutoff upstream of flexible line
Instrument Malfunction		
General equipment failure	Loss of reaction control as a result of inadequate control system performance	Install additional control systems where none are employed Redundancy of key components Upgrade components to improve accuracy, reliability, repeatibility or response time Backup the entire control system Simplify to improve operability Alter the variables that are monitored to more closely monitor the hazard Add trip systems Redesign the system to conform to present standards
	Loss of reaction control as a result of failure of an adequately designed control system	Install alarms For critical areas, install alarms that are audibly distinct Provide emegency backup systems such as emergency cooling Periodically inspect and test control systems Design so that the failure of one component will not result in a total control system failure Use only fail-safe equipment For critical areas, regularly replace portions of the system to prevent an on-line failure Use of non-interruptible power supply

(Continued)

TABLE 5-1 (Continued)

Location of Accidental Release	Cause	Potential Controls
Equipment Failure		
General equipment failure	Leak and ignition from hot oil system	Substitute with a nonflammable or less flammable material Locate furnace away from process
	Vibration	Additional supports Shortened piping (lower the vibration frequency) Pulsation dampeners Pipe loop to allow for expansion Determine if mechanical problem and repair it
	Loss of process control as a result of internal valve mechanism failure	A more reliable type of valve Periodically test valves
	Loss of process control as a result of check valve failure	Multiple check valves More sophisticated backflow protection Regularly inspect check valve internals
Vessel	Overpressure as a result of relief device failure	Size pressure relief devices using accepted practices Consider two-phase flow when sizing Install backup relief device Install a trip system as a first line of defense Design the vessel to fail in a manner that minimizes the release Enclose the vessel in a protective enclosure
Vessel shell	Corrosion	Fabricate the tank with an additional corrosion allowance Upgrade the materials of construction Use double walled construction Seal all insulation around fittings
Heat exchanger tube	Overpressure of trapped fluid	Shell and tube side pressure relief Periodic inspection of tube integrity
Drain line from storage tank	Valve failure	Add remotely operated emergency isolation valve Add a second drain valve in the line Limit the diameter of the drain line to 3/4 inch to restrict flow
Pipe break	Thermal expansion	Install a pipe loop to allow for expansion Use supports that allow for lateral movement
Bolted or riveted seams	Stress corrosion	Use welded seams

(Continued)

TABLE 5-1 (Continued)

Location of Accidental Release	Cause	Potential Controls
Welded seam	Poor fabrication	Radiographic weld testing Leak and/or pressure test equipment Use ASME code vessels
	Corrosion	Use ASME code vessels Use alternate type of weld Apply protective measures specific to the chemicals and type of corrosion involved
	Thermal shock Failure due to temperature or pressure cycles	Use ASME code vessels Set controls to restrict the rate and frequency of heating and cooling cycles Regularly replace crucial equipment before failure
Threaded joint	Corrosion	Replace with a flanged or welded joint Weld seal necessary threaded joints
Valve stem	Wear Overpressure	Tighten packing Choose a more reliable type of valve
Valve	Overpressure caused by water hammer	Limit the closing rate of the valve
Pump or compressor	Seal failure	Upgrade type of seal Install double mechanical seal with sealing fluid Enclose the pump or compressor in a ventilated enclosure Substitute with a sealless variety of pump
Design Error		
General equipment failure	Overpressure as a result of a valved off relief device	Interlock two parallel relief devices so that only one may be closed at a time
Vessel	Overpressure as a result of overfill	Restrict fillrate by limiting the size of fill lines Use identical capacity for fill lines and drain lines
	Loss of reaction control as a result of a loss of mixing	Monitor agitation directly by monitoring the mechanical drive assembly on the mixer or indirectly by monitoring temperature or flow
	Overpressure as a result of undersized relief device	Size using accepted methods Resize every time a system is used for a new service Consider the potential for two phase flow

(Continued)

TABLE 5-1 (Continued)

Location of Accidental Release	Cause	Potential Controls
Vessel (continued)	Internal tank explosion caused by a static discharge in the presence of a flammable mixture	Deflector plates at liquid entry Nitrogen blanketing Restricted feedrate Properly ground all equipment Use of explosion pressure relief
	Tank collapse as a result of emptying or cooling too quickly	Install breather vent or nitrogen blanketing Restrict the maximum possible empty or cooling rate
	Stress failure of a nozzle caused by inadequate pipe support	Added support for piping Additional reenforcement around the nozzle Wherever possible, place vessel fittings above the normal liquid level
Heat exchanger tube	Overpressure of trapped fluid	Provide shell and tube side pressure relief
Vent or vented enclosure system or vented enclosures	Internal explosion caused by ignition of a flammable mixture	Purge vents with inert gas Dilute with inert gas until the concentration is below the flammable limit Monitor for flammable mixtures
Piping	Incorrect materials of construction	Proper material specifications based on lab and pilot testing Inspect all parts to validate the material of construction and their integrity before installation during construction or maintenance
	Overpressure as a result of a reaction between process material and old material held up in a seldom used piping run	Simplify piping, eliminate dead ends or seldom used sections of pipe Install overpressure protection where mixing of incompatible materials is likely to occur
Pump or compressor	Blocked discharge, resulting in an overpressure	Provide upgraded overpressure protection Monitor flow
		Monitor temperature of the fluid in the pump case
Loading/Unloading operation	Truck overfill	Install remotely operated shutoff valves Install level monitoring controls

6. Guide to Facility Inspections

The preceding sections of this manual have broadly discussed the identification, evaluation, and control of hazards. In addition to procedures already discussed, there remains the actual inspection of physical facilities.

This section of the manual presents a brief example guide to facility inspections. General procedures including one possible approach to setting up and conducting an inspection is discussed. Specific procedures involving detailed inspection are also covered. Regulators may find this material useful as a guide to establishing their own procedures for facility inspections. Companies may find the information useful for the same purpose or for comparison with existing procedures. In either case, while the broad aspects of the procedures are applicable to most facilities that handle toxic chemicals, there may be variations in detail that are site-specific. These procedures are intended to initiate a thought process for an inspector knowledgeable about chemical processes.

6.1 BACKGROUND

Inspection of process facilities is an inherent part of reducing the probability of accidental chemical releases. Such inspections may be carried out by regulatory agencies or by companies themselves. The kind of inspection will usually differ for these two bodies; an inspection by a regulatory agency will generally be less detailed and focus on a few key items, whereas that by company will generally be more detailed and broader.

The hazard identification procedures discussed earlier can be used to identify key areas for inspection. An inspection should include all

functional parts of a physical facility where a release could conceivably
occur. These parts were generally identified in Figure 1-4, Section 1 of this
manual.

The purpose of an inspection by a regulatory agency is to ensure that
equipment and procedures for prevention and control of accidental releases are
consistent with what a company has reported to the agency, and to determine if
there may have been oversights in critical areas. The regulatory agency
inspection will primarily identify problem areas. The purpose of an inspection by a company is similar, except that the time and resources available are
likely to allow a more detailed look at the facility. The company's inspection will be much more focused on setting priorities for specific corrective
actions rather than just identifying problem areas.

6.2 GENERAL PROCEDURE

There are many ways to set up an inspection evaluation protocol and conduct
inspections. This section discusses one possible approach.

The overall procedure involves the following steps:

- Establishment of an inspection team;

- Determination of preliminary information requirements and
  acquisition of information including preparation of a facility
  questionnaire if necessary. The questionnaire will be completed by responsible parties at the facility to be inspected
  and returned to the inspection team if the team is from
  outside the facility;

- Review of the questionnaire and other preliminary information,
  including process flow diagrams, piping and instrumentation
  diagrams, operating manuals, and descriptions of maintenance

records to become familiar with the facility and identify critical areas for special attention;

- Meeting between the outside inspection team and a team of plant personnel including management to arrange a specific inspection plan, or planning meeting for the internal inspection team;

- A walk-through overview tour of the facility;

- In-depth inspections of specific areas according to the considerations discussed below in Subsection 6.2 and any other specific considerations that may be appropriate;

- A meeting between the outside inspection team and plant personnel at the conclusion of the inspection to review results, or a meeting of the internal inspection team for the same purpose;

- Preparation of a written inspection report which:

-- states the purpose and summarizes the findings,

-- identifies hazards found during the inspection,

-- discusses the risk implications of those hazards, and

-- discusses possible remedies to correct deficiencies and reduce hazards.

6.3   SPECIFIC PROCEDURES

Specific inspection procedures involve detailed examination of all parts
of the process facility where toxic materials are used, manufactured, or
stored. This inspection covers a number of specific elements which should
include, but are not necessarily limited to:

- Process characteristics and process chemistry,
- Facility siting,
- Plant layout,
- Pressure relief systems,
- Maintenance and structural integrity,
- Fire protection,
- Electrical system,
- Transportation practices, and
- Contingency plan and emergency response coordination.

Details of specific considerations in each of these areas are presented
in Appendix B of this manual. It must be emphasized that these are guidelines
only, and are representative of, but not necessarily all inclusive of, the
kinds of considerations and observations that should be made during inspections. The technical literature presents many other examples of how an
inspection protocol would be set up.

7. Costs of Accidental Release Prevention

With a commitment to and the knowledge required for accidental release prevention in place, the inevitable question is the cost of hazard identification and evaluation procedures, inspections, and control measures themselves. It is beyond the scope of this manual to exhaustively analyze the economic implications of accidental release prevention, but some general, very rough costs can be presented to provide a feel for the significance of economic issues.

This section provides cost data associated with the various hazard identification and evaluation procedures, inspections, and control technologies presented in previous sections of this manual. An example of the costs associated with the application of different combinations of controls to a specific system is illustrated. The implications of the costs of different control options is also discussed in light of policy planning.

7.1 COSTS OF HAZARD IDENTIFICATION, EVALUATION, AND INSPECTION

Attempting to assign specific costs to these activities is extremely difficult without specifics on the number, size, and type of facilities involved. As was shown in Table 4-1, formal hazard identification and evaluation procedures can be broadly classified in relative terms as low, medium, and high cost activities. Costs for any procedure depend on the complexity and size of the system being analyzed, and on the quality of the initial information on the system.

The literature from which the table was developed attempted to assign staffing and time requirements to each of the procedures listed. Careful analysis of these requirements suggests that some approximate costs can be developed as a lower bound for such activities. Results are shown in Table 7-1.

Costs for inspections are also highly variable, and again depend on system complexity, size, and quality of the initial information available. A minimal inspection might take at least four hours by one or two people, and a more detailed inspection might last two to three weeks with a team of three or four people. On this basis, estimated costs are as presented in Table 7-2.

These costs are intended as rough guidelines only, and can be expected to vary significantly in specific situations.

7.2 COSTS OF CONTROL TECHNOLOGIES

Costs of control technologies range from the costs of an individual component such as an additional thermocouple to costs of a complete alternatively designed process system for handling a toxic chemical. A fundamental concept in evaluating the costs of control technologies for accidental release prevention is that increased safety may result from increasing levels of controls. As controls are added to a system, costs will increase. It is possible to compare costs for systems with different levels of controls and evaluate the relative improvements in safety that might result. This could be done through estimates of the reduction in accident probability by quantitative fault or event tree analyses.

Costs of control technologies are addressed in more detail in a companion volume in this series on control technologies. The ensuing discussion presents an overview of control technology cost issues.

Table 7-3 summarizes cost ranges for some individual components that could be involved in enhancing the safety of a facility handling toxic chemicals, whether a storage or process facility. Costs for such enhancement could

TABLE 7-1.   ESTIMATED LOWER BOUND COSTS FOR VARIOUS HAZARD IDENTIFICATION
AND EVALUATION PROCEDURES[a]

Basis: One process unit[b]

Procedure	Staffing (personnel)	Time[c] (days or weeks)	Cost[d] ($)
Checklist	1	1d–2w	1,040 – 10,400
Safety Review	3	3d–3w	4,680 – 23,400
Ranking Procedures	3	3d–1w	4,680 – 7,800
Preliminary Hazard Analysis	3	1d–1w	1,560 – 7,800
What–If Analysis	3	3d–1w	4,680 – 7,800
Hazard and Operability Study (HAZOP)	4	2w–6w	20,800 – 62,400
Failure Mode, Effects, and Criticality Analysis (FMECA)	2	2w – 6w	10,400 – 31,200
Fault Tree Analysis (FTA)	4	2w – 6w	20,800 – 62,400
Event Tree Analysis	4	2w – 6w	20,800 – 62,400
Cause Consequence Analysis	4	1w – 6w	10,400 – 62,400
Human Error Analysis	1	1w	2,600

[a]Based on staffing and time estimates in Reference 12.

[b]"One process unit" is roughly defined as a process system consisting of from one to perhaps three major unit operations (e.g., a complex chemical reactor system, or a simple reactor, a few heat exchangers, and a distillation column).   The basis for the time and staffing for the various methods was not well defined in Reference 12.

[c]"d" = days;   "w" = weeks

[d]Based on an average loaded rate of $65/hour per staff person.

TABLE 7-2.   ESTIMATED COSTS FOR TYPICAL INSPECTIONS[a]

Team Size (Personnel)	Preparation	Time (Days) On-Site	Reporting	Cost ($)
3	0.5	0.5	1.0	3,120
3	2	5	2	14,040
3	5	15	10	46,800

[a] Preparation time, on-site, and reporting for team only based on loaded labor rate of $65/hour. Does not include plant personnel time required to assist inspection team. Does not include team travel costs or subsistence costs, if any, at site.

TABLE 7-3. COSTS OF SOME INDIVIDUAL INSTRUMENTATION AND CONTROL
COMPONENTS FOR PROCESS SYSTEM SAFETY MODIFICATIONS[a]

	Capital Cost Range ($)	Annual Cost Range ($/yr)
Flowmeters	2,500 – 5,100	400 – 800
Flow indicators	400 – 1,000	60 – 150
Check valves	400 – 600	60 – 90
Pressure sensors	200 – 500	30 – 80
Pressure indicators	200 – 600	30 – 100
Rupture disks	150 – 300	30 – 50
Relief valves	7,000 – 12,000	600 – 1,000
Temperature sensors	200 – 300	30 – 50
Indicators	1,000 – 1,700	150 – 250
Auxiliary cooling water capacity	30 – 80 per gpm	5 – 12 per gpm
Auxiliary refrigerated brine capacity	3,000 – 8,000 per ton	450 – 1,200 per ton
Load cell systems	12,000 – 16,000	2,400 – 3,200
Level detection systems	1,100 – 15,000	80 – 2,300
Flow switches	ca. 500	ca. 80
Pressure switches	ca. 500	ca. 80
Density measurement	500 – 5,000	90 – 900
pH measurement	4,000 – 5,000	700 – 900
Viscosity measurement	2,000 – 12,000	350 – 2,200

(Continued)

TABLE 7-3 (Continued)

	Capital Cost Range ($)	Annual Cost Range ($/yr)
Chemical species analyzers[b]	700 – 40,000	130 – 7,300
Controllers	800 – 6,000	70 – 500
Control valves	3,000 – 6,000	500 – 900
Complete control loops	6,000 – 15,000	300 – 1,300

[a]Basis: All costs were based on specific standard sizes which are documented
          in a companion manual in this series on control technologies. Costs
          are intended as rough guides to general magnitude only.

[b]Specific types will be found in the manual on control technologies.

involve something as simple as adding a sensor where one was not previously used, or adding an extra one for backup, to something as complex as the design and implementation of an entirely different control system.

More extensive application of control technologies could involve numerous changes in individual components, and even complete add-on subsystems such as scrubbers. Table 7-4 presents an example of how different levels or combinations of controls might affect the costs of a toxic gas storage tank system, based on the conceptual drawing in Figure 7-1. This type of analysis can be applied to any system where the evaluation of the cost implications of accidental release control measures are desired.

All of these costs are to illustrate concepts and the general magnitude of costs. Actual costs will vary widely depending on the size and type of facility, exact specifications of equipment used, and the individual organization that builds or modifies the facility.

7.3   COST IMPLICATIONS FOR POLICY PLANNING

Costs of identification, evaluation, inspection, and control technologies can be used to roughly estimate costs of various programs for accidental release prevention and control by either regulatory agencies or companies. The costs given previously in Tables 7-1 and 7-2 can be used to estimate time requirements that might be associated with some of the regulatory aspects of hazard control plans evaluations by regulation, and compliance and preparation of plans by affected companies. Cost estimates such as shown in Table 7-4 provide a measure of what specific regulatory requirements involving process equipment would cost companies, excluding, of course, many indirect costs. This methodology can also be used to provide companies with a rough measure of cost impacts of various process control options.

A complete analysis of cost implications of accidental release control is beyond the scope of this manual, but the overview presented here should be useful for preliminary planning purposes.

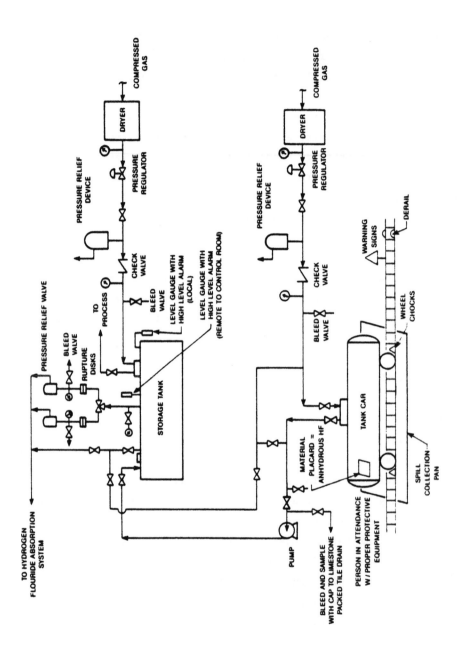

Figure 7-1.   Example of a toxic gas storage system.

TABLE 7-4. EXAMPLES OF CONTROL COSTS FOR A TOXIC
STORAGE SYSTEM[a]

Prevention/Protection Measure	Capital Cost (1986 $)	Annual Cost (1986 $/yr)
Continuous moisture monitoring	7,500-10,000	900-1,300
Flow control loop	4,000-6,000	500-750
Temperature sensor	250-400	30-50
Pressure relief		
- relief valve	1,000-2,000	120-250
- rupture disk	1,000-1,200	120-150
Interlock system for flow shut-off	1,500-2,000	175-250
pH monitoring of cooling water	7,500-10,000	900-1,300
Alarm system	250-500	30-75
Level sensor		
- liquid level gauge	1,500-2,000	175-250
- load cell	10,000-15,000	1,300-1,900
Diking (based on a 10,000 gal. tank)		
- 3 ft. high	1,200-1,500	150-175
- top of tank height, 10 ft.	7,000-7,500	850-900
Increased corrosion inspection		200-400

[a]See Table 7-5 for design basis. Details of estimating procedures are
provided in a companion manual on control technologies of this series.

TABLE 7-5. SPECIFICATIONS ASSOCIATED WITH TOXIC STORAGE SYSTEM

Item	Specification
Continuous moisture monitoring	Capacitance or infrared absorption system
Flow control loop	2-inch globe control valve, Monel® trim, flowmeter and PID controller
Temperature sensor	Thermocouple and associated thermowell
Pressure relief - relief valve	1-inch x 2-inch, Class 300 inlet and outlet flange, angle body, closed bonnet with screwed cap, carbon steel body, Monel® trim
- rupture disk	1-inch Monel® disk and carbon steel holder
Interlock system for flow shut-off	Solenoid valve, switch, and relay system
pH monitoring system	Electrode, electrode chamber, amplifier - transducer and indicator
Alarm system	Indicating and audible alarm
Level sensor - liquid level gauge - load cell	Differential pressure level gauge Electronic load cell
Diking - 3 ft. high	6-inch reinforced concrete walls, 5.2 yd^3
- 10 ft. high	10-inch reinforced concrete walls, 36.3 yd^3

8. References

1.   Lees, F.P. Loss Prevention in the Process Industries. Butterworth's, London, England, 1980.

2.   Industrial Economics, Inc., et al. Acute Hazardous Events Data Base, Executive Summary. EPA-560/5-85-029(a), U.S. Environmental Protection Agency, 1985.

3.   One-Hundred Largest Losses, A Thirty-Year Review of Property Damage Losses in the Hydrocarbon-Chemical Industries. Ninth Edition, Marsh and McLennan Protection Consultants, Chicago, IL, 1986.

4.   Kletz, T.A. Talking About Safety. The Chemical Engineer, April 1976.

5.   Chemical Emergency Preparedness Program, Interim Guidance. Revision 1, 9223.0-1A, U.S. Environmental Protection Agency, Washington, DC, 1985.

6.   Hazard Survey of the Chemical and Allied Industries. Technical Survey No. 3, American Insurance Association, 1979.

7.   Kirk, R.E. and D.F. Othmer. Encyclopedia of Chemical Technology. Third Edition, John Wiley & Sons, Inc., 1980.

8.   Weast, R.C. (ed.). CRC Handbook of Chemistry and Physics. 63rd Edition, CRC Press, Inc., Boca Raton, FL, 1982.

9.   Green, D.W. (ed.). Perry's Chemical Engineers' Handbook. Sixth Edition, McGraw-Hill Book Company, New York, NY, 1979.

10. Dean, J. (ed.). Lange's Handbook of Chemistry. Twelfth Edition, McGraw-Hill Book Company, New York, NY, 1979.

11. NIOSH/OSHA Pocket Guide to Chemical Hazards. DHEW (NIOSH) Publication No. 78-210, September 1985.

12. Battelle Columbus Division. Guidelines for Hazard Evaluation Procedures. The Center for Chemical Plant Safety, American Institute of Chemical Engineers, New York, NY, 1985.

13. Process Safety Management, Control of Acute Hazards. Chemical Manufacturers' Association, Washington, DC, May 1985.

14. Kletz, T.A. Eliminating Potential Process Hazards. Chemical Engineering, April 1, 1985.

15. Ozog, H. Hazard Identification, Analysis and Control. Chemical Engineering, February 18, 1985.

16. McNaughton, D.J. et al. Evaluation and Assessment of Models for Emergency Response Planning. TRC Environmental Consultants, Inc., Hartford, CT, February 1986.

17. Fire and Explosion Index. Hazard Classification Guide. Fourth Edition. Dow Chemical Company, Midland, MI, 1976.

18. Fisher, H.G. DIERS Research Program on Emergency Relief Systems. Chemical Engineering Progress, August 1985.

19. Chemical Emergency Preparedness Program, Interim Guidance. U.S. EPA, Revision 1, 922.3.01A.

20. National Fire Codes, 1986, National Fire Protection Association, Quincy, MA.

21. UL Handbook for Fire Ratings. Underwriters' Laboratories, Chicago, IL.

22. Perry, R.H. and Chilton, C.H. Chemical Engineer's Handbook. Fifth Edition, McGraw-Hill Book Company, New York, NY, 1973.

Appendix A: List of Acutely Toxic Chemicals

This appendix contains a listing of acutely toxic chemicals which appeared in a U.S. EPA publication addressing community preparedness for chemical emergencies (19). There are actually two separate lists entitled "Acutely Toxic Chemicals" and "Other chemicals." Some of the chemicals on the second list are also acutely toxic. Both lists appear separately in this appendix. At the time this manual was prepared word was received that three chemicals from the second list had been deleted. The modified list actually is the one used in this appendix.

11/01/85

Acutely Toxic Chemicals
Alphabetic List of Common Names and CAS Numbers

Common Name	CAS Number
Acetone cyanohydrin	00075-86-5
Acetone thiosemicarbazide	01752-30-3
Acrolein	00107-02-8
Acrylyl chloride	00814-68-6
Aldicarb	00116-06-3
Aldrin	00309-00-2
Allyl alcohol	00107-18-6
Allylamine	00107-11-9
Aluminum phosphide	20859-73-8
Aminopterin	00054-62-6
Amiton	00078-53-5
Amiton oxalate	03734-97-2
Ammonium chloroplatinate	16919-58-7
Amphetamine	00300-62-9
Aniline, 2,4,6-trimethyl-	00088-05-1
Antimony pentafluoride	07783-70-2
Antimycin A	01397-94-0
Antu	00086-88-4
Arsenic pentoxide	01303-28-2
Arsenous oxide	01327-53-3
Arsenous trichloride	07784-34-1
Arsine	07784-42-1
Azinphos-ethyl	02642-71-9
Azinphos-methyl	00086-50-0
Bacitracin	01405-87-4
Benzal chloride	00098-87-3
Benzenamine, 3-(trifluoromethyl)-	00098-16-8
Benzene, 1-(chloromethyl)-4-nitro-	00100-14-1
Benzenearsonic acid	00098-05-5
Benzenesulfonyl chloride	00098-09-9
Benzotrichloride	00098-07-7
Benzyl chloride	00100-44-7
Benzyl cyanide	00140-29-4
Bicyclo[2.2.1]heptane-2-carbonitrile, 5-chloro...	15271-41-7
Bis(chloromethyl) ketone	00534-07-6
Bitoscanate	04044-65-9
Boron trichloride	10294-34-5
Boron trifluoride	07637-07-2
Boron trifluoride compound with methyl ether (1:1)	00353-42-4
Bromadiolone	28772-56-7
Butadiene	00106-99-0
Butyl isovalerate	00109-19-3
Butyl vinyl ether	00111-34-2
C.I. basic green 1	00633-03-4
Cadmium oxide	01306-19-0
Cadmium stearate	02223-93-0
Calcium arsenate	07778-44-1
Camphechlor	08001-35-2
Cantharidin	00056-25-7
Carbachol chloride	00051-83-2

11/01/85

Acutely Toxic Chemicals
Alphabetic List of Common Names and CAS Numbers

Common Name	CAS Number
Carbamic acid, methyl-, O-[[(2,4-dimethyl...	26419-73-8
Carbofuran	01563-66-2
Carbophenothion	00786-19-6
Carvone	02244-16-8
Chlordane	00057-74-9
Chlorfenvinfos	00470-90-6
Chlorine	07782-50-5
Chlormephos	24934-91-6
Chlormequat chloride	00999-81-5
Chloroacetaldehyde	00107-20-0
Chloroacetic acid	00079-11-8
Chloroethanol	00107-07-3
Chloroethyl chloroformate	00627-11-2
Chloromethyl ether	00542-88-1
Chloromethyl methyl ether	00107-30-2
Chlorophacinone	03691-35-8
Chloroxuron	01982-47-4
Chlorthiophos	21923-23-9
Chromic chloride	10025-73-7
Cobalt	07440-48-4
Cobalt carbonyl	10210-68-1
Cobalt, [[2,2'-[1,2-ethanediylbis(nitrilomethy...	62207-76-5
Colchicine	00064-86-8
Coumafuryl	00117-52-2
Coumaphos	00056-72-4
Coumatetralyl	05836-29-3
Cresylic acid	00095-48-7
Crimidine	00535-89-7
Crotonaldehyde	00123-73-9
Crotonaldehyde	04170-30-3
Cyanogen bromide	00506-68-3
Cyanogen iodide	00506-78-5
Cyanophos	02636-26-2
Cyanuric fluoride	00675-14-9
Cycloheximide	00066-81-9
Cyclopentane	00287-92-3
Decaborane(14)	17702-41-9
Demeton	08065-48-3
Demeton-S-methyl	00919-86-8
Dialifos	10311-84-9
Diborane	19287-45-7
Dibutyl phthalate	00084-74-2
Dichlorobenzalkonium chloride	08023-53-8
Dichloroethyl ether	00111-44-4
Dichloromethylphenylsilane	00149-74-6
Dichlorvos	00062-73-7
Dicrotophos	00141-66-2
Diepoxybutane	01464-53-5
Diethyl chlorophosphate	00814-49-3
Diethyl-p-phenylenediamine	00093-05-0

11/01/85

Acutely Toxic Chemicals
Alphabetic List of Common Names and CAS Numbers

Common Name	CAS Number
Diethylcarbamazine citrate	01642-54-2
Digitoxin	00071-63-6
Diglycidyl ether	02238-07-5
Digoxin	20830-75-5
Dimefox	00115-26-4
Dimethoate	00060-51-5
Dimethyl phosphorochloridothioate	02524-03-0
Dimethyl phthalate	00131-11-3
Dimethyl sulfate	00077-78-1
Dimethyl sulfide	00075-18-3
Dimethyl-p-phenylenediamine	00099-98-9
Dimethyldichlorosilane	00075-78-5
Dimethylhydrazine	00057-14-7
Dimetilan	00644-64-4
Dinitrocresol	00534-52-1
Dinoseb	00088-85-7
Dinoterb	01420-07-1
Dioctyl phthalate	00117-84-0
Dioxathion	00078-34-2
Dioxolane	00646-06-0
Diphacinone	00082-66-6
Diphosphoramide, octamethyl-	00152-16-9
Disulfoton	00298-04-4
Dithiazanine iodide	00514-73-8
Dithiobiuret	00541-53-7
EPN	02104-64-5
Emetine, dihydrochloride	00316-42-7
Endosulfan	00115-29-7
Endothion	02778-04-3
Endrin	00072-20-8
Ergocalciferol	00050-14-6
Ergotamine tartrate	00379-79-3
Ethanesulfonyl chloride, 2-chloro-	01622-32-8
Ethanol, 1,2-dichloro-, acetate	10140-87-1
Ethion	00563-12-2
Ethoprophos	13194-48-4
Ethyl thiocyanate	00542-90-5
Ethylbis(2-chloroethyl)amine	00538-07-8
Ethylene fluorohydrin	00371-62-0
Ethylenediamine	00107-15-3
Ethyleneimine	00151-56-4
Ethylmercuric phosphate	02235-25-8
Fenamiphos	22224-92-6
Fenitrothion	00122-14-5
Fensulfothion	00115-90-2
Fluenetil	04301-50-2
Fluorine	07782-41-4
Fluoroacetamide	00640-19-7
Fluoroacetic acid	00144-49-0
Fluoroacetyl chloride	00359-06-8

11/01/85

Acutely Toxic Chemicals
Alphabetic List of Common Names and CAS Numbers

Common Name	CAS Number
Fluorouracil	00051-21-8
Fonofos	00944-22-9
Formaldehyde cyanohydrin	00107-16-4
Formetanate	23422-53-9
Formothion	02540-82-1
Formparanate	17702-57-7
Fosthietan	21548-32-3
Fuberidazole	03878-19-1
Furan	00110-00-9
Gallium trichloride	13450-90-3
Hexachlorocyclopentadiene	00077-47-4
Hexachloronaphthalene	01335-87-1
Hexamethylenediamine, N,N'-dibutyl-	04835-11-4
Hydrazine	00302-01-2
Hydrocyanic acid	00074-90-8
Hydrogen fluoride	07664-39-3
Hydrogen selenide	07783-07-5
Indomethacin	00053-86-1
Iridium tetrachloride	10025-97-5
Iron, pentacarbonyl-	13463-40-6
Isobenzan	00297-78-9
Isobutyronitrile	00078-82-0
Isocyanic acid, 3,4-dichlorophenyl ester	00102-36-3
Isodrin	00465-73-6
Isofluorphate	00055-91-4
Isophorone diisocyanate	04098-71-9
Isopropyl chloroformate	00108-23-6
Isopropyl formate	00625-55-8
Isopropylmethylpyrazolyl dimethylcarbamate	00119-38-0
Lactonitrile	00078-97-7
Leptophos	21609-90-5
Lewisite	00541-25-3
Lindane	00058-89-9
Lithium hydride	07580-67-8
Malononitrile	00109-77-3
Manganese, tricarbonyl methylcyclopentadienyl	12108-13-3
Mechlorethamine	00051-75-2
Mephosfolan	00950-10-7
Mercuric acetate	01600-27-7
Mercuric chloride	07487-94-7
Mercuric oxide	21908-53-2
Mesitylene	00108-67-8
Methacrolein diacetate	10476-95-6
Methacrylic anhydride	00760-93-0
Methacrylonitrile	00126-98-7
Methacryloyl chloride	00920-46-7
Methacryloyloxyethyl isocyanate	30674-80-7
Methamidophos	10265-92-6
Methanesulfonyl fluoride	00558-25-8
Methidathion	00950-37-8

11/01/85

Acutely Toxic Chemicals
Alphabetic List of Common Names and CAS Numbers

Common Name	CAS Number
Methiocarb	02032-65-7
Methomyl	16752-77-5
Methoxyethylmercuric acetate	00151-38-2
Methyl 2-chloroacrylate	00080-63-7
Methyl chloroformate	00079-22-1
Methyl disulfide	00624-92-0
Methyl isocyanate	00624-83-9
Methyl isothiocyanate	00556-61-6
Methyl mercaptan	00074-93-1
Methyl phenkapton	03735-23-7
Methyl phosphonic dichloride	00676-97-1
Methyl thiocyanate	00556-64-9
Methyl vinyl ketone	00078-94-4
Methylhydrazine	00060-34-4
Methylmercuric dicyanamide	00502-39-6
Methyltrichlorosilane	00075-79-6
Metolcarb	01129-41-5
Mevinphos	07786-34-7
Mexacarbate	00315-18-4
Mitomycin C	00050-07-7
Monocrotophos	06923-22-4
Muscimol	02763-96-4
Mustard gas	00505-60-2
Nickel	07440-02-2
Nickel carbonyl	13463-39-3
Nicotine	00054-11-5
Nicotine sulfate	00065-30-5
Nitric acid	07697-37-2
Nitric oxide	10102-43-9
Nitrocyclohexane	01122-60-7
Nitrogen dioxide	10102-44-0
Nitrosodimethylamine	00062-75-9
Norbormide	00991-42-4
Organorhodium complex	PMN-82-147
Orotic acid	00065-86-1
Osmium tetroxide	20816-12-0
Ouabain	00630-60-4
Oxamyl	23135-22-0
Oxetane, 3,3-bis(chloromethyl)	00078-71-7
Oxydisulfoton	02497-07-6
Ozone	10028-15-6
Paraquat	01910-42-5
Paraquat methosulfate	02074-50-2
Parathion	00056-38-2
Parathion-methyl	00298-00-0
Paris green	12002-03-8
Pentaborane	19624-22-7
Pentachloroethane	00076-01-7
Pentachlorophenol	00087-86-5
Pentadecylamine	02570-26-5

11/01/85

Acutely Toxic Chemicals
Alphabetic List of Common Names and CAS Numbers

Common Name   CAS Number

Peracetic acid	00079-21-0
Perchloromethylmercaptan	00594-42-3
Phenarsazine oxide	00058-36-6
Phenol	00108-95-2
Phenol, 2,2'-thiobis(4-chloro-6-methyl-	04418-66-0
Phenol, 2,2'-thiobis[4,6-dichloro-	00097-18-7
Phenol, 3-(1-methylethyl)-, methylcarbamate	00064-00-6
Phenyl dichloroarsine	00696-28-6
Phenylhydrazine hydrochloride	00059-88-1
Phenylmercury acetate	00062-38-4
Phenylsilatrane	02097-19-0
Phenylthiourea	00103-85-5
Phorate	00298-02-2
Phosacetim	04104-14-7
Phosfolan	00947-02-4
Phosmet	00732-11-6
Phosphamidon	13171-21-6
Phosphine	07803-51-2
Phosphonothioic acid, methyl-, O-(4-nitrophenyl...	02665-30-7
Phosphonothioic acid, methyl-, O-ethyl O-[4-...	02703-13-1
Phosphonothioic acid, methyl-, S-[2-[bis...	50782-69-9
Phosphoric acid, dimethyl 4-(methylthio)phenyl...	03254-63-5
Phosphorous trichloride	07719-12-2
Phosphorus	07723-14-0
Phosphorus oxychloride	10025-87-3
Phosphorus pentachloride	10026-13-8
Phosphorus pentoxide	01314-56-3
Phylloquinone	00084-80-0
Physostigmine	00057-47-6
Physostigmine, salicylate (1:1)	00057-64-7
Picrotoxin	00124-87-8
Piperidine	00110-89-4
Piprotal	05281-13-0
Pirimifos-ethyl	23505-41-1
Platinous chloride	10025-65-7
Platinum tetrachloride	13454-96-1
Potassium arsenite	10124-50-2
Potassium cyanide	00151-50-8
Potassium silver cyanide	00506-61-6
Promecarb	02631-37-0
Propargyl bromide	00106-96-7
Propiolactone, .beta.-	00057-57-8
Propionitrile	00107-12-0
Propionitrile, 3-chloro-	00542-76-7
Propyl chloroformate	00109-61-5
Propylene glycol, allyl ether	01331-17-5
Propyleneimine	00075-55-8
Prothoate	02275-18-5
Pseudocumene	00095-63-6
Pyrene	00129-00-0

11/01/85

Acutely Toxic Chemicals
Alphabetic List of Common Names and CAS Numbers

Common Name	CAS Number
Pyridine, 2-methyl-5-vinyl-	00140-76-1
Pyridine, 4-amino-	00504-24-5
Pyridine, 4-nitro-, 1-oxide	01124-33-0
Pyriminil	53558-25-1
Rhodium trichloride	10049-07-7
Salcomine	14167-18-1
Sarin	00107-44-8
Selenium oxychloride	07791-23-3
Selenous acid	07783-00-8
Semicarbazide hydrochloride	00563-41-7
Silane, (4-aminobutyl)diethoxymethyl-	03037-72-7
Sodium anthraquinone-1-sulfonate	00128-56-3
Sodium arsenate	07631-89-2
Sodium arsenite	07784-46-5
Sodium azide (Na(N3))	26628-22-8
Sodium cacodylate	00124-65-2
Sodium cyanide (Na(CN))	00143-33-9
Sodium fluoroacetate	00062-74-8
Sodium pentachlorophenate	00131-52-2
Sodium selenate	13410-01-0
Sodium selenite	10102-18-8
Sodium tellurite	10102-20-2
Strychnine	00057-24-9
Strychnine, sulfate	00060-41-3
Sulfotep	03689-24-5
Sulfoxide, 3-chloropropyl octyl	03569-57-1
Sulfur tetrafluoride	07783-60-0
Sulfur trioxide	07446-11-9
Sulfuric acid	07664-93-9
TEPP	00107-49-3
Tabun	00077-81-6
Tellurium	13494-80-9
Tellurium hexafluoride	07783-80-4
Terbufos	13071-79-9
Tetraethyllead	00078-00-2
Tetraethyltin	00597-64-8
Tetranitromethane	00509-14-8
Thallic oxide	01314-32-5
Thallous carbonate	06533-73-9
Thallous chloride	07791-12-0
Thallous malonate	02757-18-8
Thallous sulfate	07446-18-6
Thallous sulfate	10031-59-1
Thiocarbazide	02231-57-4
Thiocyanic acid, (2-benzothiazolylthio)methyl...	21564-17-0
Thiofanox	39196-18-4
Thiometon	00640-15-3
Thionazin	00297-97-2
Thiophenol	00108-98-5
Thiosemicarbazide	00079-19-6

11/01/85

Acutely Toxic Chemicals
Alphabetic List of Common Names and CAS Numbers

Common Name	CAS Number
Thiourea, (2-chlorophenyl)-	05344-82-1
Thiourea, (2-methylphenyl)-	00614-78-8
Titanium tetrachloride	07550-45-0
Toluene 2,4-diisocyanate	00584-84-9
Toluene 2,6-diisocyanate	00091-08-7
Triamiphos	01031-47-6
Triazofos	24017-47-8
Trichloro(chloromethyl)silane	01558-25-4
Trichloro(dichlorophenyl)silane	27137-85-5
Trichloroacetyl chloride	00076-02-8
Trichloroethylsilane	00115-21-9
Trichloronate	00327-98-0
Trichlorophenylsilane	00098-13-5
Trichlorphon	00052-68-6
Triethoxysilane	00998-30-1
Trimethylchlorosilane	00075-77-4
Trimethylolpropane phosphite	00824-11-3
Trimethyltin chloride	01066-45-1
Triphenyltin chloride	00639-58-7
Tris(2-chloroethyl)amine	00555-77-1
Valinomycin	02001-95-8
Vanadium pentoxide	01314-62-1
Vinylnorbornene	03048-64-4
Warfarin	00081-81-2
Warfarin sodium	00129-06-6
Xylylene dichloride	28347-13-9
Zinc phosphide	01314-84-7
Zinc, dichloro[4,4-dimethyl-5-[[[(methylamino)...	58270-08-9
trans-1,4-Dichlorobutene	00110-57-6

OTHER CHEMICALS

Name	CAS Number
Acrylamide	79-06-1
Acrylonitrile	107-13-1
Adiponitrile	111-69-3
Ammonia	7664-41-7
Aniline	62-53-3
Bromine	7726-95-6
Carbon disulfide	75-15-0
Chloroform	67-66-3
Cyclohexylamine	108-91-8
Epichlorohydrin	106-89-8
Ethylene oxide	75-21-8
Formaldehyde	50-00-0
Hydrochloric acid	7647-01-0
Hydrogen peroxide	7722-84-1
Hydrogen sulfide	7783-06-4
Hydroquinone	123-31-9
Methyl bromide	74-83-9
Nitrobenzene	98-95-3
Phosgene	75-44-5
Propylene oxide	75-56-9
Sulfur dioxide	7446-09-5
Tetramethyl lead	75-74-1
Vinyl acetate monomer	108-05-4

Appendix B: Example of Detailed Procedures for Hazard Evaluation Facility Inspections

B.1 Process and Process Chemistry Evaluation

Purpose

The purpose of this evaluation is to identify the most critical areas in the process facility, based on the fundamental process chemistry and the sequence of unit processes and unit operations. Process chemistry considerations also include chemical reactions that might occur with materials that might unintentionally enter the process as contaminants.

Procedures

Procedures for this part of the evaluation include the following:

- Review of chemical equations for the process to identify fundamental hazard potential associated with basic process chemistry.

- Review of written process descriptions and process procedures to identify critical process areas or procedures.

- Review of process flow diagrams (PFD) and/or piping and instrumentation diagrams (PID) for the critical process areas identified above.

- Visual inspection of process areas, units, and specific critical equipment items for oversights or deficiencies using the information discussed in ensuing subsections as a guide.

- Comparison of process characteristics of the inspected process with other processes in the plant and the chemical industry as a whole to judge relative hazard potential.

- Application of various formal hazard identification and evaluation procedures to determine qualitatively how an accidental release might occur.

Key Factors

- Materials characteristics: evaluated to determine which materials in each process are potentially the most dangerous.

  -- Comparison of physical properties at both process and ambient conditions, including boiling point, melting point, vapor pressure, viscosity, and vapor density.

  -- Flammability as characterized by flashpoint, upper and lower explosive limits, and auto-ignition temperatures. Also, the chemical compatibility of mixtures in storage areas and during handling.

  -- Acute toxicity as characterized by the health effects of exposure by inhalation or skin contact.

  -- The evaluation of reactivity according to the following parameters:
     a. Whether reactions are exothermic, endothermic or thermodynamically balanced.
     b. Potential for uncontrolled reactions due to such things as decomposition, excessive temperature, backflow, or spontaneous polymerization.

    c.   Process reactivity with water or other possible
        contaminants.

    d.   Potential side reactions and conditions under which
        they are favored.

-- Corrosiveness considered in the context of the appropriateness of construction materials.

- Range of process conditions: the hazard potential is normally increased the larger the process and hence inventory of toxic material and the more severe the process conditions.

  -- Evaluation of process capacities in terms of operating throughput and in-process inventories of toxic materials.

  -- Classification of the reaction type based on a unit operation or unit process approach, for example, halogenation, polymerization, etc. Some types may show a history of more hazards than others.

  -- Categorization of reaction and separation process temperatures as low if less than 200°F, moderate between 200 and 500°F, or high temperature if above 500°F.

  -- Categorization of process pressures reaction and separation as low if less than 100 psig, moderate between 100 and 500 psig, or high pressure if above 500 psig.

  -- Evaluation of the process conditions relative to the physical properties of the chemicals.

- Mode of processing and process configuration: certain process types and configurations may be inherently more hazardous than

others and hazards vary according to process characteristics
(e.g., reaction time, volume, and type of reaction).

-- Evaluation of the implications of a process being batch,
   semi-batch, or continuous. For example a large inventory
   batch processes may be more of a hazard than a small
   inventory continuous processes.

-- Evaluation of process configuration in terms of operational sequences and physical placement of equipment.

-- Evaluation of process complexity in terms of the number
   of process steps and the nature of individual unit
   operations.

● Thermodynamics of Key Reactions

-- Exothermic reactions: considered more hazardous than
   endothermic or balanced reactions.

-- Qualitative evaluation of the hazard potential of an
   exothermic reaction according to whether it has a weak,
   moderate, or strong exotherm.

● Types of Instrumentation and Control Systems

-- Evaluation of the control system in terms of the
   reliability and responsiveness to deviations and
   corrective actions.

-- Consideration of relative reliability of pneumatic or
   electronic systems for given application.

-- Computer control systems: considered to be safer than reliance solely on manual controls, but computer systems should have some kind of manual backup.

-- Backup for increasing reliability: emergency cooling and heating systems, redundancy of the instruments, the instrument air supply, electrical power supply, and the computer system.

-- Evaluation of alarms and emergency shutdown systems for type, complexity, location and reliability.

● Process Isolation

-- Evaluation of the ability to isolate, quench, or dump the process materials in an emergency: location of shut-off valves and whether manual or automated shut-off is used.

-- Assessment of the possible effects of a fire, explosion, or release in a nearby process on the process being inspected.

● Operator Training

-- Evaluation of operator training for routine operations by review of operator training programs, operating manuals, and observations of and discussions with selected operators, and observations of operators carrying out their duties.

-- Evaluation of emergency awareness and preparedness for operator response to emergency situations including

drills and tests in the same general manner as for
routine operations.

B.2   Facility Siting Evaluation

Purpose

The purpose of this evaluation is to assess the potential impact of a
facility's location on the frequency or severity of an accidental release and
the vulnerability of the surrounding community to an accidental release.

Procedures

- Review of surrounding community: the evaluation of plant
  location relative to the surrounding community by direct
  observation and by maps and other written information.

- Review of climatic conditions: evaluation of the potential
  for severe climatic conditions and natural disasters from
  discussions with plant personnel, by direct observation and
  from government records for the area.

- Review of municipal utility reliability: evaluation of the
  reliability of municipal utilities (electricity, water, gas,
  etc.) from discussions with plant personnel, direct observations of main supply lines and information of past utility
  failures.

Key Factors

- Location in the community: evaluation of the plant location
  within the community in terms of proximity to other businesses
  and population centers, land use and terrain features in

surrounding areas, wind patterns, and surface groundwater in the area.

-- Other businesses and population centers near the plant are potential receptors of any accidental releases, fires, or explosions in the subject facility. The potential danger varies with the specific hazard and magnitude of that hazard.

-- The exact extent of the hazard zone depends on the quantity released. In general, however, as a rough guideline, receptors within a few hundred yards of a facility may be considered sensitive receptors, while receptors beyond about 2-5 miles may be considered to be on the outer limits of typical hazard zones. It is emphasized here, however that toxic materials can be dangerous in clouds traveling many miles, and these distances are only for rough evaluation purposes.

-- An estimate of total population, including temporary population such as workers during certain parts of the day, within various distances of the plant (i.e., plant shutdown and construction personnel): the basis for estimating the number of people outside the plant who might be affected by a release.

-- Especially sensitive receptors such as schools, parks, and hospitals.

-- Land use in surrounding areas: noted in the context of the considerations just discussed, and especially prominent terrain features such as physical plants (i.e., columns and reactors), rivers, forests, and hills which could affect the dispersion of airborne chemical releases.

-- Directions and speeds of seasonal and diurnal wind patterns for the facility. This is usually shown in a diagram called a windrose. Data for windroses are available for most major metropolitan areas and some other areas from the National Weather Service.

-- The presence of surface and groundwaters in the area: noted in the context of the potential for contamination by sudden accidental releases of chemicals.

-- Evaluation of the potential impact of an accidental release or other hazard coming from an adjacent business.

-- Traffic flow patterns around the perimeter of the facility: especially points of congestion and the potential impact of traffic congestion on the movement of emergency response equipment and evacuated plant personnel.

● Climatic conditions

-- Evaluation of the potential for flood, landslide, brush fire, earthquake, severe wind or hail, subfreezing temperatures, or other climatic conditions including consideration of how such events might cause an accidental release.

-- Assessment of the need for other protective measures to reduce the potential for an accidental release resulting from climatic conditions.

- Review of municipal utility reliability

  -- Review of past reliability of municipal utilities.

  -- Consideration of unusual causes for utility failure, including downed power lines resulting from a vehicle collision with power line poles, or the loss of all utilities because of an earthquake.

  -- Evaluation of backup utilities in the event of a facility-wide utility failure.

B.3 Facility Layout Evaluation

Purpose

The purpose of the plant layout evaluation is to determine if specific features of the layout could contribute to an accidental release. A primary consideration in the evaluation of the layout is the potential for an accidental release or other accident in one section of the facility to adversely affect other sections of the facility. As far as possible, each section within a facility should be protected from the effects of accidents in other sections of the facility.

Procedures

- Review of the total facility plot plan and plot plans of individual process areas.

- Tour of facility: to complement information obtained from plot plans.

- Discussions with plant personnel: information obtained directly from plant personnel, as required.

Key Factors

● Overall Plant Configuration

-- Review of the overall configuration of the facility, as well as individual process units.

-- Evaluation of the compatibility between materials in adjacent process units, especially considering how the possible release of a chemical, fire, or explosion in one unit could affect adjacent units and lead to additional accidental releases.

-- Evaluation of the compatibility of chemicals between process units and adjacent storage areas, and between the adjacent storage areas themselves

-- Evaluation of the location of individual facility process and storage areas relative to utility and other plant areas, considering such items as ignition sources, an incident in a process or storage area affecting a critical utility system, and potential effects on other areas from incidents in process and storage areas.

-- Evaluation of the distance between various parts of the facility and property lines and of special terrain features.

-- Process areas: should be well separated from utilities, storage, office, and laboratory areas.

-- Process and storage areas for flammable materials: should be in the prevailing downwind direction from ignition sources, or otherwise located away from ignition sources to the extent possible.

-- Hazardous units: separation distances from all critical areas such as control rooms and process computer installations should at least be similar to those specified for flammable materials as given in Lees (1) for example.

-- Administrative buildings and warehouses: preferably located at the periphery of the plant.

-- Control rooms: should be protected from potential fire or explosion damage and from the adverse affects of an accidental release. Where possible, process control rooms should be located at the perimeter of the unit they control.

• Spacing of Process and Storage Areas

-- Inter- and intra-unit spacing: consideration of the distance between risk areas, and between equipment and systems within risk areas.

-- Spacing of equipment: should consider the nature of the materials, quantity, operating conditions, sensitivity of the equipment, the need to combat fires, and the concentration of personnel and hardware valuables in a given area.

-- Storage tanks: should be reasonably spaced and appropriately diked. Applicable codes and standards should be adhered to as a minimum requirement.

-- Toxic materials in processes or storage areas: special considerations of spacing and isolation.

-- Easy isolation and containment of hazardous materials in an emergency: for example, is a critical shutoff valve too close to the area of immediate impact in an accident?

● Vehicular Access and Clearances

-- Entrances and exits to various facility areas: should be adequate and free from uncontrollable obstruction in an emergency. For example, would a rail car accident on plant property block the only access road in or out so that emergency equipment could not respond to such an accident?

-- Access: there should be a minimum of at least two means of access or egress to the facility and critical areas within the facility.

-- Overhead clearances: observed for possible collapse and obstruction of access or egress in emergencies; also observed as the possible cause of a chemical release incident due to a collision with vehicular traffic.

● Security

Security considerations are included in the facility layout evaluation because of the possibility of deliberate or accidental sabotage. The facility property should be fenced and access limited through gates under ready observation or direct control of facility security personnel.

B.4   <u>Pressure Relief System Evaluation</u>

<u>Purpose</u>

The purpose of this evaluation is to evaluate the adequacy of pressure relief systems designed to prevent rupture of vessels, pipelines, or equipment which would result in the uncontrolled release of toxic, explosive, or flammable materials.

<u>Procedures</u>

- Design and Procedures Review

  -- Review of the need for and extent of pressure relief systems by examining process flow diagrams, process and storage conditions, and process instrumentation diagrams.

  -- Review of maintenance and engineering records on relief systems.

  -- Questioning of appropriate plant personnel about current installations, practices, and procedures.

- Field Inspection

  -- Visual inspection of relief systems in a convenient priority sequence determined by a consideration of hazard potential, location in the plant, and inspection schedule.

  -- Examination of process equipment for the presence of protection and its overall adequacy.

-- Examination of individual relief devices and systems for proper configurations and specifications.

-- Checking of nominal pressure and temperature ratings on vessels and other equipment against actual use conditions.

-- Examination of the physical condition of pressure relief equipment.

Key Factors

● Evaluation of the appropriateness and applicability of the relief system includes the following considerations:

- Relief systems: should be in place and functional on all equipment where it is required by codes and standards, and other equipment where the hazard for rupture from overpressure exists.

-- Safety relief valves: should be provided on the discharge side of positive displacement pumps, between positive displacement compressors and block valves, between back-pressure turbine exhaust flanges and block valves, and on any equipment where liquid can be blocked in and later warmed, or where chemical reactions, external fire, overfilling, or other process malfunction could result in equipment internal overpressure.

-- Vacuum relief devices: should be used where vacuum drawn on equipment, if blocked in, could cause equipment collapse from external pressure.

-- Relief devices: should be of the proper type and
   specifications for the application. State-of-the-art
   equipment should be used where possible on equipment
   containing large inventories of toxic materials.

-- Specifications: including sizes, construction materials,
   relief set pressure, set pressure tolerances, and service
   temperature range. These must be compatible with and
   specific to process conditions.

-- Consideration should be given to the possibility of
   solids formation which could plug relief device inlets,
   outlets, and working mechanisms, (e.g. polymerization of
   monomer vapors from condensation on cold surfaces).

● Sizing

-- Sizing: based on the maximum relief rate after consideration of four relief situations:

   a. Fire exposure
   b. Reaction/decomposition overpressure
   c. Maximum fill rate
   d. Thermal expansion

-- Relief devices: should be sized using accepted procedures of the American Petroleum Institute (Recommended
   Practice for the Design and Installation of Pressure
   Relieving Systems in Refineries, Part I - Design, API,
   1973) the National Fire Protection Association (NFPA 30
   and NFPA 68), the American Society of Mechanical Engineers (ASME) Boiler and Pressure Vessel Code, Section
   VIII, consistent with type of service. Sizing for

reaction/decomposition cases can be done using methods developed by the Design Institute for Emergency Relief Systems (DIERS), which was sponsored and published by the American Institute of Chemical Engineers (AIChE) (18).

- Configuration

    -- Evaluation of the possible hazards of manifolding, such as discharge of incompatible materials.

    -- Overall installation: should provide for ease of removal, inspection, testing, and replacement of the relief devices.

    -- Relief devices: should not be blocked by shut-off valves upstream or downstream unless a fail-safe system is provided with parallel relief such that both relief lines cannot be out of service at the same time, or some other means is used to protect the shut-off valves from improper closure. The former is sometimes accomplished using a 3-way, 2-port valve upstream of dual relief valves such that only one can be isolated at a time. Protection from shut-off is accomplished using breakable seals or even locks on shut-off handles to indicate or prevent unauthorized closure.

    -- Outlets not leading into common manifolds or flares: should be directed in a safe manner. This means that they should not be directed toward personnel or equipment where the discharge could cause fire, explosion, serious contamination, or other accidents. This is especially important in the ignited pressure relief of flammable materials where impingement of flame on vessels or other equipment could have serious consequences.

-- Vents: should normally terminate outside of buildings and be at a height that minimizes exposure hazards.

-- Discharge piping: should be supported independently of the relief valve to withstand dynamic forces involved when relief valve opens.

-- Drain connections, weep holes, or rain guards: should be provided on relief discharge piping where rain, snow, or condensation could accumulate and plug the discharge. Low inertial covers may be provided in some cases.

-- Rupture disks used in series with safety relief valves: a pressure gauge or other pressure indicator should be provided between the rupture disk and relief valve to indicate disk integrity. A pressure reading indicates that the rupture disk must be replaced.

-- Blockage: auxiliary devices which might plug, such as check valves, or flame arresters should not be installed on relief system piping. Pipe plugs or caps should not be present on relief valves.

-- Backpressure effects: should be considered for effect on relief valve operation, especially when several valves discharge into a common system.

-- Location: Safety relief valves should be located as close as practical to the equipment they protect to minimize pressure drop and valve "chattering."

-- Inlet piping: Should not be less than same nominal pipe size as relief valve inlet.

-- Discharge piping: Should not be less than same nominal pipe size as relief valve outlet. Should be run as directly as possible with minimum changes of direction.

-- Common headers: Should have a cross sectional area at least equal to the total of the connected valves.

-- Pressure reducing stations: For steam, air gas, etc., should be fitted with safety relief valves on low pressure side. Safety relief valve should be sized to carry bull load of reducing valve and its bypass.

● Inspection and Testing

-- Safety relief valves: should be periodically inspected for structural integrity and for signs of corrosion or plugging. Materials that undergo polymerization, or are extremely corrosive may require especially frequent inspections.

-- Safety relief valve repair firms, including in-house testing units, should have a current certificate of authorization from the National Board of Boiler and Pressure Vessel Inspectors and have a "VP" stamp.

-- Pressure indicators on rupture disk installations: should be periodically checked for indications of rupture disk activation or leakage.

-- Safety relief valves: should be tested for opening at the required set pressure and for proper reseating.

-- Inspection frequencies: should follow a fixed schedule commensurate with the type of service and risk associated with a system failure. Safety relief valves should be inspected at least once in five years and as much as annually or more in high hazard applications.

-- Records of inspections and testing: should be maintained.

B.5 Maintenance and Structural Integrity Evaluation

Purpose

The purpose of this evaluation is to assess the adequacy of maintenance procedures and the structural integrity of equipment.

Procedures

Discussions with maintenance personnel are held and written maintenance procedures and records are examined. The physical facility as a whole and individual equipment items believed to be most hazardous are examined during a site tour. Maintenance equipment and facilities are also observed.

Key Factors

● Maintenance Organization and Scheduling System: the overall maintenance organization and scheduling system is reviewed with key maintenance personnel by examining written procedures and records.

-- Staffing levels should be appropriate for the size of the facility. Changes in staff levels are compared to changes in the physical facility. For example, has staffing kept pace with and is it of the appropriate type for recent plant modifications or additions?

-- Examination of staff qualifications and skill levels relative to the hazard potential of the specific operation.

-- Evaluation of the duties and responsibilities of contract versus in-house maintenance. In-house supervision or auditing of contract maintenance should be sufficient to ensure satisfactory contractor performance.

-- Review of systems for work order initiation and implementation are reviewed.

-- Review of procedures for maintenance scheduling and ranking to ensure that high hazard areas are properly recognized.

-- Evaluation of the extent and adequacy of preventive maintenance. The frequency of such maintenance is considered in light of the kinds of equipment and systems involved, their potential inherent reliability, and the consequences of failure.

● Inspection, Testing, and Monitoring Program (ITM Program)

-- Determination of the presence or lack of an inspection testing and monitoring program for process equipment and instrumentation.

-- Evaluation of the ITM program in terms of four areas:

    a. Vessels, piping, and other process equipment,

    b. Rotating equipment, and

    c. Instrumentation.

    d. Utility systems (steam, water, air, electrical, etc.)

-- Factors considered for a formal ITM program:

    a. Severity of service (defined by operating temperatures and pressures and corrosiveness of materials used in the equipment),

    b. Hazard potential of specific chemical and equipment used, and

    c. Size of process facility and ITM program resource requirements.

-- Visual inspections of the presence and extent of corrosion, cracks, and improper installation such as inadequate foundations, missing supports, excessive vibration, etc.

-- Testing: may include determination of wall thickness, corrosion rates, the presence of cracks or pinholes, pressure tests, and temperature tests.

-- Testing methods: include ultrasonics, radiography, liquid penetrant, magnetic particle, eddy current, acoustic emission, visual leak testing, and others. Special tests may be required for vessels lined with glass, rubber, or other polymeric materials.

-- Assessment of the need for continuous monitoring methods or programs.

-- Inspections for rotating equipment: include vibration analysis and monitoring, unusual sounds, and leakage around rotating parts, in addition to inspection items used for other equipment.

-- Instrumentation ITM: depends on how important the instrumentation is to the process and the kinds of instruments involved. For example, thermocouples may be more reliable than resistance temperature devices.

- Maintenance Record Keeping

-- Engineering drawings and design specifications on equipment should be retained and readily available.

-- Records of inspections, tests, repairs, and modifications should also be available.

-- Equipment records should include a detailed safety check list for inspections, testing, and maintenance. Such check lists should cover special precautions and procedures to be taken before, during, and after maintenance work.

-- Maintenance work order and scheduling records should be maintained.

- Physical Condition of Equipment

-- Visual examination of the physical condition of equipment for signs of excessive corrosion, structural weathering, and physical flaws such as cracks or other physical damage.

-- Examination of the integrity of insulation, especially of
storage tanks and other equipment containing hazardous
materials where accumulated moisture under damaged
insulation could cause external corrosion.

-- Examination of instrumentation and control equipment to
ensure that housings and enclosures are in place to
prevent dirt, moisture, and corrosion from impairing the
accurate functioning of the equipment.

-- Examination of control valves for worn or sticking stems,
and of other valves for excessive corrosion, indicating
they might be inoperable in an emergency.

-- Examination of the integrity of foundations and structural steel supports for evidence of cracks, subsidence,
and corrosion.

-- Examination of rotating equipment for evidence of vibration, leaks, and unusual or abnormal noises.

● Design Specifications and Plant Practices

-- Design specifications for pressure and temperature: must
be adequate for the intended service, and plant practices
must be consistent with pressure and temperature specifications. Unfired pressure vessels as a minimum standard
should be ASME Code, Section VIII Divisions 1 or 2
constructed and stamped. Other equipment should be
designed and constructed in accordance with recognized
codes and standards as a minimum requirement. Minimum
standards may not be adequate for toxic materials.

-- Safety controls for overpressure and overtemperature: should be present and in good working order.

-- Type and location of sensors: as far as possible, a sensor should directly measure the variable that must be controlled.

-- Where backup systems are present: both the backup and the primary system should be in good working order.

-- The most effective backup system: one that functions on a different principle than the primary system.

-- Materials of construction and corrosion allowances: must be appropriate for the service at hand.

-- Proper construction and installation of equipment: including provision for ease of inspection and maintenance.

-- Piping systems: should be designed with allowance for stresses and movement due to thermal expansion, and systems should be properly supported and guided.

-- Piping systems: should be uncluttered, with valves and lines labelled or easily identifiable.

-- Systems: should be designed so that the failure of one valve or sensor does not result in an accidental release. All instrumentation should be fail-safe.

-- Valves and fittings: should be appropriate to their intended service. Threaded fittings are inappropriate for most piping handling hazardous materials.

-- Backflow protection: must be present in lines where backflow is a hazard. A single backflow device is rarely sufficient and some devices provide more reliable protection than others.

-- Flexible hoses: should be used only where necessary. Some type of operational audit system, or physical systems should be in place to prevent using hoses in situations they are not designed for. Hoses should be inspected regularly for signs of wear or abuse.

-- Freeze protection: should be provided where required, especially in cold water lines, instrument connections, and lines in dead-end service such as piping at standby pumps.

-- Lubrication and cooling systems for process machinery: should be in good working order. Oil filters should be used for lubrication to critical components.

B.6 Fire Protection Evaluation

Purpose

The purpose of the fire protection evaluation is to assess the potential contribution of fire or explosion to an accidental release. The fire protection system is evaluated for its ability to control or extinguish a fire, limit its extent, and limit the ensuing damage. The fire protection system must especially protect facilities containing toxic chemicals. A fundamental principle is to maintain operations outside the explosive range of flammable materials.

Procedures

- Review of Drawings

    -- Examination of the plant process diagrams and layout drawings to identify areas where combustible, flammable, or explosive materials are used and stored.

    -- Review of drawings of the fire protection systems to identify the location of the water supply, the distribution system, sprinkler systems, fire monitors, hydrants, and special fire fighting equipment.

- Review of Written Procedures

    -- Review of emergency response plans to evaluate lines of authority, communications, and general procedures.

    -- Documentation of fire protection team personnel assignments and training procedures.

- Discussions with Personnel: interviews with personnel responsible for fire protection concerning history, current procedures, problem areas, and future plans.

- Site Inspection: inspection of fire protection systems in light of the many considerations listed below under Key Factors.

- Comparison of specific practices with applicable fire codes, discussion of good practices that may not be explicitly covered by codes, and evaluation of practices in the context of site-specific considerations.

- Selected National Fire Protection Association Codes with special significance for chemical process plants include, but are not restricted to, the following (20):

Code	Title
11	Foam Extinguishing Systems
12	Carbon Dioxide Systems
12A & B	Halon Systems
13	Sprinkler Systems Installation
14	Standpipe and Hose Systems
15	Water Spray Fixed Systems
16	Foam-Water Sprinkler and Spray Systems
17	Dry Chemical Systems
19B	Respiratory Equipment for Firefighters
20	Centrifugal Fire Pumps
22	Water Tanks
24	Private Fire Service Mains
30	Flammable and Combustible Liquids Code
385	Tank Vehicles for Flammable and Combustible Liquids
386	Portable Shipping Tanks
43A	Storage of Liquid and Solid Oxidizing Materials
43C	Storage of Gaseous Oxidizing Agents
493	Intrinsically Safe Apparatus
496	Purged Enclosures for Electrical Equipment
50	Bulk Oxygen Systems
50A & B	Gaseous and Liquid Hydrogen Systems

Code	Title
54	National Fuel Gas Code
58	Liquified Petroleum Gases, Storage and Handling
59	Liquified Natural Gas, Storage and Handling
61A	Manufacturing and Handling Starch
63	Industrial Plants Dust Explosions
654	Plastics Industry Dust Hazards
66	Pneumatic Conveying Systems
69	Explosion Prevention Systems
71-72	(Signaling Systems and Fire Detectors)
76A	Essential Electrical Systems
231	General Storage Indoor
231C	Rack Storage of Materials
512	Truck Fire Protection
1961-1963	(Fire Hose and Connections)
1	Fire Prevention Code
13A	Sprinkler Systems Maintenance
27	Private Fire Brigades
291	Fire Hydrants Uniform Makings
329	Underground Leakage of Flammable and Combustible Liquids
497	Electrical Installations in Chemical Plants
68	Explosion Venting Guide
70B	Electrical Equipment Maintenance
80A	Protection from Exposure Fires

Key Factors

- Water Supply and Distribution

    -- Water sources: must be of adequate capacity, quality,
       and reliability. Supply pressure and a backup supply of
       adequate capacity are fundamental considerations.

    -- Redundancy of supply: may be advisable in some high risk
       situations. Water bodies may require pretreatment, such
       as filtration and chlorination to remove dirt and debris
       and control organisms that could plug the system.
       In-plant reservoirs or reserve supplies should typically
       provide at least 4 hours of coverage. This depends on
       the availability of other sources and the nature of
       potential fires.

    -- Distribution system: should consist of a looped or
       gridded network of large-diameter pipe, feeding all of
       the fire protection systems and equipment requiring
       water. Underground piping or appropriate freeze protec-
       tion should be used, depending on climate. Where above
       ground portions of the system are run, they should be
       secure from mechanical, fire, and explosion damage, and
       freezing weather.

    -- Pumps: must provide adequate pressure and volume for the
       plant requirements.

    -- Additional pumps may be required because of plant expan-
       sion or modifications.

--   Pumps should be automatically started, but in either case startup must be reliable and secure in an emergency.

--   Adequate protection must be provided to ensure that lines, pumps, valves, and discharge devices do not freeze and impair the fire water supply.

--   Pump suction supply: normally sized to provide maximum flow rate for a minimum of four hours.

--   Evaluation of the sizing of the distribution system in light of the size of the facility and nature of the operation: normally a system is not sized to cover simultaneous fires in all areas. However, sizing of various portions of the system should account for the actual layout of a specific plant since in some facilities simultaneous demands on a fire protection system may be greater than at others. A minimum diameter for underground mains is 6 inches.

--   System pressures: sprinkler and water spray systems should normally require between 50 to 100 psig. Monitors, large hose systems and some foam systems may require 100 to 150 psig. Fire trucks supplied by a main can normally use 20 to 50 psig.

--   Fire pumps: should have capacities of 150% delivery requirement at 65% of rated head.

--   Fire pumps: should preferably have automatic start controls and a backup drive in the event of electrical power failure. Diesel drives are usually preferred, but if electric pumps are used, backups should be steam or

diesel; again diesel is usually preferred. Electric
pumps should be UL listed.

-- Adequacy of coverage considers but is not limited to the
following: fire protection for indoor and outdoor
storage of flammable liquids in drums, adequate spacing
and fire protection for flammable materials storage
tanks, fire protection in warehouses, fire protection of
cooling towers.

-- Sectional control valves in the underground fire mains
should divide the grid system into sections, limiting the
area subject to a single impairment. The number of
sprinkler risers and hydrants out of commission during
any change or repair should be specified depending on the
size of the area to be covered. With any one section
shut off, at least one water supply should be available
for the remainder of the system.

-- The distribution system should ensure protection of
structures and tankage.

● Sprinkler and Deluge Systems

-- Sprinkler and deluge systems: used for localized and
broad area protection.

-- Automatic sprinkler protection: necessary for all
buildings containing combustible construction, or flammable and combustible materials.

-- Evaluation of adequacy in terms of areas covered, the
density of coverage (flowrate per unit area), the

physical conditions of the system components, and
frequency of testing.

--   Sprinkler waterflow alarms: should give an audible local
alarm and automatically transmit water flow signals to a
central supervised location for any flow of water through
the sprinkler piping.

--   Automatic detectors: should be used where the quantity
of combustibles is limited or sprinklers are not compatible with the hazard to be protected.

● Hydrants and Monitors

--   Hydrants provide hose connections to the fire protection
water system. Monitors provide a fixed, quick response
discharge point for fire protection water streams.

--   Sufficient hydrants having at least two 2.5 inch hose
streams should be provided at any point in the property
where fire may occur regardless of wind direction. This
will require spacing hydrants 225 to 250 feet apart at
plant with ordinary hazards. At plants having highly
combustible occupancies, the spacing may need to be
reduced to 100 to 150 feet. For facilities with noncombustible buildings and non-hazardous occupancies,
hydrant spacing may be extended to 250 to 300 feet. For
average conditions locate hydrants 50 feet from the
building or equipment protected.

--   Fire protection monitors: should be used to protect
equipment containing flammable liquids that is not in a
building or structure protected by automatic sprinklers.

Their principle advantage is to provide a quick stream of water which can be operated by one man while hose lines are being laid. Spacings of 200 to 250 feet.

--   Assessment of the possibility of blockage of streams from fire monitors. Such blockage can occur when plant modifications are made. For example, an additional tank being put in behind an existing tank protected by a monitor. The new tank is blocked from the monitor stream by the existing tank. Portable monitors should be used in such cases.

● Building Protection and Portable Fire Extinguishers

--   Evaluation of fire protection for buildings in terms of type of construction and contents.

--   Small hose stations: located inside buildings so that every square foot of floor area is within 20 feet of a hose nozzle attached to not more than 75 feet of 1.5 inch woven jacketed rubber lined hose or equivalent. The nozzles should be the combination spray and solid stream with shutoff.

--   Small hose: preferably attached to risers independent of the sprinkler system if hose streams are considered needed when sprinklers are not operating. If this cannot be arranged, small hose may be attached to 2.5 inch or larger pipe on a wet pipe system.

--   Portable fire extinguishers: available in sufficient number, located where they are readily accessible, and in good operating condition.

-- Fire walls, partitions, and barricades: provided to separate personnel areas, high value property, critical process units, and critical utility and auxiliary units.

● Foam Systems and Other Special Protection

-- Special fire hazard protection: includes foam systems, carbon dioxide, dry chemicals, and explosion suppression systems.

-- Fire extinguishing agents: compatible with process materials.

● Fire Proofing and Structural Protection

-- Fire proofing as a surface coating material: gunite or other synthetic material. A UL standard is for two hours of protection (21). However, hourly rating should be chosen as appropriate for situation.

-- Fire proofing: used on structural steel and on walls of vessels in a chemical process area, applied by spraying or spread coating onto the structure. It should be present in all areas where equipment may be exposed to fire. It should be used on all main load-bearing structural members that support either process piping or equipment within hazardous areas. Fireproofing on vessel walls is not a common practice.

-- Evaluation of the adequacy of fire protection in terms of type, thickness, coverage, and integrity of the coating. Gaps or peeling are a basis for rejection.

-- Fireproofing should extend at least 30 feet above potential pool fires.

-- Fire proofing should be provided for valve operators for all emergency safety devices.

● Mobile Fire Apparatus

-- Mobile fire apparatus: may consist of fire trucks or hand carts.

-- Fire apparatus: located in areas protected from but accessible to plant areas where fires are likely.

-- Equipment: should be in sound condition and subjected to periodic tests.

-- Evaluation of the adequacy of the equipment in terms of capacity and presence of the right kinds of equipment for the types of fires likely to occur in the specific facility. If ponded water is a source of supplementary water, an inlet system needs to be in place to ensure that blockage of suction hose can not occur.

● Alarm Systems

-- Alarm Systems: consist of various combinations of sensors and alarm devices and should be appropriate to the intended fire hazard. Sensors may be based on detection of flammable vapors or heat, for example. The selection would depend on the specific situation.

-- Evaluation of sensor locations relative to locations of flammable materials and ignition sources.

-- Evaluation of the physical condition of sensors and alarm systems.

-- Evaluation of system adequacy in terms of location, coverage, and sensitivity.

-- Flow alarms: should be provided on sprinkler systems to indicate their activation.

● Fire Emergency Response Organization

-- Lines of authority and communications for fire emergencies: should be clearly defined and readily available to all plant personnel.

-- Procedures, individual assignments, and emergency numbers: should be clearly posted in operator control room and other areas where personnel are likely to be in an emergency situation.

-- Training procedures: should be clearly defined and written. Training program should include both formal classroom type instruction as well as field drills with equipment. For high hazard areas good practice suggests simulated incidents as well. Training frequency should be consistent with the risk associated with a given process area.

-- A cadre of assigned individuals should form the core of a fire fighting team. The size of this team will depend on the size of the plant and the level of risk and must be adequate to at least contain an incident, if outside help is available. If outside help is not available the team must be of sufficient size and training to fully control an incident on its own.

● Maintenance of Fire Protection Equipment

-- A regular program of scheduled inspections, tests, and preventive maintenance on fire protection equipment should be adhered to.

● Plant Layout Considerations

1. Adequate spacing, diking, and drainage: should be provided for process equipment and tanks of flammable and combustible materials.

2. Flammable storage pumps, compressors, and other equipment should be specific distances from ignition sources. Location near toxic materials should be avoided where possible.

3. Adequate drainage: should be provided to avoid large concentrations of flammable materials in the event of spills. It should be pitched to drain away from high hazard structures with a minimum of 1% grade.

B.7 Electrical System Evaluation

Purpose

The purpose of the electrical system evaluation is to determine to what extent the electrical system could contribute to an accidental release through design, operation, or reliability deficiencies. All portions of the electrical system should be installed in accordance with the National Electric Code or stricter standards. Individual components should comply with the acceptability criteria of recognized testing organizations.

Procedures

The overall electrical system is reviewed through discussions with plant personnel, by a plant inspection tour, and by evaluation of written data and drawings.

Key Factors

- Reliability

    -- Total power requirement and sources of electrical power to the plant: evaluation of the reliability of the sources (public utility system, private system, cogeneration system), including the history of outages, and examination of the physical source of power into the plant from outside power sources or from generating facilities within the plant for location and physical condition.

    -- Evaluation of locational factors: proximity to flood-prone areas or areas within exposure zones of fire, explosions, or frequency of lightening.

    -- Evaluation of the nature and extent of redundancy backup power system: focusing primarily on critical areas, including major power sources such as diesel generators for a whole process area, and battery backup units for individual processes or specific critical control elements.

-- Evaluation of historical reliability: including the frequency and duration of outages at the plant.

-- Determination of whether the facility has considered implications of power outages for the process unit as a whole as well as individual components such as instrumentation for specific intervals ranging from say, one minute to two hours.

-- Evaluation of voltage variations in light of possible effects on sensitive equipment. Could power surges damage a critical component in a control system such as software on a disk drive in a process control computer?

-- Evaluation of the configuration of the distribution system in terms of the adequacy of loops or independent circuit to different process area. Within a process area, electrical load blocks should correspond to process load blocks.

-- Electrical system: should be physically protected to minimize exposure to fire, corrosion, and mechanical damage; should be simple in schematic and physical layout to minimize human error in isolation load transfer; and should be accessible for ease of repair and maintenance.

-- Electrical system: should have adequate instrumentation for monitoring and the efficient diagnoses of failures, and protection by fuses and circuit breakers should be adequate.

-- Bonding and grounding: should be provided to protect personnel and protect systems from static buildup and lightening.

-- Maintenance and testing: the nature, frequency, and adequacy for power transformers, circuit breakers, relays and other devices, whether by the utility company, outside contractors, or the company, is evaluated.

● National Electric Code (NEC) Compliance

-- The electrical system in a chemical plant must conform to the specifications for hazardous locations as specified in Section 501-5(a) of the National Electric Code (NEC).

-- Process and storage areas are viewed in terms of the NEC hazardous location class, division, and grouping system. For example, Class I - Division II - Group D describes areas where flammable liquids and gases are handled, but are normally confined within closed systems, and where chemicals are typical organics. Other common categories for chemical plants include chemicals in Groups B and C. Class II areas, which is for atmospheres containing combustible dust, and Class I - Division I service, which is for atmospheres which continuously or intermittently contain high concentrations of flammable gases or vapors.

-- Key principles to be checked: isolation of the electrical system components and containment of the flame front should ignition occur inside of equipment. In practice this is accomplished by sealed conduits, circuit breakers, lights, motors, and switches.

-- Evaluation of proper pressurization and venting of closed
   areas.

-- Inspectors: should be alert to the use of any portable
   electrical equipment not conforming to NEC codes.

-- Evaluation of equipment maintenance conditions by observations of specific components of the electrical system
   such as: outside electric lines, insulators, support
   structures, switchgear, distribution panels, circuit
   breakers, lighting, grounding systems, motor starters and
   control centers, generators, transformers, relays, and
   lightening arresters.

B.8 Transportation Practices Evaluation

Purpose

The purpose of this evaluation is to assess the potential for a transportation incident causing an accidental release on plant property or elsewhere.
Areas of concern include: loading and unloading procedures, adequate design
of vehicles to handle the materials they transport, inspection and maintenance
practices, and practices regulating the movement of vehicles within the plant.

Procedures

-- Review of written guidelines and procedures pertinent to
   in-plant transportation practices.

-- Inspection of loading and unloading areas, along with
   equipment and selected vehicles that happen to be present
   at the time of the visit.

-- Discussions with both supervisory and operating personnel associated with transportation.

Key Factors

● Review of General Transportation Procedures.

-- Determination of vehicle type and of the rationale for the use of these vehicles. The compatibility of the vehicle with the type of service is evaluated in terms of materials being hauled in adherence to Department of Transportation requirements as well as special procedures and requirements established by the individual plant and company.

-- Evaluation of construction materials, and vehicle tank specifications in terms of pressure ratings, temperature ratings, wall thicknesses, valving, and on-vehicle sensors and instrumentation relative to the kinds of materials being shipped.

-- Evaluation of special practices, such as refrigerated and insulated vehicles for temperature sensitive products from a "what-if" mentality.

-- Brief visual inspections of vehicles may be made at the time of the inspection to look for malfunctioning, poorly maintained, or incorrect equipment, such as inappropriate pressure relief devices.

-- Review of plant procedures for routine inspections.

-- Consideration of extra precautions taken for pressurized tank vehicles and high hazard materials.

-- Review of certification procedures, usually through discussions with plant personnel. Such procedures would cover verification of the previous contents of the vehicle, cleaning after the last load, and similar considerations.

-- Review of numbers, frequency of shipments, and sizes of shipments.

● Review of Loading and Unloading Operations

-- Examination of the use and scope of loading checklists.

-- Methods of overfill protection: common procedures for overfill protection include level sensors, automatic shut-off actuated by quantity totalizers, scales and load sensors, and reliance on operators. For hazardous materials an overfill protection system should have a back up.

-- Spill control measures: provisions should be available for both containment and cleanup, and where highly volatile materials or gases are involved, the plant should have a contingency plan for responding to air borne releases.

-- Availability of automatic shut-offs actuated by abnormal conditions and remote shut-offs: in case equipment can not be reached during an emergency.

-- The responsibilities of the plant personnel and the
driver: should be clearly defined and understood. For
high hazard loading and unloading operations a plant
representative should always be present.

-- Review of labeling, stenciling, placarding, and other
informational practices for compliance with DOT requirements and a plant's own special requirements.

-- Review of controls on in-plant routing and stationing of
vehicles. Potential hazards related to these factors
include the potential for vehicular collisions with
equipment in tight areas and vehicle accidents caused by
poor locational practices. As an example of the latter,
a fully loaded tank trailer parked off the pavement on
soft ground, could overturn as a result of the wheels
sinking on one side and the vehicle falling over.

-- Definition of procedures for the management of rail car
traffic on the plant premises. Special provisions for
protection against derailment incidents are noted.

-- Review of procedures for dealing with liquid heels in
tank vehicles. It is common practice not to sample or
analyze the heels in vehicle tanks dedicated to a single
service. Plant procedures for avoiding cross-
contamination of products in transportation vehicles and
receiving racks are evaluated.

-- Evaluation of equipment condition and its use. For
example, do hoses show signs of significant abrasion and
where? Are couplings appropriate for the type of service? What precautions are taken to maintain cleanliness

in filling and unloading equipment where contamination could be hazardous? Is equipment being used as it was intended, or has jerry-rigging occurred as products have changed?

-- Evaluation of precautions used for static electricity grounding.

-- Check valves and/or other precautions: should be in place to prevent backflow and siphoning conditions form occurring.

-- Color coding or other means of designating multiple lines and spouts: desirable at multi-material loading and unloading facilities.

-- Are flow rate limiters in use?

-- Is there thermal expansion relief for blocked-in valve lines?

-- Evaluation of procedures to prevent drive-away with the attendant breakage of lines and accompanying chemical releases.

-- Recalibration of gaging and metering equipment: should be carried out periodically.

• Off-Site Risks

-- Review of driver qualification, training, and certification procedures.

--   Review of accident histories for clues as to potential
        causes and impacts of future incidents. Changes in
        equipment or procedures occasioned by previous incidents
        are noted.

--   Review of routing procedures in the context of DOT
        regulations, special local rules, and other considera-
        tions that may be specific to the plant and the materials
        it handles.

        a.   DOT regulations
        b.   Other considerations

B.9   Contingency Plan and Emergency Response Coordination

Purpose

The purpose of the emergency response evaluation is to ensure that
adequate procedures and equipment are in place to reduce the effects of an
accident on people and property both within and outside the plant. Plant
personnel and, where necessary, personnel from local emergency response
agencies should be trained to participate in plans for controlling plant
emergencies during large windstorms, earthquakes, floods, power failure,
fires, explosions, and accidental releases.

Procedures

- Review of Written Procedures: written procedures are reviewed
  and the accessibility of these procedures to plant personnel
  are observed. Written procedures are evaluated in terms of
  comprehensiveness and specificity to the peculiarities of the
  individual facility. Recognition of the most significant
  hazards is noted.

- Discussions with Plant Personnel: the interviewing of selected plant personnel to evaluate their perceptions, to gauge
  their knowledge and attitudes toward emergency response, and
  to obtain additional factual information for use in the
  evaluation.

Key Factors

- Evaluation of contingency plans for dealing with various
  emergencies in terms of the following:

  -- The plan should be comprehensive and cover fire, explo-
     sion, and chemical releases.

  -- The plan should be specific. It should clearly designate
     responsibilities for individual unit personnel as well as
     plant personnel involved in fire fighting teams, medical
     teams, evacuation teams, etc. It should also address
     specific high hazard situations such as incidents in
     specific units or process areas, and specific kinds of
     incidents such as the accidental large release of a toxic
     chemical.

  -- Plans should be up to date. A process plant is rarely a
     static entity. Changes and modifications made over the
     years may affect process hazard potential. Corresponding
     changes in emergency response plans should also be made
     and clearly dated to allow evaluation of the appropriate-
     ness of the current plan.

  -- Evaluation of the availability of the plan to plant
     personnel: this includes physical distribution of the

plan and the way it is stored and treated by plant
personnel.

-- Responsibilities for personnel: should be clearly
defined. The definition of responsibilities for the
evening and night-time work shifts is especially important, since staffing on these shifts is usually less than
in the day time.

- Personnel Training

-- Personnel training programs: should include written
materials and include both formal "class room" instruction as well as field drills.

-- Instruction and drills: commonly cover routine fire
fighting and some times non-catastrophic spills. For
areas with high hazard potential, specific drills for
dealing with potential catastrophic incidents are important.

-- Evaluation of operator awareness: by questioning them
about what incidents they consider to be the greatest
hazards in their areas and how they might respond to such
incidents.

- Emergency Communications Systems

-- Communications systems available for dealing with emergencies: may include telephones, radios, signals, and
alarms.

-- Evaluation of the effectiveness and reliability of the communications system. Because telephone lines may be out of service in an emergency, radio communication is an important backup.

-- Definition of communications responsibilities and plans for relaying information in times of emergency.

-- Consideration of alternatives to telephone and radio communications, such as area wide alarm signals.

- Emergency Response Equipment

-- Emergency response equipment availability: may include air packs, chemical suits, medical packs, and mobile tool kits.

-- Evaluation of the effectiveness and reliability of the equipment. The total available supply of emergency breathing air is critical.

- Coordination with Outside Agencies and the Community

-- Evaluation of the availability of support facilities, equipment, and personnel. Support facilities include hospitals, emergency aid stations, and fire stations. Equipment includes fire vehicles, ambulances, and specialized tools. Numbers and skills of support personnel are noted. Because chemical plant operations can sometimes be esoteric to the outside community, the hazard potential of a facility may be reduced if outside authorities and emergency services are properly informed and are familiar with the plant and its operations.

Local fire departments should be aware of the methods and equipment necessary, to fight a chemical fire for each chemical in use. The plant's program in this regard is evaluated.

-- Consideration of plant participation in joint training activities with the community.

-- Consideration of the proximity of support facilities and response times. Response times longer than 15-20 minutes are reaching the extreme of utility for an emergency response.

-- Consideration of accessibility to the plant and various areas within the plant, especially to how emergency access and egress may differ from normal entrance and exit patterns.

-- Review of plans for emergency notification of the community and for community evacuation. While a community evacuation plan is beyond the control of the plant, recognition of the need for such a plan and steps the plant may have taken to have the community develop such a plan are noted.

-- Mutual aid from neighboring industry: is it available?

-- Evaluation of general community relations and of historical relations by discussions with plant personnel. This may be important in engaging community support in emergencies, as well as in securing the plant against possible sabotage.

Part II

Prevention and Protection Technologies

The information in Part II is from *Prevention Reference Manual: Control Technologies for Controlling Accidental Releases of Air Toxics,* prepared by D.S. Davis, G.B. DeWolf, and J.D. Quass of Radian Corporation for the U.S. Environmental Protection Agency, August 1987.

Acknowledgments

This manual was prepared under the overall guidance and direction of T. Kelly Janes, Project Officer, with the active participation of Robert P. Hangebrauck, William J. Rhodes, and Jane M. Crum, all of U.S. EPA. In addition, other EPA personnel served as reviewers. Sponsorship and technical support was also provided by Robert Antonpolis of the South Coast Air Quality Management District of Southern California, and Michael Stenberg of the U.S. EPA, Region 9. Radian Corporation principal contributors involved in preparing the manual were Graham E. Harris (Program Manager), Glenn B. DeWolf (Project Director), Daniel S. Davis, Nancy S. Gates, Jeffrey D. Quass, Miriam Stohs, and Sharon L. Wevill. Contributions were also provided by other staff members. Secretarial support was provided by Roberta J. Brouwer and others. Special thanks are given to the many other people, both in government and industry, who served on the Technical Advisory Group and as peer reviewers.

1. Introduction

Accidental toxic chemical releases, fire, and explosion are the three types of major accidents in chemical process facilities. Toxic chemical releases may occur as the result of fire or explosion, but may also occur in their absence.

This manual, which addresses the prevention of and protection from accidental releases of toxic chemicals, is part of a set of reference manuals whose purpose is to summarize the major concepts of release hazard control so that the probability and consequences of accidental toxic chemical releases can be reduced. The volumes in the series include:

Prevention Reference Manual - User's Guide

Prevention Reference Manual - Control Technologies
      Volume 1: Prevention and Protection Measures
      Volume 2: Mitigation Measures

This series of manuals is intended to assist both regulatory and industry personnel in developing approaches to reducing accidental chemical releases and their consequences. In addition to highlighting the various technical aspects of the problem, the manuals also briefly address costs.

1.1 FUNDAMENTAL CONCEPTS

The purpose of preventing toxic chemical releases is to prevent harm to the health of human beings and other life and to the environment. An accidental toxic chemical release is the final event in a sequence of events leading to the release. In the terminology of formalized fault tree and event tree analyses, two of several formal methods for analyzing causes and effects in event sequences, that are discussed in the PRM-User's Guide, an event sequence begins with the initiating, or primary, event and passes along a chain made possible by enabling events. The chain is broken by preventing either the initiating event or the enabling events from occurring. Selection of the best of several possible options that could prevent a particular accidental toxic release depends on correctly identifying individual events and event chains, on knowing the relative probability of the events, and on the skill and knowledge of the individuals charged with solving the problem.

Figure 1-1 illustrates the place of prevention in the overall sequence of an accidental chemical release. Prevention is the second barrier between the realization of a toxic chemical release hazard and its consequences. It follows proper identification of the hazards. Protection, reduction of the quantity of material released, and mitigation, reduction of the magnitude of consequences, are two additional barriers. Community response is also shown as a barrier since community response measures can reduce the ultimate adverse consequences of a release. Evacuation is an example of a community response measure. Since the total prevention of an accidental release implies a zero probability for that release, and since this can never be assured, prevention in practice means reducing the probability of a release to within acceptable limits. At present, such acceptable limits, which are related to acceptable risk, are subjective and are not quantitatively defined.

Preventing accidental releases means preventing the loss of primary containment. The three general classes of equipment that contain chemicals are:

- Vessels;

- Piping; and

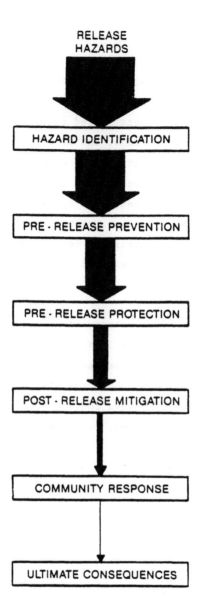

Figure 1-1.   The role of various accidental release control measures in
reducing the consequences of an accidental release.

- Process machinery.

In the context of this manual, vessels include all major process equipment: tanks, columns, heat exchangers, etc. Piping includes pipe lines, valves and fittings, and instruments fitted into pipe lines. Process machinery includes pumps, compressors, fans, and other such material moving equipment, either rotating or reciprocating. To ensure containment, a process must operate in proper sequence under acceptable conditions of temperature, pressure, and composition. The equipment must operate within design limits defined by specifications for pressure, temperature, externally applied mechanical stresses, chemical compatibility (corrosion, erosion), and within the physical limits imposed by the process system as actually installed and operated.

An accidental toxic chemical release can occur when there has been a physical breakdown of equipment or a loss of process control that leads to a breach in primary containment. Process modifications that enhance control and/or physical process steps and procedures that reduce the probability of equipment failure or a loss of process control will reduce the probability of a release. The overall problem, then, requires consideration of:

- Process design;

- Physical plant design;

- Operation; and

- Supervision or management.

The first two elements address the process, its chemistry, and hardware; the second two elements address the human factor.

1.2   ORGANIZATION OF THE MANUAL

Following this introduction, Section 1, the remainder of the manual addresses the above elements. Section 2, which covers process design considerations, addresess the basic functional and operational characteristics of any chemical process system, highlights major hazards associated with various characteristics of systems, and discusses the control measures that reduce those hazards. These include specific approaches for preventing deviations from intended operating conditions that could directly or indirectly cause an accidental toxic chemical release.

Section 3 covers physical plant design and considers how inappropriate design practices for the physical facility and its hardware contribute to accidental chemical releases. Adhering to proper design and equipment specifications for vessels, piping, valves, process machinery, and instrumentation can reduce the probability of an accidental toxic chemical release. Fundamental principles of configuration, layout and other design considerations associated with process hardware are discussed.

Section 4 discusses procedures and practices such as management policies, operator training, maintenance practices, and other topics. Some experts in loss prevention believe that the most significant decreases in the probability for accidental toxic chemical releases can be made in these areas.

Section 5 discusses protection systems, which include scrubbers, flares, and secondary containment systems. These devices, which can provide an emergency control system for process units handling some toxic chemicals, reduce the quantity of toxic chemical reaching the environment once primary containment has been breached.

A selection of hazard controls depends partly on the economics of their application. Example control costs are presented in each section for various process and equipment control technologies. These technologies involve individual hardware and procedural components, as well as multicomponent combinations for controlling an accidental release hazard. Since numerous

combinations of individual controls are possible, costs are presented for
individual hardware and procedural components that can be combined into a full
control system.

Since costs in this manual are intended to be a starting point for
initial planning, cost bases selected for each component are for specific
sizes and overall specifications representative of that class of component.
For example, for piping, one control option might be to use Schedule 80 rather
than Schedule 40 pipe in some process unit. Costs are presented for three
sizes of both Schedule 80 and Schedule 40 carbon steel pipe, two-inch,
four-inch, and six-inch, which are common sizes encountered in many process
plants handling toxic chemicals. While other sizes may be encountered, a
first estimate of the relative costs of changing the schedule size can be
estimated using one of these pipe sizes as an example.

The costs shown in this manual were obtained from various process plant
estimating references and are presented in 1986 dollars. For each component,
both capital and total annual costs are presented. Capital costs represent
direct installed equipment costs, exclusive of indirect costs such as engineering or contractor's fees. Total annual costs include only labor where a
procedure is involved, or only maintenance and capital recovery where a
hardware component is involved. Maintenance costs for all hardware are
assumed to be a flat two percent of the direct installed equipment cost.
Capital recovery has been calculated at a 10 percent cost of capital over an
assumed 10-year component life.

Costs for various controls applied to new plants can be expected to
differ substantially from the same controls applied to existing plants. The
relative differences in costs will vary according the type of controls, but,
in general, retrofitted equipment is usually more costly than the same equipment installed initially. Costs in this manual apply to new plants. Retrofit
costs can be roughly estimated as about fifty percent higher. Costs of
introducing changes in procedures and practices can also be expected to vary,
but generalizations about these costs cannot be as easily made as for equipment.

2. Process Design Considerations

A first step in hazard control is planning the initial process design. This section describes the basic considerations of initial process design, as well as subsequent process modifications, and identifies operational aspects of a process where upset and loss of control could lead to a toxic chemical release. Process design considerations involve several fundamental principles applied to the process materials, process variables, and equipment. These principles are addition, substitution, deletion, and redundancy (or duplication). Each process chemical, operation, or piece of equipment should be viewed in terms of how changes based on these qualities may lead to a reduction in the probability of an accidental release. For example, substitution might involve the replacement of a toxic chemical with one less toxic. Duplication or redundancy might involve the use of a second thermocouple to measure a critical temperature.

Modifications, then, of fundamental process chemistry, operations, and equipment can reduce the severity of operating conditions, the quantities of toxic materials, the complexity of the process, and can enhance process control. Such modifications involve a number of process components:

- Control characteristics of the process;

- Process thermodynamics and chemistry;

- Flow measurement and control;

- Pressure measurement and control;

- Temperature measurement and control;

- Quantity measurement and control;

- Mixing; and

- Composition measurement and control.

Process hazards associated with each area that can contribute to accidental releases and the control measures that can reduce those hazards are discussed in the following subsections. Such measures may include changes in both hardware and operational procedures. Hardware changes may involve individual equipment items or entire systems. Costs of hardware components likely to be involved in process modifications are presented. Since process modifications can involve combinations of components, costs are presented on a component or subsystem basis, allowing the user of this manual to select those components that apply to any specific modification for which the user wants an approximate estimate of costs.

2.1 CONTROL CHARACTERISTICS OF THE PROCESS

Good process control, including the interaction of human operators with the process system, is basic to successful process operations. Process control is achieved by manipulating the variables of flow, temperature, pressure, composition, and quantity.

2.1.1 Process Control and Release Prevention

Adequate control of the process is fundamental to accidental release prevention, since loss of control might begin or propagate the sequence of events leading to loss of containment and accidental release. Loss of containment through loss of control occurs when some process variable exceeds the physical limits of the containment system. Equipment failure, operational failure, or both can cause loss of control. The interaction of human operators with the process is not explicitly addressed here, but certain aspects of the human factors, such as operator training, are discussed in Section 4.

Process control means maintaining process variables within prescribed limits. Process variables are either controlled directly or indirectly. Some indirectly controlled variables are maintained through control system action on directly controlled or manipulated variables. The physical characteristics of a process unit, the process chemicals, and the operating conditions determine the control characteristics of the particular system.

The following essential functions of a process control system may be modified to enhance control and reduce the probability of accidental releases:

- Measurement;

- Normal control; and

- Emergency or protective control.

These functions of a control system are the same whether manual, analog, or computer control is used. In any control system, the conceptual component configuration for controlling a variable is as shown in Figure 2-1. The connecting arrows indicate information flows between components. The dotted lines indicate that both direct or indirect measurement and control of variables is encountered in process systems. A sensor for the measurement device monitoring the measured variable transmits its signal to a controller, where a comparison or decision is made relative to specified process conditions. A signal is then sent from the controller to the actuator of a final control element, which adjusts the manipulated variable. In chemical process systems the most common manipulated variable is flow and the most common final control element is the actuator/control valve. The actuator is the device that physically operates the valve. The manipulated variable controls change in the measured variable. For example, a temperature might be the measured variable and flow of a cooling medium might be the manipulated variable. In some situations, the measured and manipulated variable may be the same: the flow rate of a fluid stream, for example. In a completely manual system, a human operator responds to a measurement and acts as both the controller and

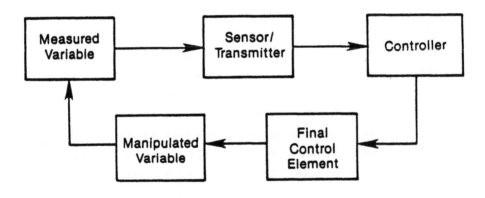

Figure 2-1.   Generalized information flow for process control loop.

actuator of the final control element. In an analog system the response occurs through electrical, electronic and/or mechanical components. A computer system combines the components of an analog system with the capabilities of logic and computations software algorithms.

The primary variables for any chemical process unit are:

- Flow;

- Pressure;

- Temperature;

- Composition; and

- Quantity.

Typically, these variables are all controlled by manipulating the flow of some process stream.

The specified value of the controlled variable in the process system is referred to as its set-point. Deviations from this set-point can occur in response to various influences known as process loads.

Process loads may arise from disturbances in the following:

- Feed stream quality and rates;

- Service stream (e.g. steam, cooling water) quality and rates;

- Product stream quality and rates (especially where there is recycle);

- Equipment malfunctions;

- Process operating conditions;

- Ambient conditions; and

- Material properties and behavior.

Deviations in variables can cause deviations in the process. Deviations from acceptable conditions are the fundamental source of process hazards. A valve that is open when it should be closed or an overpressure leading to equipment failure or emergency venting are two examples of direct causes of accidental releases. The specific role of primary variables in process hazards is explored further in Sections 2.3 through 2.8. Certain additional general concepts of process control as they relate to hazard reduction are discussed below.

2.1.2   System Response

The overall control characteristics of a process depend on the response characteristics of both the process being controlled and on the control system itself. The response characteristics of a process are known as the process dynamics. The interaction of a control system with a process creates a new set of dynamic responses that may differ from the dynamic responses of the process alone. A key concept of process control and dynamics is that any corrective action taken will not result in an instantaneous response and might not result in the return of a deviating variable to its set point. The response of a system to a control action can result in a continued deviation, both in terms of duration and magnitude, or to new deviations, if a control system is not properly designed. Either the duration or magnitude of deviation may lead directly or indirectly to a loss of containment and to an accidental release. The relationship between deviations and specific causes of events leading to releases is the primary theme throughout other parts of this manual.

Any assessment of the probability of an accidental chemical release depends, therefore, on the extent to which deviations in magnitude <u>and</u> duration can be tolerated before a loss of containment occurs. Once this is known, the limits of control are defined, and appropriate initial design or modifications that enhance control can be undertaken.

2.1.3 Control System Integrity

The success of process control depends on the integrity of the control system. Integrity means that the control strategies and hardware are adequate, reliable, for the job at hand. Judging the adequacy of a control system requires examination of its components and specifications. An evaluation cannot fully determine adequacy, in terms of controller settings and response to system upsets, without understanding the intent of the basis for design, the design, or observing the control system in operation under actual process upset conditions. Partial insight can be achieved by examining the control system configuration on a piping and instrumentation diagram, the actual hardware used, by discussions with personnel familiar with operating the process, and by formal methods of hazard identification and evaluation such as hazard and operability studies and fault tree analysis, among others. Such methods should be used in control system design (1). From these approaches the control system's adequacy may be inferred.

If control strategy is correct, any increase in the reliability of a control system depends on improving the quality, specifications, and maintenance of the physical components, and on duplication of (redundancy) components where the potential hazard warrants. This may ultimately be a subjective decision; no specific criteria can be given here for that decision. The formal hazard evaluation method mentioned earlier can be a guide, however. For example, duplication could be applied to an entire loop, or only to the components most likely to fail. This judgement must be made on a case-by-case basis.

High points in design philosophy have been well-stated by one author (1). There should be a clear design philosophy and proper performance and reliability specifications for the control and instrumentation. The design philosophy should deal with, among other things, the characteristics of the process and of the disturbances to which it is subject, the constraints within which the plant must operate, the definition of the functions which the control system has to perform, the allocation of these between the automatic equipment and the process operator, the requirements of the operator and the administration of fault conditions. The philosophy and specification should cover measurements, displays, alarms, control loops; protective systems; interlocks; special valves (e.g., pressure relief, non-return, emergency isolation); the special-purpose equipment; and the process computer(s).

A process that appears to require a complex control system should be re-examined to determine process changes that would reduce the complexity.

The control system as a whole and the individual instruments should have the turn-down capabilities necessary to maintain good measurement and control at low throughputs.

Transient situations, such as startup, shutdown, and restarting after a trip or restarting an agitator, tend to be particularly hazardous.

2.1.4   Control System Components

Major control system components discussed in this section include:

- Sensing and measurement;

- Controllers;

- Final control elements;

- Switches and alarms;

- Emergency shutdown and interlock systems; and

- Computer control.

Certain aspects of control systems are general to the system as a whole or apply to all kinds of instrumentation used. Some important aspects are (1):

1. The control loops should have fail-safe action as far as possible, particularly on loss of instrument air or electrical power to the control valves. The action for other equipment should also be fail-safe where applicable. Control that adds material or energy should receive special consideration for integrity.

2. The ways in which common cause failure can occur and the ways in which the instrument designer's intentions may be frustrated should be carefully considered.

3. Instrumentation which is intended to deal with a fault should not be disabled by the fault itself. And if the process operator has to manipulate the instrumentation during the fault, he should not be prevented from doing so by the conditions arising from the fault.

4. The instrument system should be checked regularly and faults repaired promptly. It should not be allowed to deteriorate, even though the process operator compensates for this. The process operator should be trained not to accept instrumentation unrepaired over long periods.

5. Ease of detection of instrument faults should be an objective in the design of the instrument system. The process operator should be trained to regard detection of malfunction in instruments as an integral part of his job.

6. Instruments which are required to operate only under fault conditions, and which may therefore have an unrevealed fault, require special consideration.

7. Important instruments should be checked regularly. The proof test interval should be determined from a reliability assessment where possible. The checks should include measurements, alarms, control loops, etc.

8. Tests should correspond as nearly as possible to the expected plant conditions. It should be borne in mind that an instrument may pass a workshop test, but still not perform satisfactorily in the plant.

9. Valves, whether control or isolation valves, can leak even when closed. Control valves in particular may not give a tight shutoff. More positive isolation may require measures such as double block and bleed valves.

10. Valves, particularly control valves, also tend to stick, giving rise to conditions that do not always emerge from simple applications of the fail-safe philosophy. Two dangerous conditions are a feed valve jamming in an open position, or a cooling water valve jamming in a closed condition.

11. Practices which process operators tend to develop in their use of the instrumentation should be borne in mind, so that these practices do not invalidate the assumptions made in the reliability assessments.

12. The possibility of human error should be fully taken into account. The reliability of the process operator should be assessed quantitatively where possible. Human factor principles should be applied to reduce human error.

It is also necessary to pay careful attention to the details of the individual instruments used (1). Some important features are:

1.   Instruments are a potential source of failure, either through a functional fault on the instrument or through a loss of containment at the instrument.

2.   Use of inappropriate materials of construction can lead to both kinds of failure. Materials should be checked carefully in relation to the application, bearing in mind the possible impurities in the bulk chemicals. It should be remembered that the instrument supplier usually has only a very general idea of the application.

3.   Instruments containing glass, such as sight glasses or rotameters, can break and give rise to serious leaks and should be avoided, where possible, with toxic chemicals.

4.   Instruments may need protection against the corrosiveness of the process fluid. Examples of protection are the use of inert liquids in the connecting lines on pressure transmitters or of chemical diaphragm seals on pressure gauges.

5.   Sampling and connecting lines should be given careful attention. Purge systems are often used to overcome blockage in sampling and connecting lines. Freezing is another common problem, which can be overcome by the use of steam or electrical heat tracing.

6.   Temperature measuring elements should not normally be installed bare, but should be protected by a thermowell. A thermowell is frequently exposed to quite severe conditions, such as erosion/corrosion or vibration and should be carefully designed.

7.   Pulsating flow is a problem in flowmeters such as orifice plate devices and can give rise to serious inaccuracies. This is a good example of a situation where duplication of identical instruments is no help.

8. Pressure transmitters and regulators are easily damaged by over-
    pressure.

9. Complex instruments such as analyzers, speed controllers, vibration
    monitors and solids weighers are generally less reliable than other
    instruments. This requires not only that such instruments receive
    special attention, but that the consequences of failure be analyzed
    with particular care.

10. Different types of pressure regulator are often confused, with
    perhaps a pressure reducing valve being used instead of a non-return
    valve or vice versa. It is specially necessary with these devices
    to check that the right one has been used. Also, bypasses should
    not be installed across pressure regulators.

11. Selection of control valves is very important. A control valve
    should have not only the right nominal capacity but also appropriate
    rangeability and control characteristics. It should have any
    fail-safe features required, which may include not only action on
    loss of power but also a suitable limit to flow when fully open. It
    should have any necessary temperature protection (e.g. cooling
    fins). Bellows seals may need to be provided to prevent leaks. The
    valve should have a proper mechanical balance for the application,
    so that it is capable of shutting off against the process pressure.
    It should be borne in mind that any valve, but particularly a
    control valve, may not give completely tight shutoff, and also that
    a badly adjusted valve positioner can prevent shutoff.

12. Instruments should not be potential sources of ignition and should
    conform with the electrical area classification requirements.

A further discussion of the loss prevention aspects of control and instrument
systems is given by Hix (2).

Sensing and Measurement--

Sensing and measuring both controlled and manipulated variables is fundamental to the proper operation of any control system. A wide variety of instrument types and vendors is available. Sensing and measuring devices include devices for normal process control, as well as trips and alarms for emergencies, and monitoring and detecting equipment for impending equipment failure and chemical leakage. Once a process system is constructed, its successful operation depends on the effectiveness of the sensing and measuring devices on which process control decisions are based.

Components for this function of the control system vary in sensitivity, range, accuracy, and reliability. Component selection must consider the severity of operating conditions, including chemical attack.

The variable of interest should always be measured directly where possible, rather than being inferred from some other measurement. The measurement should be made at the right location. If the variable is critical for process safety, the same measurement should not be used for control and for an alarm or trip.

Controllers--

The controller is the decision-making and command component of a process control system. The controller receives information from the process sensing and measuring equipment, compares this information with its set point, and takes appropriate action to minimize deviations. In a manual system, the human operator acts on the information. In a computer-supervised control system a computer acts on the information. Traditional analog control systems operate between these extremes.

Controllers come in a variety of types and operate in any or in all three common control modes: proportional, integral, and derivative; or in other specialized modes. Details of such features are beyond the scope of this manual but are widely discussed in the literature of process control (e.g.,

References 3, 4, 5). Pneumatic controllers have been available for many years
and are still in wide use. Electronic controllers have grown in popularity
over the years and include traditional analog electronic controllers as well
as newer digital types.

The numerous changes in controller technology in recent years have
created opportunities for both increased safety and hazards. Opportunities
for safety exist because of the versatility, accuracy, and response capabilities of the newer generation of controllers. Opportunities for hazard result
from the large number of available control systems, making inappropriate
selection, application, installation and operation a greater possibility than
before. Proper controller selection is an important step in accidental
release prevention.

Final Control Elements--
A process control loop is completed with the final control element. In
most cases this is a fluid control valve. In other cases it may be another
device, an electric heater, for example. Again, many types and vendors exist.
For a control valve, proper sizing, trim, and actuator selection are critical
in selection since these factors determine how well the valve will perform its
control function under both normal and abnormal operating conditions.

Switches and Alarms--
Switches and alarms are important devices for process upset and emergency
conditions. Alarms are devices that warn of an impending or existing abnormal
process condition; switches are action devices that divert a flow, close a
valve, or perform some other emergency function such as tripping an alarm.
Necessary characteristics are sensitivity to the condition being monitored and
hardiness in hostile environments, especially because such devices are typically used only intermittently.

The most common of these devices respond to pressure, flow, level, and
temperature. Details are described in the sections of this manual specific to

each of these process variables, but certain general features are discussed here.

Switches may be indicating or non-indicating. An indicating switch has a scale readout that allows the process operator to make changes in the switch's set-point. A non-indicating switch is not adjustable by the process operator. This type may be appropriate in a process where the switch's function is crucial in controlling the potential for an accidental release and undocumented changes in the settings are to be avoided. The danger with this type, however, is that if process conditions are changed, the switch setting may also have to be changed, but there is no indication of the switch setting to verify that the necessary change has been made.

A switch should be applied to the parameter showing the most significant change for the condition to be alarmed to enhance switch sensitivity. For example, pressure switches are sometimes used in pump discharge systems to indicate a pump failure, or to start a spare pump. However, with a low-head pump system, the difference in pressure between the shutoff setting and operating pressure may be small, making switch operation unreliable. The change in flow rate between normal operation and shutoff condition may be significant, making a flow switch the better choice. Switches differ in their susceptibility to damage from overrange. This should be considered when selecting a switch for a process prone to overrange upsets.

The philosophy of the alarm system should relate the variables alarmed, the number, types and degrees of alarm, and the alarm displays and priorities to factors such as instrument failure and operator confidence, the information load on the operator, the distinction between emergency alarms and mere status indications, and the action the operator should take.

Emergency Shutdown and Interlock Systems--
These systems may use a combination of many instrument components to achieve a safe process shutdown condition. Their success depends on proper

design and operating strategy, as well as on the proper selection of system hardware. An emergency shutdown system is used primarily for the shutdown of process equipment and for the operation of block, dump, and control valves. An interlock system presents a change in the operation of a pump, valve, or other equipment unless certain conditions are met. In the design of these systems, redundant components and circuits are normally recommended since these systems are safety systems and high reliability is a fundamental requirement. This is especially important in the temperature and level control of exothermic reactions or when heat sensitive materials are handled.

Instrument Air and Electrical Backup Systems--

Backup instrument air or electrical systems also should be provided for these systems. The backup electrical system for instruments and control may be part of the same system for general electrical backup, or it may be an independent system. Overall electrical backup can be achieved with diesel-generator sets. Sometimes the needs of critical process control instrumentation are best met by localized battery backup. These systems are often supplied as part of standard control instrumentation packages when the latter are purchased as a complete system, but can also be provided as packages for individual control loops, if required.

Where pneumatic systems are used, backup instrument air supplies may be system-wide or local. System-wide backup may require tying a main compressor into both the basic plant power supply and into the plant electrical backup system to ensure compressor availability during power outages. Another approach is to provide a large inventory of compressed air storage sufficient to bring a process system to a safe shutdown. Sometimes backup is provided by tying the instrument air into a liquid nitrogen system so that the nitrogen can be substituted for air on an emergency basis. Local backup of instrument air can be achieved by having a portable cylinder or emergency tank of either compressed air or nitrogen tied into the instrument air system for emergency use.

Computer Control--

The application of computer- and microprocessor-based control has increased rapidly over the last decade. There are at least two fundamental
reasons for such growth: 1) reduced costs and a simultaneous increase in
capability, and 2) hardware developments that permit more effective automated
distributed control. An example of the former is the advent of microcomputers
that offer a small operation true computer control for the first time. An
example of the latter is the "smart" controller, a stand-alone controller that
can be programmed for its various control functions and settings. A discussion of many more specific developments, including the use of coaxial cables
to replace multiple cable runs between parts of a control system and a central
computer, video screens to replace individual meters and displays in control
rooms, and other such developments, is beyond the scope of this manual, but
the reader may find ample information in the many books on process control and
in technical periodicals (e.g., References 6 and 7).

The new instrumentation and control technologies offer an increase in the
quality and quantity of data that can be handled at a given time. Two very
desirable features are that priority scanning and rate of change monitoring
can be employed. Such systems free the human operator from many routine and
complex monitoring and evaluation functions, allowing him to focus on critical
process deviations or upsets. Sequencing and the automatic shutdown features
of such systems can make the system freer from operator error.

When computer systems are used, the ability to revert to manual control
or ordinary analog control may be desirable in many processes for safety
reasons. Such backup systems may be provided for the most critical parts of
processes handling toxic chemicals.

2.1.5 Effectiveness of Control Systems

The effectiveness of a process control system is its ability to minimize
the magnitude and duration of deviations in process variables. Maintenance of

process variables at or near the set point is accomplished by direct manipulation of the flow of mass or energy into or out of the process unit. These manipulations alter the values of other variables. Loss of control, which can initiate the chain of events leading to an accidental release, can occur when manipulations fail in sequence, time of application, or magnitude. A fundamental loss of control can occur from a failure of components in the control system, including physical components and human error, or from faulty logical design of the control system. This latter failure is most likely to occur when deviations lead the control system to operate near or beyond the extremes of its design capabilities, in which case the design fault may be relative to the range of applicability rather than absolute. Proper design of the system and proper selection of individual components are therefore essential to ensure control system reliability and effectiveness.

The effectiveness of process control depends on the following key items:

- There are a limited number of fundamental manipulated variables for controlling any chemical process system.

- The controllability of a process depends on the interaction of both the process control system and the dynamics of the specific process.

- Critical variables should be measured directly. For example, a pressure indication should not be used as an indirect indication of temperature in an exothermic chemical reaction.

- The control system must be compatible with the level of training and skill of both operating and maintenance personnel.

- Duplication or redundancy of individual system components can increase the reliability of individual parts of the system.

- A redundant component must be completely independent of its duplicate.

- A definitive evaluation of process control for the critical variables in a process once a system has been installed is not easily accomplished by an evaluator and a judgement of its reliability must be inferred from limited information.

The reliability of the components that make up any control system has been studied for many common process equipment and instrumentation components (1, 8). One measure of reliability is the mean time between failures (MTBF), or the failure rate or frequency, the number of failures that occur per unit of time. Table 2-1 presents some values for selected components. Additional values are presented in other sections of this manual for specific equipment items. Such values can be used in conjunction with standard methods of quantitative systems analysis, fault tree analysis for example, to estimate overall system reliability.

Changes in control strategy or control system hardware may be an appropriate process modification to reduce the probability of a failure leading to an accidental toxic chemical release. However, a secondary hazard can be introduced if the effects of the changes on the process and its operation are not carefully reviewed beforehand. For example, although more sophisticated instrumentation and control schemes may be desirable in theory, in practice a lack of adequately trained or skilled operating or maintenance personnel may make such systems more hazardous. Direct process effects may include unforeseen changes in process dynamics.

TABLE 2-1.   FAILURE RATES OF SELECTED PROCESS SYSTEM COMPONENTS

Component	Failure Rate (faults/yr)
Control valve	0.25 - 0.60
Differential pressure transmitter	0.76 - 1.73
Variable area flowmeter transmitter	0.68 - 1.01
Thermocouple	0.088 - 0.52
Pneumatic controller	0.29 - 0.38

Source: Adapted from Reference 1. More detailed information may be found
in Reference 8 and other sources.

The following performance criteria should be considered when selecting
components for process control systems:

- Mechanical or electrical;

- Range;

- Accuracy (repeatability);

- Precision;

- Operating environment and materials of construction;

- Sensitivity; and

- Ability to withstand excursions from specifications or
operating conditions.

In summary, the process control system, which is understood to mean both
the human operator and the control hardware, must be one of the key elements

examined as a possible basis for process modifications to a process system. The specific modifications required must be made on a case-by-case basis but, in general, should be directed toward increasing the reliability and responsiveness of the control system to deviations and corrective actions for those variables most critical in preventing a loss of control. In the sections that follow, some of the specific fundamental principles of a chemical process, and pertinent process variables are explored in the context of possible process changes that could achieve the above objectives.

2.2 PROCESS CHARACTERISTICS AND CHEMISTRY

Numerous formal hazard identification and evaluation procedures are available which include examining chemical process characteristics and chemistry, as summarized in the PRM-User's Guide, that are amply discussed in the technical literature (9, 10). Fundamental process changes can often reduce inherent hazards by reducing in-process inventories of toxic materials and by reducing the severity of operating conditions. Some factors important in examining chemical processes for potential hazards are (1, 11):

- Inventory – the quantity of hazardous material in the process;

- Energy content – the inherent energy content specific chemical species as well as process temperature and pressure conditions;

- Time factor – maximum rate of a potential release, and warning time available for emergency counter measures;

- Substitutability – whether a less hazardous material can be substituted, or whether a substitution may result in less severe operating conditions; and

- Complexity - the number and kinds of process steps; the operational sophistication required for the process.

Hazard reduction through changes in these broad process characteristics involve other factors discussed below.

2.2.1 Process Materials

Evaluation of the properties of the materials in the process must include:

- Raw materials;

- In-process materials, such as reaction intermediates; and

- Product materials and process wastes.

The more known about all the materials in a process the better, but in practice, all potentially significant properties and even the identity of some intermediate process materials may not be known. Table 2-2 lists properties of significance for process hazards (12). For release prevention, the properties of toxic materials that should always be known are:

- Boiling point;

- Vapor pressure at ambient and process conditions;

- Specific heat;

- Heat of vaporization;

- Acute toxicity;

- Corrosiveness and reactivity properties;

TABLE 2-2.   PROPERTIES OF MATERIALS TO BE CONSIDERED IN CHEMICAL
PROCESS HAZARD EVALUATION

GENERAL INFORMATION REQUIRED

Corrosivity
Purity (Specifications of Grade
    Used)
Formula (Chemical Structure)
Contamination Factors
    (Incompatibility)
Quantity of Material Anticipated
Color
Hygroscopicity
Molecular Weight
Appearance
Odor
Physical State
Solubility
Synonyms

FLAMMABILITY INFORMATION REQUIRED

Flash Point
Fire Point
Flammable Limits
Specific Gravity
Vapor Density
Vapor Pressure
Heat of Vaporization
Boiling Point
Ignition Temperature,
    Autoignition Temperature
Spontaneous Heating
Dielectric Constant (Static
    Hazard)
Melting Point
Flow Point
Percent Volatiles
Extinguishing Media
Special Fire Fighting Procedures
Unusual Fire and Explosion Hazard
Gases Released During
    Decomposition
Heat of Fusion

REACTIVITY (INSTABILITY)
INFORMATION REQUIRED

Differential Thermal Analysis (DTA)
Impact Test
Thermal Stability
Detonation with Blasting Cap
Drop Weight Test
Thermal Decomposition Test
Lead Block Test
Influence Test
Self Acceleration Decomposition
    Temperature
Card Gap Test
Thermal Stability (Under
    confinement) JANAF
Critical Diameter
Limiting Oxygen Value
Hazardous Decomposition Products
Incompatibility
Conditions Contributing to
    Hazardous Polymerization
Conditions Contributing to
    Instability
Shock/Friction Sensitivity
Decomposition Temperature
Specific Heat
Gas Evolution
Adiabatic Temperature Rise
Heat of Reaction

Source: Reference 12.

- Flammability; and

- Explosive limits.

The boiling point defines the physical state of the material in the process, whether it is a liquid or gas at ambient and process conditions, for example. The vapor pressure defines the volatility of the material when it is below its boiling point, and how much pressure buildup might be expected from overheating below the boiling point. The specific heat and heat of vaporization allow estimation of volatilization rates when the release occurs as a liquid.

Toxicity data alerts the designer, operator, or evaluator of a chemical process facility to what materials are most hazardous to health, and suggests where to focus attention when taking measures for release prevention. Parameters that should be considered include the IDLH, LC50, and others defined in standard chemical toxicity references.

Toxic materials that are also corrosive require extra care in specification of materials of construction and also in monitoring equipment conditions. Corrosive materials that are not the primary toxic materials in a process are also important; however, when they are involved in processes where toxics are present. Corrosion can contribute to equipment failure and to a toxic chemical release.

Unrecognized reactivity in a process can cause a release hazard. If certain materials are accidentally mixed, or unexpected contamination occurs, a runaway thermal reaction or violent evolution of gases can result. Both may cause overpressure or spillage of a toxic chemical.

Flammability and explosive limits are important because fires and explosions can cause an accidental release as well as be caused by a release of a flammable material. A small leak could result in a major release if ignited.

Other properties may be significant for other reasons. One example is density. In a two-phase non-miscible liquid system, the denser liquid will settle out in vessels and tanks if agitation is not provided. Another example is viscosity, which can be a significant variable affecting process behavior in flow, heat transfer, and mixing parts of a process. A special situation where this might contribute to a hazard is a loss of temperature control where emergency cooling might not respond as rapidly as expected because a viscous layer near a cool heat transfer surface reduced the heat transfer coefficient to a lower value than anticipated in the original process design. These are only some examples. Evaluation of process materials may have to take into account properties beyond those that at first appear obvious.

2.2.2 Process Mode

The mode of process operation can affect the safety of a process. Three basic modes of process operation are: 1) batch, 2) semibatch, and 3) continuous operation. Table 2-3 lists some considerations that apply to each of these modes.

2.2.3 Reaction Thermodynamics

The fundamental thermodynamic characteristic of concern for reaction safety is the exothermicity of a reaction. This is the direction and rate of energy release. An exothermic reaction releases heat and an endothermic reaction absorbs heat. Exothermic reactions are usually of concern because they can lead to thermal runaway reactions and the attendant adverse consequences. The hazard of a specific exothermic reaction depends on both the magnitude of the heat of reaction and on the rate at which the heat is released or absorbed. The magnitude is determined primarily by the specific reaction and by the quantities of chemicals involved. The rate depends on these variables as well as on the temperature at which the reaction takes place. The primary process consideration when exothermic reactions are present is provision of adequate cooling. Temperature control is discussed more completely in Section 2.5.

TABLE 2-3.   CONSIDERATIONS FOR VARIOUS PROCESS MODES

	Batch	Semibatch	Continuous
Residence Time	Long	Long	Short
Feed System	Intermittent – short duration	Intermittent – long duration	Continuous – long duration
Startup – Shutdown	Frequent	Frequent	Infrequent
In-process Inventory*	Large	Large	Small
System Response			
Temperature	Slow	Slow	Fast
Pressure	Fast	Fast	Fast
Composition	Fast	Fast	Slow
Flow	Not applicable	Long	Fast
Fluid Dynamic Stability	Stable	Stable	Subject to Fluctuation

*At same annual throughput rate.

Endothermic reactions are not as hazardous as exothermic reactions, but do carry some hazard potential. Systems with endothermic reactions require heating. The hazard lies in the potential failure of the heating system in a way that would result in too much heat input to the process. This could cause overpressure through flashing, boiling, or expansion, and weaken or damage equipment.

2.2.4 Process Control System

Some fundamental concepts of control systems were discussed in Section 2.1. Types of control systems include:

- All manual,

- Manual/analog,

- Automatic/analog, and

- Automatic/digital, and

- Computer control.

The type of system used must be appropriate to the process. In general, reliability increases with sophistication, assuming proper design and installation.

2.2.5 Process Type

The type of process is characterized in terms of unit processes and operations. Certain unit processes and unit operations may be inherently more hazardous than others. The basic unit processes that make up most of the chemical industry have been defined in a classic work by Groggins (13) and more recently in a comprehensive study (14). Unit processes defined by the latter are presented in Table 2-4. Recognizing the class into which a process fits will highlight the specific hazard features associated with that class.

TABLE 2-4. UNIT PROCESSES IN THE CHEMICAL PROCESS INDUSTRIES

Acylation	Halogenation
Alkaline Fusion	Hydration
Alkylation	Hydrocracking
Amination	Hydroforming
Aromatization	Hydrogenation
Calcination	Hydrolysis
Carboxylation	Ion Exchange
Caustization	Neutralization
Chlorination	Isomerization
Combustion	Nitration
Condensation	Oxidation-Reduction
Coupling	Polymerization
Cracking	Pyrolysis
Diazotization	Reforming
Double Decomposition	Sulfonation
Electrolysis	
Esterification	
Fermentation	

The most common unit operations of chemical engineering are listed in Table 2-5. Again, certain operations that may inherently be more hazardous than others can be considered for substitution in process modification.

2.2.6 Severity of Process Conditions

The severity of process conditions is defined by the levels of pressure and temperature for each step of the overall process, in reactors as well as separation equipment. The effects of pressure and temperature from a process hazard perspective are discussed in Sections 2.4 and 2.5. In general, the higher these variables, the more hazardous the process and the greater the accidental release potential. Pressures in excess of 1,000 psig may be considered extreme. Pressures above 400 psig and temperatures above 500°F may be considered high.

While severity is usually interpreted in terms of high pressure and temperature, extremely low conditions can also be hazardous. As discussed in Sections 2.4 and 2.5, these conditions require special precautions in design and operations. Conditions near ambient pressures and temperatures generally will result in safer operation than temperatures and pressures either much above or below ambient. Since operating conditions are usually determined by the fundamental chemistry of the process, changes in these conditions will require research into the corresponding changes in process chemistry. Important factors include the physical states of reactants, products and rate processes, fundamental reaction kinetics and rates of mass transfer. These factors of the process can be examined to determine where changes might reduce severity.

The history of methanol synthesis is an example of a situation where changes in the fundamental process reduced the severity of operating conditions. For many years the synthesis of methanol required high temperature and very high pressure. Research resulted in fundamental change to a new process that allows methanol to be synthesized at much lower temperatures and pressures (15); thus operation hazards were reduced.

TABLE 2-5.   UNIT OPERATIONS IN THE CHEMICAL PROCESS INDUSTRIES

Absorption	Extraction
Adsorption	Filtration
Agglomeration	Heat Transfer
Centrifugation	Leaching
Crystallization	Mixing
Deionization	Membrane Transport
Desorption	Size Reduction
Distillation	Sublimation
Drying	Fluids Transport
Evaporation	Solids Transport

2.2.7 Process Complexity

Process complexity is characterized by the number of unit operations, streams, and variables that must be monitored and controlled in a process. In general, increased complexity increases the process hazard by creating more opportunities for unforeseen process interactions and more opportunity for errors in design, construction, and operation. Process complexity can be roughly estimated by considering the number of equipment items and the number of chemicals involved. A special consideration is how much recycling of process streams is in the process.

2.2.8 Process Modifications

Potential process modifications include any changes in the previously discussed areas that could reduce the toxicity, severity, sensitivity, or complexity of the process. Such modifications are limited by the fundamental constraints of process chemistry. The number and extent of possible modifications is probably greater the earlier in the development of a process they are made. An existing process offers less opportunity for fundamental change than a new process, but even here a change of catalyst, for example, might result in less severe, hence less hazardous, operating conditions.

The effects of modifications on traditional process costs will be process and plant specific. The effects will depend on equipment and operational features of the modified process. Though such costs cannot be estimated generically, some insight about the potential research and development costs of modifications can be gained by talking with technical and managerial personnel. It has been estimated that it costs about $150,000/year to maintain the average technical person in a large corporate environment. This translates into about $75/hour of researcher time. If it is assumed that this reflects the cost of the average technical professional/managerial person who would be involved in a program to modify the basic process characteristics and chemistry, the costs of such a program can be estimated from estimated levels

of effort for the program. Figure 2-2 presents a range of costs for such
programs based on average staff level commitment and program duration. Though
these costs exclude lost production while modifications are being made or the
capital investment required to physically modify a chemical process system,
they illustrate order of magnitude costs that show that fundamental process
modifications for process safety may equal the cost of new product/ process
development and implementation.

Assume that a typical chemical product may take 5 or more years of effort
and $10,000,000 to bring to market, excluding marketing costs. If a
modification required only 10% of that effort, costs would be on the order of
$1,000,000.

2.3   FLOW MEASUREMENT AND CONTROL

Flow measurement and control of both primary process fluids and service
fluids such as steam, air, nitrogen, and heating and cooling media is at the
heart of all process system operations. This section explores the role of
fluid flow in process events that may lead to accidental toxic chemical
releases, and flow hazard controls.

2.3.1   Flow Hazards

Process hazards associated with flow involve deviations that cause the
flow to be too low, too high, reversed, or fluctuating. Flow changes may be
caused by pressure changes in some prime mover for a fluid, or by changes in
the resistance to flow in piping or other equipment. Special hazardous
consequences of flow deviations can be identified for various individual
equipment items and processes. Some consequences are obvious, while others
are not. Some causes and hazards of loss of flow control are discussed below.

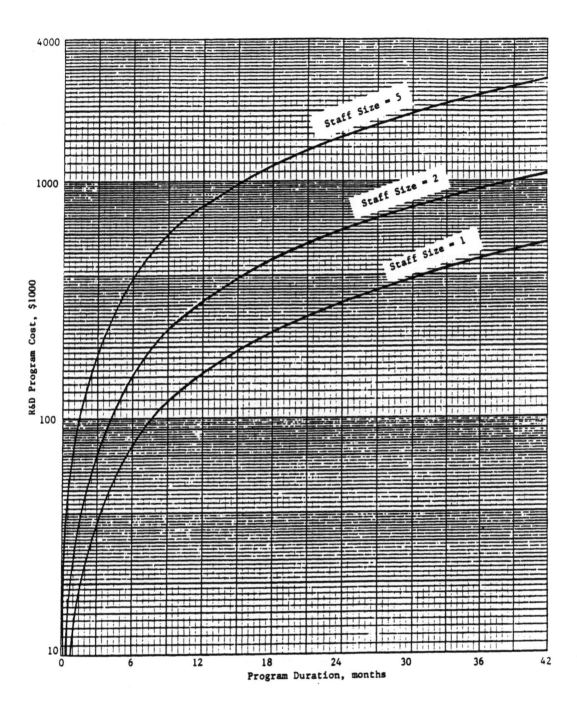

Figure 2-2. Estimated typical program costs for our R&D effort to
develop process modifications.

Low Flow--

Flow can be interrupted or reduced by factors related to the process, equipment, or operations. These factors may act independently or in combination and include:

- Blockage;

- Vapor lock;

- Cavitation in pumps;

- Mechanical failures of prime movers; and

- Leakage.

Blockage can occur from physical obstructions such as dirt, tramp materials in a line, or deposits which build up over time. These include products of corrosion and erosion. Polymerizable fluids may cause blockage or restrictions to flow. Freezeup is another possibility with some materials.

Certain equipment especially prone to blockage includes all piping, especially small diameter piping and tubing, heat exchangers, packed beds, and filters. Support screens and perforated plates in process equipment are also prone to blockage. Control valves, which usually have orifices much smaller than line size, are especially prone to restrictions and blockage, as are orifices in differential pressure flow sensors. Control valves, being automatic devices, can sometimes jam in a full or partially closed position, due to such conditions as dirt buildup on the stem, or mechanical failure of the activator.

Vapor lock can occur when a gas or vapor collects in a line or other process equipment space so that the vapor pressure exceeds the available pressure head available for flow. Liquid flow through the line or equipment may be restricted or stopped.

Cavitation in pumping equipment can occur when a liquid near its boiling point or contains a dissolved gas. If the suction pressure to a pump falls low enough, the dissolved gas may be suddenly released or the liquid may boil, trapping a vapor bubble in the pump that causes the pump to lose suction and flow.

Mechanical failures in prime movers, such as pumps and compressors, are obvious causes of low flow. These mechanical failures include electrical outages as well as actual breakage of mechanical components.

Leakage upstream in a fluid system can cause deficient flow downstream. Causes of leakage are discussed in the subsections on equipment in Section 3 of this manual.

Many possible hazards are associated with low or no flow, depending on the process, but the following are typical:

- Overheating caused by inadequate cooling to a process. The consequences of excess temperature are discussed in Section 2.5. Some consequences include runaway chemical reactions or physical damage to equipment.

- Overcooling caused by inadequate heating to a process. The consequences of deficient temperature are discussed in Section 2.5. Some consequences include an accumulation of unreacted material that could cause a runaway reaction on reheating, a freezeup in a critical line, or collapse of an atmospheric vessel due to vapor condensation and unrelieved vacuum.

- Process upsets caused by incorrect feed rate. In some exothermic reactions a portion of the cooling relies on a low temperature feed. An increase in the feed temperature or loss of one feed and not the other can lead to an excess of one reactant. This can sometimes lead to an excess exotherm and a runaway reaction. Too low a feed rate to distillation columns can cause excess heating from a reboiler if the control system does not respond properly, causing overpressurization and a relief valve discharge.

- Low flow in a pump can cause overheating and pump failure.

Excess Flow--

A primary cause of excess flow is an open valve which should either be partially or completely closed. The following are some typical hazards associated with excess flow:

- Excess flow of a cooling medium can cause the same problems as deficient flow of a heating medium, discussed previously.

- Excess flow of a heating medium can cause the same problems as deficient flow of a cooling medium, discussed previously.

- As with low flow, incorrect reactant ratios can result, causing excess exotherms, or excess boilup and pressures in heated process equipment.

- Overfilling of vessels can result if excess flow is not detected and shut off in time.

- Unwanted siphoning of liquids is another example of excess flow; or flow where none is intended.

Reverse Flow--

Reverse flow can occur when downstream pressure exceeds upstream pressure. A common cause of reverse flow is a pump failure, or failure of a directly pressurized fluid transport system. Reverse flow is not uncommon, even in systems protected by check valves that may fail to close or seal properly. Some typical hazards of reverse flow are the following:

- Reverse flow can lead to unintended chemical or physical reactions. An unexpected exothermic reaction in a storage tank from backflow of process reactants is an example of a chemical reaction hazard. Overpressurization through unintended flashing of liquid is an example of a physical reaction hazard. This could occur, for example by reverse flow of a liquid into a heated process vessel. Unintentional contact of water and hot oil in refinery systems has sometimes caused steam explosions.

- Reverse flow through pressure relief vent headers can cause obstruction of vent lines, contamination by leakage through unseated pressure relief valves on other equipment connected to a common header, and excessive back pressure on relief devices, with all the attendant hazards that may result from these effects.

Fluctuating Flow--

Fluctuating flow can be caused by a variety of conditions. Reciprocating prime movers such as piston pumps inherently generate a fluctuating flow, but processes using such devices are either designed to accommodate the fluctuations, or have special dampers to reduce the fluctuations. Fluctuating flow may be a hazard when it is unintentional. Some common unintentional causes of fluctuating flow include cavitation in centrifugal pumps and improperly tuned automatic process control loops. Some hazards of fluctuating flow include:

- Severe pulsations which can damage piping and equipment.
  Even moderate pulsations over time may lead to fatigue
  failure in piping or other process hardware.

- The overall controllability of a process is degraded,
  making it inherently less safe.

- Depending on the oscillatory time period of fluctuations,
  fluctuating flow can lead to the same problems as low or
  excess flow, depending on the relative process system
  dynamics in response to heating or cooling, for example.

Hammer Blow--

The phenomenon of hammer blow, also referred to as "water hammer," is the
propagation of a pressure shock wave through a liquid in a pipe line or other
equipment. It occurs when flow is abruptly shut off. An example is when a
control or solenoid valve is suddenly closed. It can also occur with manual
valves, such as ball valves, that can be closed quickly. The pressure shock
can be destructive to piping and other equipment in liquid-handling systems.
The primary hazard is piping or equipment rupture that could directly or
indirectly cause the release of a toxic chemical. A more insidious aspect of
this phenomenon is that even if its effects are not immediately evident, its
repeated occurrence may cause stress fatigue and eventual piping or equipment
failure.

Three conditions that can lead to "water hammer" are (16):

- Hydraulic shock;

- Thermal shock; and

- Differential shock.

Hydraulic shock occurs when there is an abrupt interruption of flow. The conversion of the kinetic energy of the moving fluid to a physical pressure force is proportional to the square of the velocity, and the resulting pressure shock wave can be considerable. Hydraulic shock can occur from an abrupt valve closure, which includes check valves in pump systems when pumps are started or stopped. This phenomenon can be alleviated by providing a water hammer arrester in the line or by providing slower closing valves. Special consideration needs to be given to this effect when the line handles toxic materials directly, or in other process lines whose rupture could indirectly lead to a toxic release. This phenomenon must also be avoided in automated process systems where actuator action can cause abrupt valve closures. Manual systems may be less prone to this effect, but may not be entirely immune.

Thermal shock can occur when bubbles of vapor trapped in condensing liquids suddenly collapse, causing rapid liquid influx and shock waves to occur within the liquid. This can also occur in condensing two-phase systems where gas becomes entrapped between slugs of liquids. Each of these forms of hydraulic shock can damage equipment.

Differential shock can occur from two-phase flow in piping where the gas phase creates a plug of liquid from wave action, resulting in a high gas side pressure drop from the upstream to downstream side of the plug. The plug accelerates and gains enough momentum that when a change of direction is encountered, at an elbow for example, the inertia prevents the plug from changing direction as rapidly as the gas and a shock occurs. This shock can be great enough to cause damage to the piping.

2.3.2 Technology of Flow Control

The flow measurement and control system is intended to prevent flow deviation from leading to accidental releases. Initial process design or subsequent process modifications must provide for proper selection, implementation, and operation of these systems. Adequacy and reliability are the two evaluation criteria for these systems.

Measurement and Control--

Flowmeters are used for sensing and measuring fluid flow. A controller
responds to this measurement and commands the final control element to appropriate action. The primary control element for fluid flow is the valve,
actuated manually for manual control or through a mechanical actuator system
for automatic control. The reliability of the fluid flow system depends on
the reliability of the individual components.

A distinction is made between sensing and measurement elements. Sensing
refers to detection; whereas measurement refers to converting flow sensing to
a numerical quantity expressed in appropriate units of measurement. This is
an important distinction when considering the appropriateness of various
components in a flow measurement system, since different combinations of
sensors and final measurement devices may encountered. Proper sensing and
measurement of flow is the basis for controlling all other process variables:
temperature, pressure, composition, and quantities.

The primary types of flow meters in common use in chemical process plants
include the following:

- Differential static pressure;

- Momentum transfer (target, turbine, and positive displacement meters);

- Variable area displacement;

- Vortex shedding; and

- Magnetic field.

Each type of meter is more appropriate for some applications than others. For
example, static pressure-based meters require small holes in either probes,
flanges, or pipe walls, which can become clogged by solids. Such meters are

obviously inappropriate when suspended solids are present in fluids. Improper selection of a flowmeter can lead to a process hazard.

The controller, responding to the information received from the flow measurement system, operates either a prime mover or a control valve. Flow control requires only proportional and integral controller action. The characteristic noise found in flow systems precludes derivative control action. A further characteristic of flow control is that there may be some transportation lag, but, in general, a flowing fluid responds quickly to the control action.

The final control element is usually the control valve. Sometimes, however, flow may be controlled indirectly by changing a pump setting or a gas pressure on a pressurized fluid feed tank.

Important criteria in the selection and operation of control valves include the proper size, trim, and action. Control valves and a pressurized feed tank system usually allow a quicker control response than a pump, with the exception, perhaps, of a complete shut off. This should be born in mind when evaluating flow control options for high-hazard systems. An automatic control valve should have a fail-safe action. It should fail appropriately open or shut in the event of a loss of electrical power or instrument air.

Emergency Trip Systems--

Flow switches may be used to sense either the initiation or cessation of fluid flow. These are primary devices in alarm and shutdown systems for chemical processes based on a flowing stream.

There are two categories of flow switches: variable-area devices and velocity-sensing devices (17). Low cost, simple switches based on both of these operating principles are available. These switches are suitable for applications where accurate settings are not required, such as lubricating-oil flow to rotating machinery.

More accurate switches may be required in critical applications, such as detecting flow deviations before a process is endangered. Here variable-area or rotameter type switches or orifice type differential pressure flow switches would be appropriate. Differential type switches can be direct acting or can transmit a pneumatic or electronic signal to the switch mechanism, alarms, or relays.

Variable-area switches have good rangeability, but usually cannot be used for flows as large as differential switches can handle. The latter are more limited at low flow ranges because of the square-root dependency of the signal or flow. They may become unreliable if the flow drops too rapidly below the maximum design flow. These types of switches are sensitive to the range selected for their application. Many other types of switches are available. For example, a pressure switch can be used as a flow switch by sensing pressure in a system when flow is present.

When integrated into a total emergency trip system, flow switches actuate valves and relays to close and open lines, and turn motors on and off. As will be discussed in other portions of this document, emergency systems controlling fluid flows are the basis for emergency control of pressure, temperature, quantity, and composition.

In general, exceeding the operating range of these types of switches may cause no permanent harm. However, where the reliability of the switch is really critical, the range must be selected with careful consideration of the likelihood of overrange service and the risk of harm caused by it. More details on the role of flow switches in process emergency control can be found in various literature sources (e.g., Reference 17).

Reverse Flow Prevention--

Reverse Flow prevention can be achieved by specific equipment, equipment configurations, and operational procedures. Equipment includes check-valves, sensing, automatic shutoff valves, and positive displacement pumps (rather than centrifugal pumps). Equipment configurations include relative elevation differences or traps between equipment items. Procedural approaches include direct shutoff with existing valves, or diversion of flow. The procedural methods depend on the sensing of and corrective response to reverse flow, and on previous provision for diversion in the design or construction of a facility.

The technological basis for reverse flow prevention is, therefore, the installation of the appropriate equipment. The most common device is the check valve. Various types of check valves are shown in Figure 2-3. Typical applications of the various types are given in Table 2-6 (18). While check valves provide some protection, they can fail by sticking open or not reseating properly. Sole reliance on check valves may be undesirable in critical service.

The principle of reverse flow prevention using positive displacement pumps is illustrated by comparison between a positive displacement pump and a centrifugal pump, as shown in Figure 2-4. Because the inherent clearances within a positive displacement pump are less than a centrifugal pump, reverse flow may be reduced when a positive displacement pump is not running. When it is running, an increase in upstream pressure is met by an increase in pumping head and maintenance of forward flow. In a centrifugal pump, the flow falls off sharply as upstream pressure increases and in extreme cases reverse flow can occur even if the pump is running. An idle centrifugal pump offers very little resistance to backflow.

Reverse flow is prevented through elevation differences when the feed vessel is sufficiently higher than the host vessel, ensuring that no condition in the host vessel would be sufficient to overcome the fluid head differential between the vessels.

The trap principle of reverse flow prevention is based on interruption of flow with an intermediate vessel. During normal flow, the fluid is repumped for continued transport. Other arrangements may exist. The basic principle is interruption of stream continuity so that flow is not readily reversed.

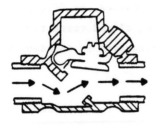

SWING CHECK VALVE

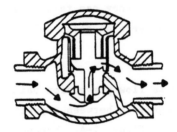

HORIZONTAL LIFT CHECK VALVE

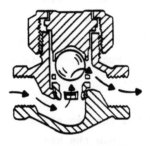

BALL CHECK VALVE

Figure 2-3.   Check valve types.

TABLE 2-6.   TYPES OF CHECKVALVES

Type	Mechanism	Application	Limitations
		General: Prevent back-flow in lines.	
Swing check	Flow keeps swing gate open, while gravity and flow reversal close it. Tilting-type is pivoted at center and insures closing without slamming. Outside levers and weights are used on standard swing checks greater sensitivity for changes in flow is required.	Where minimum pressure drop is required-Best for liquids and for large line sizes.	Not suitable in line subject to pulsating flow. Same styles operate only in a horizontal position.
Piston check	Flow pattern as in globe valve. Flow forces piston up and reversal and gravity returns it to seat.	Good for vapors, steam, and water. Suitable for pulsating flow.	Many designs are for horizontal service only. Not common in sizes over 6 in. Not recommended for service which deposit solids.
Ball check	A lift-type check consisting of a ball with guides.	Stops flow reversal more rapidly than others. Good for viscous fluids which deposit solid residue that would impair operation of other types. Vertical or horizontal installation is possible.	Not common in sizes over 6 in. Not suitable for lines subject to pulsating flow.

Source: Adapted from Reference 18.

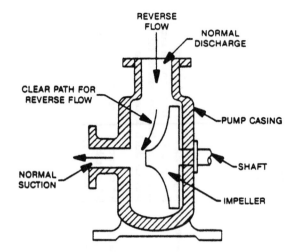

A. ILLUSTRATION OF REVERSE FLOW THROUGH CENTRIFUGAL PUMP

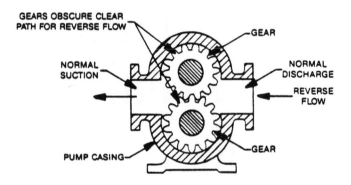

B. ILLUSTRATION OF PREVENTION OR REDUCTION OF REVERSE FLOW IN A GEAR PUMP

Figure 2-4. Examples of backflow through pumps.

Blockage Prevention--

Prevention of blockage means that any valves or pumps must be prevented from failing in a closed position when they should remain open. Lines, ports, and orifices must be kept clear of debris.

There are numerous potential sources of debris in chemical processes. Especially during startup especially, debris is present from pipe tape, ordinary dirt, welding drippings, corrosion scale, and the nuts, bolts, and other items inadvertently left in lines and equipment. Vigilance on the part of the startup personnel is the first defense against flow blockage occurring from such sources. Technological solutions include screens, strainers, filters, and traps.

When in-line screens, strainers, and filter devices are applied to prevent blockage, a secondary hazard is associated with the use of such devices; if not properly maintained they can become the source of blockage. The appropriateness of use must be evaluated on a case-by-case basis. Also, a filter can be provided with a bypass activated by a high upstream pressure on high pressure drop across the filter.

Two insidious causes of flow blockage are the buildup of corrosion products and the accumulation of polymer materials from the side reactions of various process fluids. Another is the attrition and breakage of solid materials in processes such as catalyst pellets and packing using in packed beds. Technological solutions here include the methods listed above, as well as changes in the characteristics of the solid materials used in the process.

Blockage or flow reduction is also a possibility under certain circumstances with slurries, high viscosity fluids, and melted materials at low temperatures. This can occur, for example, in coolers, or due to very low ambient temperatures. Solidification or buildup in slurries or freezing materials can reduce flow or cause complete blockage. The increase in viscosity can sharply reduce the flow rate for fluids.

Siphoning Prevention--

Siphoning can be prevented through proper design to ensure that if piping from a lower elevation to a higher elevation is submerged at the higher elevation end, appropriate steps have been taken to prevent backflow siphoning. Another technique is an interlock system wherein closure of a valve or cessation of pumping or failure of flow to stop will activate an alarm.

2.3.3 Control Effectiveness

The effectiveness of flow control is evaluated in terms of performance, limitations, and reliability. Proper selection is important to ensure that these criteria are met. Some guidelines for flowmeter selection are summarized in Table 2-7 (19).

The performance of individual components and complete systems depends on the interaction of physical and design choices made by the designer and fabricators. Sensors are limited by how sensitive they are to pressure and temperature beyond their design limits, and resistance to corrosion or erosion by process fluids. Controllers vary by type and individual vendor.

The effectiveness of emergency shutdown systems depends on individual components and on the overall system design. A key factor here is the response time required to achieve safe flow shutoff and the integrity of that shutoff.

Reverse flow prevention depends on the functioning of check valves and on the prevention of overpressurization of downstream vessels. Siphoning prevention relies on proper process design and construction and on other safeguards such as interlocks and alarms.

TABLE 2-7. FLOWMETER SELECTION GUIDE

Flowmeter Element	Recommended Service	Turn-Down Ratio	Typical Accuracy, Percent	Viscosity Effect	Relative Cost
Orifice	Clean, dirty liquids; some slurries	4 to 1	± 2 to ± 4 of full scale	High	Low
Wedge	Slurries and viscous liquids	3 to 1	± 0.5 to ± 2 of full scale	Low	High
Venturi tube	Clean, dirty, and viscous liquids, some slurries	4 to 1	± 1 of full scale	High	Medium
Flow nozzle	Clean and dirty liquids	4 to 1	± 1 to ± 2 of full scale	High	Medium
Pitot tube	Clean liquids	3 to 1	± 3 to ± 5 of full scale	Low	Low
Elbow meter	Clean, dirty liquids, some slurries	3 to 1	± 5 to ± 10 of full scale	Low	Low
Target meter	Clean, dirty, viscous liquids, some slurries	10 to 1	± 1 to ± 5 of full scale	Medium	Medium
Variable area	Clean, dirty, viscous liquids	10 to 1	± 1 to ± 10 of full scale	Medium	Low
Positive Displacement	Clean, viscous liquids	10 to 1	± 0.5 of rate	High	Medium

(Continued)

TABLE 2-7 (Continued)

Flowmeter Element	Recommended Service	Turn-Down Ratio	Typical Accuracy, Percent	Viscosity Effect	Relative Cost
Turbine	Clean, viscous liquids	20 to 1	± 0.25 of rate	High	High
Vortex	Clean, dirty liquids	10 to 1	± 1 of rate	Medium	High
Electro-magnetic	Clean, dirty, viscous conductive liquids and slurries	40 to 1	± 0.5 of rate	None	High
Ultrasonic (Dobbler)	Dirty, viscous liquids and slurries	10 to 1	± 5 of full scale	None	High
Ultrasonic (Time-of-travel)	Clean, viscous liquids	20 to 1	± 1 to ± 5 of full scale	None	High
Mass (Coriolis)	Clean, dirty, viscous liquids, some slurries	10 to 1	± 0.4 of rate	None	High
Mass (Thermal)	Clean, dirty, viscous liquids, some slurries	10 to 1	± 1 of full scale	None	High
Weir (V-Notch)	Clean, dirty liquids	100 to 1	± 2 to ± 5 of full scale	Very low	Medium
Flume (Parshall)	Clean, dirty liquids	50 to 1	± 2 to ± 5 of full scale	Very low	Medium

Source: Adapted from Reference 19.

Overfill protection relies on proper quantity measurement, as discussed in Section 2.6 of this manual, as well as an overall good flow control and proper shutoff.

The effectiveness of any system depends on its reliability. The fundamental elements of reliability include measurement, attenuation, and shutoff. The primary components of measurement are sensor and measurement devices. Attenuation is achieved by valve action, as is shutoff. System reliability depends on the reliability of the individual components and the system hardware and software. Software components include dead bands, dead time, and other dynamic response characteristics. And, as pointed out in the discussion on process control, the ultimate control of flow depends on the interaction of the control system with the process flow characteristics as much as on the characteristics of the control loop itself. Table 2-8 presents some typical reliability information on flow control components expressed as typical failure rates.

If a hazard evaluation discovers that the flow of specific streams is critical to preventing conditions that could lead to an accidental release, the entire flow sensing, measurement, and control system needs to be examined critically for its integrity. If necessary, the system needs to be modified to function more reliably.

2.3.4 Summary of Control Technologies

Table 2-9 summarizes major hazards or hazard categories associated with flow, and the corresponding control technology for both new and existing facilities. Numerous individual control technologies or procedural changes can be inferred for each category, based on the proceeding discussion of this section.

TABLE 2-8. TYPICAL FAILURE RATES OF FLOW CONTROL COMPONENTS

Component	Failure Rate Failures/Year
Flowmeters	
D/P cell and transmitter	0.76-1.73
Magnetic	2.18
Flow controller	0.29-0.38
Control valve	0.25-0.60
Flow switch	1.12
Flow indicators	0.026
Check valve (backflow prevention)	1.10×10^{-4}
Control loop	1.73

Source: Adapted from Reference 1.

TABLE 2-9.   MAJOR HAZARD AND CONTROL TECHNOLOGY SUMMARIES

Process Variable	Hazard	New Facility	Existing Facility
Flow	• Low Flow - Blockage	• Flow, pressure, and temperature monitoring • Screening or filtration • Materials of construction • Equipment selection • Construction supervision • Design for polymerization inhibitors	• Add new flow, pressure, and temperature monitoring • Add screening or filtration • Change materials of construction • Change equipment type • Enhance maintenance • Add polymerization inhibitors
	- Vapor lock	• Design of temperature control system • Equipment selection	• Change temperature control system • Change equipment type • Change process chemistry
	- Cavitation	• Design of temperature control system • Equipment selection	• Change temperature control system • Change equipment type • Change process chemistry
	- Mechanical failure in prime mover	• Equipment selection	• Change equipment type

(Continued)

TABLE 2-9 (Continued)

Process Variable	Hazard	New Facility	Existing Facility
	– Leakage	• Equipment selection	• Change equipment type
	• Excess Flow	• Flow, pressure, and temperature monitoring	• Add new flow, pressure, and temperature monitoring
		• Flow limiters	• Add flow limiters
		• Automatic flow shut-off trips	• Add automatic flow shut-off trips
		• Emergency bypass diversion	• Add emergency bypass diversion
	• Reverse Flow	• Check valves	• Add check valves
		• Pressure control	• Add pressure control
		• Emergency bypass diversion	• Add emergency bypass diversion
		• Equipment selection	• Change equipment type
	• Fluctuating Flow	• Use pulsation dampeners	• Add pulsation dampeners
		• Process control system design	• Change process control system
		• Equipment selection	• Change equipment type
	• Hammer Blow	• System piping design	• Change piping
		• Equipment selection	• Change equipment type

2.3.5 Costs

Both rough, order-of-magnitude capital and total annual costs of the various technological approaches to process modifications for flow have been estimated. Table 2-10 summarizes estimated costs.

2.3.6 Case Example (1)

A serious accident occurred at a plant where ethylene oxide and aqueous ammonia were reacted to produce ethanolamine. Ammonia backflowed in the ethylene oxide line, reaching the ethylene oxide storage tank. The ammonia went past several non-return valves in series, and past a positive displacement pump through a relief valve that discharged into the pump suction line. The ammonia reacted with 7,940 gal. of ethylene oxide in the storage tank, which ruptured violently. The resulting vapor exploded, causing widespread damage and destruction.

2.4 PRESSURE CONTROL

Pressure is the driving force for flow, and is therefore the driving force in any chemical release. Control of pressure is central to the prevention of accidental chemical releases. This section discusses pressure measurement and control considerations.

2.4.1 Pressure Hazards

Overpressure and underpressure can increase the probability of an accidental chemical release. Either can cause physical failure of process or storage equipment. Overpressure can cause the opening of a relief device, thereby allowing a toxic chemical to enter the environment. Potential consequences range from a small release of hazardous materials through leakage to a large, sudden release through the total and rapid failure of containment. Generally, overpressures are considered more hazardous than underpressures,

TABLE 2-10.   COSTS OF COMPONENTS ASSOCIATED WITH PROCESS MODIFICATIONS FOR
              FLOW SYSTEMS

Basis: 4 inch diameter pipe line. Flow rate 250 gpm Pressure 50 psig.	Capital Cost Range ($)	Annual Cost Range ($/yr)	References
	(1986 Dollars)		

Flowmeter (D/P cell and
    transmitter, magnetic,
    and turbine)
	2,500 - 5,100	380 - 780	5, 20, 21

Flow switch	-	-	21
Flow indicators	380 - 1,000	60 - 150	5, 20, 21, 22
Check valve	380 - 570	60 - 90	5, 20, 21, 22
Controllers - Single loop, PID - Simple interactive, PID - programmable, PID	800 - 1,600 $1,600 - 3,000 2,000 - 6,000	69 - 138 138 - 260 173 - 519	4, 24, 25
Control Valve	3,000 - 6,000	450 - 900	23, 26, 27
Control loop - Conventional	6,000 - 12,000 3,000 - 15,000	900 - 1,800 260 - 1,298	Composite

but process underpressure can also result in equipment failure by collapse or by causing leakage or backflow of incompatible materials. Process design or modification must include consideration of both over- and underpressure.

Structural failures caused by overpressure will often occur initially at a seal or joint and secondly in cracking or rupture of vessels, piping, and casings of other equipment (e.g., pumps). Potential sources of process leaks caused by overpressure include flanges, valve stems, pipe joints, welded or riveted vessel seams, and pump or compressor seals. Structural failures in vessel and pipe walls and equipment casings may occur as limited cracks or total failure.

A second, indirect group of hazards are process effects from pressure deviations from specified operating conditions. Since pressure is the driving force for flow, pressure deviations can lead to flow deviations with their attendant hazards, as discussed in Section 2.3 on Flow Control. Pressure also can affect process chemistry. Disturbances in pressure for a pressure sensitive process can sometimes result in the formation of unstable or incompatible by-products that could contribute to the pressure disturbance and result in overpressure and accidental release.

Initial process design and process modifications associated with pressure must consider all factors contributing to pressure deviations and loss of pressure control. These include process, equipment, and operating considerations.

Three related primary process events can lead to loss of pressure control: thermal expansion, excess material generation, and flow restriction.

Thermal expansion may be caused by a loss of temperature control or by excess heat due to a high reaction rate. Loss of temperature control may be caused by a malfunction in cooling and heating systems, or inadequate heat transfer due to inadequate mixing or fouling of heat transfer surfaces. Loss of reaction rate control may be caused by a loss of flow, temperature, composition control, or inadequate mixing.

Excess material generation may occur from runaway reactions or from uncontrolled flashing of liquids from sudden contact with high temperatures. If a reaction generates more moles of material than it consumes, then an accelerated reaction rate may generate enough excess material to result in an overpressure. Uncontrolled flashing could occur if a liquid were inappropriately introduced into a high temperature process.

Flow restrictions that occur as a result of fouling, freezing, valve closure, or other physical blockage may lead to an overpressure upstream of the restriction.

In many instances, the process failures listed above are caused by a mechanical failure in equipment. These failures may be the result of exceeding design capabilities, improper design, poor maintenance, defective equipment, fatigue failure, or corrosion.

Since not every form of mechanical failure can be prevented, insufficient design preparation to minimize the impact of mechanical failure can also be listed as a factor that contributes to loss of pressure control.

A loss of pressure control may occur as a result of fundamental design flaws, poor operating and maintenance practices, or insufficient operator training, especially in response to non-routine operating conditions.

2.4.2 Technology of Pressure Control

The pressure measurement and control system helps prevent accidental releases resulting from pressure deviation. Initial design or subsequent process modifications must provide for proper selection, implementation, and operation of these systems. Pressure relief for either excess over- or excess under-pressure is a fundamental aspect of pressure control.

Measurement and Control--

Pressure measurement relies on both sensing and final measurement devices. Control relies on how a system responds to pressure to take corrective action. Many types of pressure sensors and measuring devices are available, each having some unique feature in the mechanism it uses to sense and measure pressure. Probably the most common type of device in the process industries is the Bourdon gage, which is described in numerous texts and handbooks. A second category of devices is electronic. The most common type in this category is probably the strain gage sensor coupled with an appropriate signal conditioning and read-out instrument. The choice of device depends on the needs of the process and the operating environment. The actual pressure device must often be protected from direct contact with the process fluids. The device mechanism, combined with the method by which it is protected from process fluids or even external environments will determine the device's overall performance.

The pressure controller is the next device to be considered in a pressure control system. Details of these devices are discussed in numerous texts and handbooks in the general technical literature. Pressure control is achieved through control of the flow of a fluid (discussed in Section 2.3 on flow control).

Preventing the realization of pressure hazards requires the proper selection of pressure measurement and control equipment in both initial design and subsequent process modification. Proper selection must consider accuracy, precision, and reliability. A change in a pressure sensing and measurement element may be an appropriate process modification to enhance process safety.

Emergency Pressure Relief Systems--

An emergency pressure relief system is the final step in preventing overpressure or under pressure when the basic control system fails. Under certain circumstances a pressure relief system will be designed to vent directly to the atmosphere. This type of system is often used for emergency venting of explosions or where nonhazardous materials are involved. Its appropriateness where toxic materials are involved is debatable, unless an explosion or other effects of overpressure are a greater hazard. For toxic materials, a total containment system designed to capture and store the released material for later treatment and disposal, or to capture and treat the material as it is vented to render it "nonhazardous" before it is released, may be desirable. These systems are composed of a pressure relief device (a rupture disc, safety-relief valve or combinations), a vent header system, and a catch vessel or a final treatment device (usually a flare or a scrubber). The treatment devices are discussed in more detail elsewhere in this document; descriptions of the remaining components in a total containment system are presented below.

Rupture discs--A rupture or bursting disc is a non-reclosing pressure relief device composed of a pressure sensitive disk or membrane held in place by a support structure. A disc subject to reverse pressure may be fitted with an additional support designed to prevent reverse pressure rupture of the disc. A typical rupture disc arrangement is illustrated in Figure 2-5.

Types of rupture discs include domed discs, composite discs, reverse domed discs, and flat discs. Materials of construction include metal, plastics, resin-impregnated graphite or some combination of these. The rupture disc is frequently constructed with a corrosion- resistant coating or liner. A temperature shield such as an insulating flock may be installed upstream of the disc to protect it from high temperatures; however, this device would not be appropriate for rupture discs discharging into a downstream pipe or manifold system where constrictions would trap the flock material and cause a pressure venting discharge restriction.

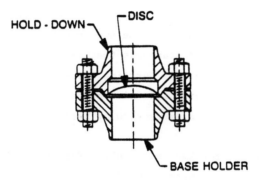

Figure 2-5.   Typical rupture disc installation.

Discs may be piercing or non-piercing. A piercing disc has an assembly that contains a knife edge against which the disc presses as it ruptures to ensure a clean, complete, full-dimensional break.

Rupture discs may be used alone or upstream of a safety relief valve (discussed below). A rupture disc is often placed upstream of a relief valve to prevent leakage through a relief valve that may not reseat effectively, or to protect the valve against corrosion or even plugging from protracted polymerization of monomer vapors. This latter phenomenon has been known to occur.

Sizing and selecting a rupture disk should follow accepted design procedures, such as those recommended by the Design Institute for Emergency Relief (DIERS) of the American Institute of Chemical Engineers (AIChE), American Petroleum Institute (API) or the American Society of Mechanical Engineers (ASME). A new set of standards that is a fairly comprehensive guide for both manufacturers and users of bursting disks has been assembled in Great Britain. An overview of rupture disks with numerous references is presented in a comprehensive book on loss prevention by Lees (1).

The Design Institute for Emergency Relief Systems (DIERS) has assembled a manual and a computer program package useful to a safety-relief system specialist; however it is not geared toward the novice. This document is particularly important when considering the effect of two-phase, vapor-liquid flow on the sizing of a relief device. Two-phase flow occurs when a pressurized vessel that contains liquid is rapidly depressurized. The occurrence of two-phase flow during emergency relief almost always requires a larger relief system (two to ten times the area), compared to that required by vapor venting. Some of the information generated by DIERS has been used by Fauske to develop a nomograph for sizing an emergency relief device (28).

Safety-relief valves--The term "pressure relief valve" is a generic term applied to various types of valves used to relieve pressure. API RP 520 gives additional definitions within the safety relief valve category (30). A safety valve is characterized by rapid full opening, or pop action, and is normally used for steam, air, gases, or vapors. Relief valves are used primarily for liquid service and are characterized by valve openings proportional to the rise in pressure over the opening pressure. A safety-relief valve can be used for either liquid, vapor or gas service. In practice, the terms safety valve, relief valve, and safety-relief valve are often used interchangeably.

A conventional safety-relief valve is illustrated in Figure 2-6. For this device, the bonnet (as shown in the diagram) may be vented either to the atmosphere or to the discharge side of the valve. Backpressure will decrease the set pressure for a valve with a bonnet vented to the atmosphere and will increase the set pressure for a valve with a bonnet vented to the discharge side of the valve. To compensate for the effects of backpressure, a balanced safety-relief valve may be used. For this type of valve, backpressure in the discharge line will have very little effect on the set pressure of the valve. The balanced safety-relief valve is often the most appropriate type of valve for a total containment system.

A final type of valve is the pilot-operated safety-relief valve, which is designed to achieve good mixing of the substance being discharged and air. A pilot-operated safety valve is often used when a flammable material must be diluted below its flammable limit with air before it can be safely released.

The API, ASME, and the National Fire Protection Association (NFPA) have each issued guidelines and standards for the sizing and installation of safety-relief valves (29,30,31). Criteria for sizing include fire exposure, fill rates, thermal expansion, and reaction/decomposition venting. DIERS manual also applies to safety relief valves. William A. Scully has prepared a brief discussion of some of the common malfunctions of safety-relief valves and how these malfunctions can be corrected (32). D.M. Papa discusses the effects of back pressure on a safety relief valve (33). The pitfalls of

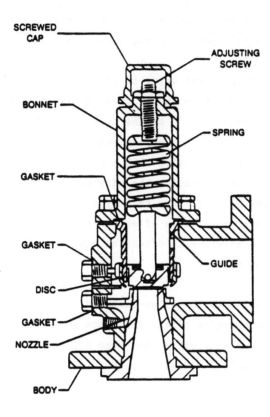

Figure 2-6.   Cross-section of a typical pressure relief valve.

sizing a relief valve when the equipment (specifically, a distillation column) will be used for more than one process is discussed by Bradford et. al. (34). An overview of pressure relief valves, with numerous additional references, is presented by Lees (1).

An important consideration in relief valve installations is to keep them from being isolatable while still providing for their maintenance while a process unit is·on-line. A twin-valve configuration with a two-way, three-port block valve, which blocks only one relief valve at a time, is desirable. When a standard block-valve is used upstream of either a single or two relief valves, it should be locked or sealed open to prevent unintentional or unauthorized closure.

Vent headers--The vent header is the pipework that delivers material from the discharge side of the relief device to a point where it can be rendered nonhazardous. These systems are often complex, having several relief devices using one common header system. For toxic materials, it might be well to consider addition of a header to the relief device system that leads to some form of secondary containment or treatment. However, depending on the situation, even with a toxic material, a discharge directly to the atmosphere still might be a safer condition than risking a restricted relief discharge. The appropriate configuration must be evaluated case by case.

Sizing the vent header for an emergency relief system is not a simple matter. For safety, the system must be sized to accommodate more than the flow from the largest single source. For economy, the system must usually be sized for some flow less than the sum of all sources. In approach presented by Fitt, overpressure due to fire, electrical power failure, instrument air failure and cooling water failure are considered (35). Fitt calculates the effect of each of these failures for three different time periods and sizes the pressure relief header for the maximum load that results from failure of one of the items. A method for sizing relief headers is also given in API RP 520 (29).

Pressure trip systems--A pressure trip system may be used instead of an emergency relief system. A pressure sensor activates a switch when the pressure is approaching an unacceptable level. The switch then shuts a valve to isolate a part of a process from the source of pressure. As an example, a pressure switch in downstream equipment may stop the flow of steam to a feed preheater or to a distillation column reboiler.

Because of the limitations discussed for relief systems, a pressure trip system should be several times more reliable than a pressure relief valve for the same application. Calculation of this reliability should be based on the calculated proportion of the time for which a system will be in the failed state, referred to as the fractional dead time. This method considers both the fault-rate of the individual components and the frequency of proof testing.

2.4.3 Control Effectiveness

The control effectiveness of pressure control is evaluated in terms of performance, limitations, and reliability.

Measurement and Control--

Table 2-11 lists a variety of pressure sensing and measurement devices and pressure ranges over which they may be used. In addition to a pressure range, each device has a variety of other performance limitations, including acceptable sensitivity, operating temperatures, reproducibility, and materials of construction. The range of values for these parameters is quite large and a device is routinely available for almost every temperature, pressure range, and operating environment likely to be encountered in typical chemical processing applications.

Because of the large number of choices, care must be taken to select an appropriate device for each application. Each pressure device can only be expected to perform satisfactorily within its design specifications. The

TABLE 2-11. PRESSURE SENSING AND MEASURING DEVICES (Adapted from Reference 19)

Applicable Pressure Ranges

mm Hg absolute (1 mm Hg = 133 Pa)

"H₂O (1"H₂O ≅ 250 Pa)

PSIG (PSIG = 6.9 kPa)

Scale markings: 10^{-14} 10^{-10} 10^{-6} 10^{-3} 10^{-1} 1 50 200 600 -300 -200 -100 -10 -5 -1 -0.1 +1 +5 +10 +100 +200 +300 4 7 11 10^2 10^3 10^4 10^5 10^6

Category	Type of Device
Bellows	Abs. Press. Motion Balance
	Abs. Press. Force Balance
	Atm. Press. Ref. Motion Bal.
	Atm. Press. Ref. Force Bal.
	Aneroid Manostats
Bourdon	Conventional Bourdon
	Spiral Bourdon
	Helical Bourdon
	Quartz Helix
Diaphragm	Abs. Press. Motion Balance
	Abs. Press. Force Balance
	Atm. Press. Ref. Motion Bal.
	Atm. Press. Ref. Force Bal.
Electronic	Strain gauge
	Electronic Transmitters
	Capacitive Sensors
High-Pressure Sensors	Dead Weight Piston Gauge
	Bulk Modulus Cell
	Manganin Cell
Manometers	Inverted Bell
	Ring Balance
	Float Manometer
	Barometers
	Visual Manometers
	Micromanometers
	Cartesian Divers
Pressure Repeaters	D/P Cell
	Standard Diaphragm
	Button Diaphragm
Ionization	Hot Cathode
	Cold Cathode
Thermal	Thermocouple
	Thermopile
	Resistance Wire
Mechanical	McLeod
	Molecular Momentum
	Capacitance

degree to which a device can withstand excursions beyond its operating range
varies. Some sensors may be permanently damaged by even brief excursions out
of their operating range. The sensitivity of devices to such excursions
should be taken into account during initial design or subsequent
modifications.

Because there may be no external indication that a pressure device is
giving an incorrect reading, redundancy of pressure sensing and measurement
should be used in all critical processes. For example, a common failure in a
Bourdon gage is to jam at some false reading. It may be advisable to use two
different types of devices if the potential cause of device failure is
sensitivity to process pressure excursions. An appropriate solution might be
to use a cruder, more robust type of device to roughly indicate the pressure
as a back-up for the more sensitive sensor actually used for process control.

Information on the reliability of a specific device can be supplied by
the vendor, although this information may not itself be reliable and must be
used with caution. A number of statistics on reliability have been assembled,
some of which are presented later in this section.

Emergency Pressure Relief Systems--

Rupture discs--A rupture disc is generally preferred over a relief valve
when pressure rise may be so rapid as to virtually constitute an explosion,
when even minor leakage cannot be tolerated, or when the potential for
corrosion and/or blockage would limit the effectiveness of a valve.

The set pressure for a rupture disc is usually specified by a range, that
can vary anywhere from +/-2% to +/-25% of the mean bursting pressure (34).

The reliability of rupture discs varies with type. The simplest form is
a flat disc made of a material weaker than the vessel it is designed to
protect. Because it is virtually impossible to accurately predict the

bursting pressure for this type of disk, a large margin is required between
the operating pressure and the bursting pressure. An exception is the brittle
bursting disc, often made of resin-impregnated graphite, which is often used
for low pressure and corrosive applications.

Several types of domed discs are available. One variety is a thin piece
of dome-shaped metal. Because they are very thin, their bursting characteristics are altered by even slight corrosion, elevated temperatures, or
minor deformations of the surface. Pressure fluctuations near the set pressure of these discs can result in premature rupture as a result of fatigue.
Generally these discs cannot be operated above 70% of their rated bursting
pressure, and even less at elevated temperatures (28). They will collapse
under a comparatively small back pressure and are generally installed with a
backpressure support. An improvement in domed design is a disc made of
thicker material radially scored to purposely weaken the surface. Because
most of the disc is made of heavier material, it is less susceptible to
fatigue. These discs can generally be operated at up to 80% of their design
burst pressure (1).

The inverted domed disc has advantages over either of the domed disks
described above. One variety is composed of an inverted dome with a knife
edge or some other variety of puncturing device positioned next to the disc.
When the bursting pressure is reached, the dome inverts and is punctured.
Another variety is so constructed that when the disk inverts it pops free of
its support and is caught immediately downstream. Both smooth and scored
surface inverted dome disks are available. The danger with inverted domed
discs, however, is that the disc may not be punctured when it inverts. These
discs can operate at pressures up to 90% of their design capacity and can be
manufactured to relieve within 2% to 5% of their design capacity (35).

Rupture discs are potentially hazardous because they are nonresealable
and they provide total depressurization. A rupture disc gives no external
indication that it has blown. Once a disc has blown, the vessel it was

protecting is no longer isolated from whatever else may be present in the containment system, and it is possible for material to backflow into the vessel from the containment system. This would be hazardous if the materials were incompatible. Vessels normally at atmospheric pressure are particularly susceptible to this since a loss of pressure from the vessel would not indicate a ruptured disc. An undetected ruptured disc in an atmospheric vessel could result from a previous overpressure, damage during installation, or corrosion.

Because a rupture disc provides sudden, total pressure relief, it may result in two-phase, vapor-liquid flow from a pressurized vessel when liquid is initially present. This may be hazardous if the disk or the containment system downstream is not sized for such a release. Flares and scrubbers for these types of systems are not usually designed to handle the large quantities of liquid that could be released. Any incompatibilities between the liquid released and the materials within a downstream containment system might be quite serious because of the quantity of liquid that could be present. A knockout vessel of some type is needed when a two-phase release is possible. In such cases, and when a heated material whose melting point is above ambient temperature is involved (such as molten sulfur), particular care must be taken. Lines leading up to, and including the catch vessel must be heated to prevent plugging when the liquid cools.

A seemingly trivial, but actually very dangerous hazard associated with the use of rupture discs is that they can be easily installed upside down. The consequences of installing a disk upside down will depend on the type of disc, but in most cases the device will relieve at a substantially higher pressure than intended. All disks should have a tag indicating the proper direction for installation; these tags should be regularly inspected.

Other disadvantages of rupture discs are that once installed they cannot be non-destructively tested, wrong discs can be installed, and multiple discs nested together can be installed. Avoidance of these latter two occurrences requires strict adherence to proper, installer training and auditing procedures.

Safety-relief valves—A properly designed safety-relief valve will:
1) open automatically at a pre-adjusted set pressure using the energy of the fluid, 2) open fully to its rated flow capacity at a pressure typically not more than 10% above the set pressure, 3) shut flow off completely when the pressure falls to a preset reseat pressure (minimally between 3% to 5% below the set pressure).

Both undersizing and oversizing a valve can impair its effectiveness. Undersizing a valve presents the obvious hazard of not being able to relieve pressure fast enough. An oversize valve may chatter (rapid opening and closing), resulting in excessive wear or damage to the valve. An oversized valve may reseat improperly, resulting in leakage into or out of the vessel.

It is common to place a safety-relief valve downstream of a rupture disc. The rupture disc will prevent minor leakage of process materials and the safety-relief valve will allow the process to be resealed when the pressure drops. Sometimes a rupture disk of corrosion-resistant materials is used to protect a safety-relief valve made out of less exotic materials and expensive materials. Care must be taken to ensure that the presence of the blown rupture disk does not affect the anticipated performance of the safety-relief valve.

The dangers associated with pressure-relief valves are the possibility of isolation with shutoff valves either upstream or downstream, two-phase flow through the valve, and the potential for failure of the valve because of human error or mechanical failure. Another hazard is having other foulable fittings (e.g. check valves, flame arresters) between a relief valve and a vessel or between a relief valve and safe discharge. These can also cause isolation.

Since the safety-relief valve will reseat when the pressure drops, the potential for two-phase flow is not as high as it is with a rupture disk. However, two-phase flow is possible while the valve is open and rapid release is occurring. In the case of a runaway reaction, a large portion of the vessel contents may empty before the pressure falls below the reseating pressure. As with a rupture disc, a safety-relief valve will be grossly undersized for a two-phase release if it was sized taking only the vapor release into account. Much research has been and continues to be conducted in this area by the Design Institute for Emergency Relief Systems (DIERS).

Human error can result in the failure of a safety-relief valve in several ways. Unlike a rupture disk, a safety-relief valve has internal settings that can be adjusted to alter the performance of the valve. Incorrect adjustment of these settings during installation or routine maintenance could increase the potential for overpressure. The physical orientation of the valve is important during installation. A valve should be installed vertically, or if a type is designed for such service horizontally, with the discharge pointing down. In addition, the valve should not be required to support the weight of the discharge piping.

The hazards associated with any valve are the same with a safety-relief valve. Seals may leak, the valve may plug, or corrosion and abrasion may destroy the integrity of the valve. These hazards may be more pronounced for a safety-relief valve since it may go for very long periods of time without being called into service. For this reason, regular inspection and testing of the valve is essential to ensure its proper function when required.

If a safety-relief valve is located downstream of a rupture disc, then a pressure-detecting device should be placed in the line between the disc and valve. A premature failure of the rupture disk would otherwise go undetected and the safety-relief valve would be exposed to the conditions the rupture disc was installed to prevent. Also, since the disc relies on a differential pressure to rupture, a leak could reduce this differential and cause the pressure to rise well above the set pressure before the disk gave way.

Vent header--The effectiveness of a vent header depends on its proper sizing. The primary danger with vent headers in a total containment system for pressure relief discharges is that the vent header may not be sized to accommodate a worst-case event of simultaneous discharge of all or several relief valves connected to the header system. A second hazard is the potential for mixing incompatible materials in the vent header system or cross-contaminating the contents of one vessel with material vented from another vessel.

Pressure trip systems--

A trip system may be preferable to sole reliance on a pressure relief system when a toxic material is involved. Because of its mode of operation, no hazardous materials are discharged in the event of an overpressure. If properly applied, a pressure trip system may be more reliable than a pressure relief system. However, the system has no backup if not used in conjunction with pressure relief. If a safety-relief valve fails to open at its set pressure, it may open at some higher pressure before the structural limits of the equipment it is protecting are reached. A pressure trip system can never substitute for a pressure-relief valve installed to protect against overpressure due to fire.

The reliability of a trip system is a design criterion. Generally, a pressure trip system is designed to be at least ten times as reliable as the pressure relief valve it is intended to complement.

The reliability of a pressure trip system can be improved by a variety of methods that may be used to improve the reliability of any control system. Individual system components may be replaced with more reliable (and usually more expensive) components. Redundancy of hardware and software in various forms of backup systems may be used.

For critical applications a combined system may be configured as follows:

- First line of prevention - pressure control loop;

- Second line of prevention - pressure trip system; and

- Third line of prevention - relief valve discharging to header with protection technology system (e.g., scrubber).

The pressure for activating the system increases at each step.

Reliability of Pressure Control Components--
The reliability of pressure control depends on the reliability of the individual components that comprise the pressure control system. The reliability of a full multicomponent control system also depends on the architecture of the specific system. Table 2-12 presents reliability data for some individual components.

2.4.4 Summary of Control Technologies

For both new and existing facilities, Table 2-13 summarizes major hazards or hazard categories associated with pressure and the corresponding control technology or procedural categories. Numerous individual control technologies or procedural changes can be inferred for each category based on the preceding discussions of this section.

2.4.5 Costs

Table 2-14 presents the costs of components associated with process alternatives for pressure measurement and control systems. This table presents a list and costs of typical components. There are many individual variations of these component groupings, since other possible components are not included here. These costs give an order-of-magnitude basis for

TABLE 2-12.   TYPICAL FAILURE RATES OF PRESSURE CONTROL COMPONENTS

Component	Failure Rate Failures/year
Pressure Transducer/Transmitter	0.76-1.73
Pressure Indicator	0.026-1.41
Pressure Switch	0.34
Control Valve	0.25-0.60
Pressure Controller	0.29-0.38
Pressure Control Loop	1.73

Source: Adapted from References 2,8,19,37.

TABLE 2-13.   MAJOR HAZARD AND CONTROL TECHNOLOGY SUMMARIES

Process Variable	Hazard	New Facility	Existing Facility
Pressure	Overpressure	• Flow, pressure, and temperature control system design	• Change flow, pressure, or temperature control system
		• Emergency trip system for flow shutoff or automatic venting	• Add emergency trip system for automatic flow shutoff or automatic venting
		• Design to avoid flow blockage	• Change system to avoid flow blockage
		• Provide pressure relief for thermal expansion	• Add pressure relief for thermal expansion
		• Provide general pressure relief	• Add or change pressure relief system
	Underpressure	• Flow, pressure, and temperature control system design	• Change flow, pressure, or temperature control system
		• Emergency trip system for flow shutoff or automatic vacuum break	• Add emergency trip system for automatic flow shut off or automatic vacuum break
		• Mechanical vacuum breaker	• Add mechanical vacuum break
	Fluctuating Pressure	• Control system design	• Change control system design
		• Equipment selection	• Change equipment type

TABLE 2-14.   COSTS OF COMPONENTS ASSOCIATED WITH PROCESS MODIFICATIONS FOR PRESSURE MEASUREMENT AND CONTROL SYSTEMS

Basis: 4 inch diameter piping. Pressure range 0-300 psig	Capital Cost Range ($) (1986 Dollars)	Annual Cost Range ($/yr)	References
General purpose pressure transducer	200-500	33-81	40
Indicators	200-600	33-99	40
Computer interface system	2,000	330	40
Pressure gauge	50-250	8-43	40
Control valve	3,000-6,000	450-900	3,20,21
Controller	2,000	326	24,25,26,27
- conventional	6,000-12,000	910-1,800	
- via process control computer	3,000-15,000	260-1,298	
Control loop			Composite
- simple, single loop, PID	800-1,600	69-138	
- simple, interactive, PID	1,600-3,000	138-260	
- programmable, PID	2,000-6,000	173-519	
Rupture disks	150-225	28-42	20,23,28
Relief valves	7,000-12,000	609-952	20,23,27

estimating costs of process modifications involving the pressure measurement
and control system.

2.4.6 Case Examples (1)

An explosion occurred in a vinyl chloride pump on the "recovered" vinyl-
chloride monomer system (RVCM), seriously injuring an operator. The line
sections before and after the pump had been removed and the explosion occurred
in the idle vented pump about an hour later. Investigation showed that the
entire RVCM system was contaminated with vinyl chloride polyperoxide, an
unstable material. There had been earlier abnormal occurrences where the RVCM
gas compressors had not shut down at low pressure because a pressure switch
had failed. RVCM liquid had accumulated in the storage tanks for 20 days and
the acidity level in the vinyl chloride feed had been high.

A failure of the control system in a chlorine cellroom caused a backpressure to develop. When operating personnel tried to shut the system down
manually, failure of these controls caused a serious chlorine release. One of
two compressors taking hydrogen from a low pressure gasholder continued to
operate and failure of a low level trip created a negative pressure, allowing
air to leak in and cause an explosion in the compressor cooling coils. The
hydrogen/air mixture in the holder diffused into the catalyst purification
unit, a high temperature developed, and another explosion occurred. The
gasholder trip did not operate because the timer was bypassed by a "jumper".

2.5 TEMPERATURE CONTROL

Like flow and pressure, temperature is one of the primary variables in a
chemical process. The temperature of a system strongly influences which
chemical reactions are preferred and the rate at which they occur. Phase
changes are determined by temperature. The volume of gases, and to some
extent liquids, depends on temperature. For these reasons, a loss of pressure
control, discussed in the preceding section, will often be the result of a

loss of temperature control. The inability to control temperature can
therefore indirectly or directly lead to conditions that cause an accidental
release. Proper design of a temperature measurement and control system is
therefore essential to preventing accidental releases.

2.5.1 Temperature Hazards

For some chemical reactions a loss of temperature control may result in a
runaway reaction with excess generation of heat and materials that could lead
to an overpressure and an accidental release.

A chemical reaction system will generally have a specified temperature
range in which it must operate to successfully produce the desired product.
This range is usually set by balancing reaction rates with economics to
achieve an acceptable reaction rate for the primary reaction, while minimizing
competing reaction rates, the size of reaction vessels, and heating and
cooling systems. The consequences of operating above or below the set operating temperature will be a function of the specific reaction system.

High temperatures are usually more hazardous than low temperatures, but
both must be considered in evaluating temperature's contribution to process
hazards. Often a low temperature will slow down or quench the reaction. This
may be hazardous if an unreacted feed stream is incompatible with downstream
operations. High temperatures may result in a number of undesirable consequences. At higher temperatures, reaction rates are likely to increase, resulting in excess heat and material generation that may lead to an overpressure. A high temperature may result in the formation of unwanted by-products because of thermal decomposition of the product or increased reaction
rates for competing reactions. Thermal decomposition often results in the
generation of more moles of material than are consumed; this could result in
an overpressure. The consequences of unwanted by-product formation due to the
competing reactions will depend on the nature of the by-products formed. The
presence of lighter molecular weight by-products would result in an increased

system pressure and could result in an overpressure. A reaction may form
by-products that react with downstream materials or that are corrosive to the
materials of construction. Both of these events could contribute to the
potential for an accidental release.

Increases in temperature result in thermal expansion, which can result in
increase in pressure and an accidental release. A loss of temperature control
in any process involving heating or cooling may lead to an accidental release.
Two-phase processes are more sensitive to temperature fluctuations than
single-phase processes. Gas phase processes are more sensitive to temperature
fluctuations than liquid phase processes. However, thermal expansion can be
very destructive in a sealed liquid-full system; a system with no vapor space.
Sometimes heating or cooling results in a phase change: freezing, condensing
or vaporization. Phase changes may lead to overpressure or underpressure.
The expansion of water on freezing is an example of one consequence of cold
temperature.

Most physical properties vary with temperature. The performance of any
physical operation will be affected. The consequences of a temperature
fluctuation must be evaluated on a case-by-case basis. Sometimes a loss of
heat to a stream will decrease the solubility of a component in the stream to
the point where it drops out of solution. Solidified material could clog
lines and would affect the chemical properties of the stream; this could
ultimately lead to an overpressure and an accidental release. Partition
coefficients are temperature dependent and processes such as liquid-liquid
extractions will not function properly as temperatures vary.

A loss of temperature control may lead to an accidental release due to
material failure. Structural or sealing materials lose physical strength as
the temperatures increases. At an elevated temperature, a construction
material may stretch, bend or crack under load conditions that were acceptable
at lower temperatures. Metals experience fatigue if subjected to repeated
heating and cooling. This can lead to eventual failure under previously

acceptable conditions. Plastic linings used to protect metals from corrosion may fail at excessive temperatures. Expansion joints are often more sensitive to high temperatures than other process piping and may fail if temperatures are elevated above design specifications. Extremely cold temperatures may result in brittle failure of metals, especially if the metal is subjected to high stress under these conditions. Rapid changes in temperature can also be destructive.

A hazard in heat transfer equipment is the possibility of accidental mixing of a heat transfer fluid with a process fluid. Heating and cooling are usually accomplished via heat exchangers or various forms of jacketed equipment. A leak may develop in a vessel or pipe wall that allows the jacket fluid to enter the vessel, or vice versa, depending on which pressure is higher. A special danger arises when the process side is at a higher pressure than the jacket side. The process side of a jacket of a vessel frequently has a pressure rating higher than the jacket. A leak could cause high pressure process material to rupture the jacket or exit through the jacket's relief valve. If the jacket is at a higher pressure than the process, then the heat transfer fluid will contaminate the process, which could result in an accidental release if the heat transfer material is incompatible with the process materials.

Flashing may occur when a hot liquid with a high boiling point is mixed with a lower boiling point liquid. This flashing may result in a rapid overpressure. The accidental mixing of a heat transfer fluid with a process fluid, as was discussed above, may sometimes be the cause of such an event. A number of accidents have occurred when a hot oil is transferred to a tank that has been steam cleaned and not thoroughly cleared of condensation. The condensate has flashed and resulted in an overpressure.

Process temperature control is closely related to flow control. Reaction rates and the heat generated by a reaction is often governed by flow

control. The rate of heat transfer depends on the flow rate of a heating or
cooling medium. Therefore, a loss of a flow control can often result in a
loss of temperature control.

Process Considerations--

A process-related loss of temperature measurement and control occurs when
the temperature measurement and control systems are functioning according to
design standards but the heat generated by the process exceeds the capacity of
the cooling system.

Chemical reactions either generate or require heat. Anything in a
chemical reaction system that alters the rate of reaction alters the rate of
heat generation and/or consumption and affects the ability to control
temperature. As an example, loss of flow control of a reactant to an
exothermic reaction may result in the generation of excess heat. If the heat
generation rate exceeds the ability of the temperature control system to
remove it, the result is a loss of temperature control.

Failure to achieve proper mixing may result in a loss of temperature
measurement and control. Improper mixing of reactants may result in localized
hot spots. In a jacketed process vessel, for example, if there is poor mixing
then there will be very poor heat transfer between the jacket and the reactor
contents. Temperature measuring probes will not measure a representative bulk
temperature when there is inadequate mixing.

Equipment Considerations--

A failure of flow control may cause a failure of temperature control. As
mentioned above, in most cases, temperature control is achieved through
control of the flow of either the process materials or of a heat transfer
fluid. In some cases heating is accomplished by direct heating with gas, oil,
or electric heating elements. These are often used to heat a hot oil, steam,
or some other heating fluid; thus flow control is still the ultimate method
for heat transfer and temperature control.

As discussed above, a leak in the wall that divides process materials from heat transfer fluids will sometimes result in an accidental release. This is particularly a problem because these surfaces have fluid on each side and are therefore difficult to inspect for signs of erosion, corrosion, or defect.

Since not every form of mechanical failure can be prevented, insufficient design preparation to minimize the impact of mechanical failure can also be listed as a factor contributing to loss of temperature measurement and control.

Operating Considerations--

A temperature control system will only be as accurate as the standard to which the measuring devices are calibrated; therefore, improper calibration or maintenance of the system can contribute to failure in the control system.

A lack of chemical and thermodynamic data will contribute to the potential for a loss of temperature control. To design a proper control system, plant engineers should have a sufficient understanding of the consequences of temperature excursions. Operators should have training sufficient to respond to deviations in normal operating conditions. Often plant personnel do not know what side reactions could occur at various temperatures and do not have enough information to quantify the temperature dependence of all the reaction rates. This detailed information is often lost with the transfer of information between research and development, or it was never obtained in the rush to commercialize a new process. A total understanding of a complex chemical process may not be possible, but a lack of chemical and thermodynamic data will always leave plant management in the position of viewing temperature control as a mystery.

2.5.2   Technology of Temperature Control

The first line of prevention of an accidental release as a result of a
loss of temperature control is the temperature measurement and control system.
Process modifications must consider the capacity and reliability of these
systems. The second line of prevention is a temperature trip system designed
to regain control of the process once temperature control has been lost. The
ability to design a system that will adequately control temperature will
depend on the chemical and thermodynamic data available; adequate data will
potentially lead to better design.

Measurement and Control--

Temperature measurement relies on both sensing and final measurement
devices. Control relies on the corrective actions that a system takes in
response to changes in temperature. A variety of temperature sensors and
measuring devices are available. Each variety has some unique feature in the
mechanism it uses to sense and measure temperature. Temperature measuring
devices organize the output of the temperature sensing device and/or convert
the output to a meaningful temperature readout.

Several commonly used temperature sensing devices are discussed below.
Additional discussions on a wide variety of temperature sensing devices may be
found in numerous sources (e.g., Reference 39). Common varieties of temperature sensing devices used in the process industries include: thermal expansion
devices, thermocouples, thermistors, resistance thermometers, and a variety of
solid state sensors.

One group of temperature sensing devices rely on thermal expansion of
materials, usually metals. The thermal expansion of these devices is often
linear with temperature, making an additional temperature measuring device
unnecessary. Additionally, the output from these devices is visual and they
are generally not used for remote sensing. Bimetallic thermometers, filled
thermal elements, and glass stem thermometers are types of temperature sensing

devices that rely on thermal expansion to measure temperature. These devices
are common in older plants or in situations where a rugged local backup to an
electronic temperature sensing device is desirable.

A thermocouple is a simple temperature device composed of two homogeneous
wires of dissimilar metallic composition joined at one end to form a measuring
junction and at the other end to a measuring device. The measuring device
provides an internal connection between the two wires so that they form a
closed path through which current may flow. When heat is applied to the
junction of the two dissimilar metals, a small electromotive force (emf) is
generated and current flows through the thermocouple loop. In commercially
available thermocouples the emf is linearly proportional to the temperature at
the measuring junction. Thermocouples are rugged and inexpensive. They are
best suited for applications where moderate accuracy over a wide temperature
range is required and are very common in the chemical process industry.

Thermistors are a form of solid-state sensor that use a ceramic semiconductor which measures the change in the resistance of the semiconductor as
temperature varies. The resistance is nonlinear and must be converted to a
linear temperature scale by a temperature measuring device. Thermistors are
usually best-suited for applications where accuracy and rapid response are
required over a narrow temperature range; they tend not to be as well-suited
for measuring temperatures over a wide range.

A resistance thermometer is a device that uses the change in resistance
and temperature of a conductor to measure temperature. These devices have
conductors made of platinum, nickel, or a nickel/iron mix. They are accurate
and are available over a fairly wide temperature range. They are physically
compact and are well-suited for use in microprocessor-based products.

A variety of solid state sensors are available that measure temperature
induced changes in resistance, frequency, current, or voltage. Optical

sensors are available for measuring the temperature of a flame from a distance.

The devices mentioned above have certain applications to which they are
well-suited. It is best to consult with the vendors of each of these devices
to determine the particular applicability of a specific sensor. The key to
making an appropriate selection is to understand what process conditions may
contribute to an accidental release and what sensor would be best suited to
measure those conditions. Some processes must operate in a very narrow
temperature range and therefore require a sensor with high sensitivity and
accuracy over a narrow range. A process may require a narrow operating range
for producing the desired product but will be hazardous only if it operates
well above or below that range. In such a case it may be appropriate to
install two sensors; one with a narrow range for controlling normal operations
and one with a wide range for monitoring upset conditions. For some processes
a loss of temperature control will develop slowly. In such situations response time may not be as important as accuracy. For other processes a loss
of temperature control will develop rapidly, and in these situations a rapid
response time may be more important than accuracy.

Temperature Trip and Emergency Cooling Systems--
A temperature trip system is essentially the same as a pressure trip
system. This type of control system monitors temperature and is activated
when temperature has exceeded a specified limit or a specified rate of increase. When activated, the system responds in an attempt to shut the process
down by eliminating the source of heat and often by providing a source of
cooling. Usually the goal is to prevent an overpressure. Any decision about
whether temperature or pressure measurement should be used as the basis for
the trip system should be based on which would provide the most immediate and
accurate indication of an impending system failure. Once activated, a trip
system will usually disrupt the process, forcing at least a partial, temporary
shutdown. The system should be designed to trip only when necessary. An

intimate understanding of the chemistry and thermodynamics of the process is essential for proper design of this type of system.

A trip system may be designed to shut off the supply of heat or it may be designed to supply an emergency supply of cooling. Emergency cooling may be accomplished by a number of methods. External cooling may be supplied by flooding a heat exchanger or vessel jacket with cooling water. Internal cooling may be supplied by adding chilled solvent to the reaction melt. The solvent would not only cool the reaction but the extra dilution could slow the reaction rate. A somewhat drastic measure would be to intentionally poison the reaction by adding a material that is known to effectively stop its progress. The more drastic measure of automatically dumping the contents of a reactor into a diked, open area or into a supplementary containment vessel has been applied in some cases; however, dumping into an open diked area may not be appropriate where toxic or flammable materials are involved.

If a temperature trip system is designed to replace a pressure relief device, then the reliability of the trip system must be ten times that of the pressure relief device for the same application. Calculation of this reliability should be based on the calculated fractional dead time (the proportion of the time for which a system will be in the failed state). This method considers both the fault-rate for the individual components and the frequency of proof testing. If a temperature trip system is designed to supplement an emergency relief device, then the designed reliability must be based on the consequences of a loss of temperature control. In a situation where the overpressure may be so rapid as to constitute an explosion, or where a loss of temperature control is likely to lead to an accidental release, it is probably appropriate to design the system to have a reliability again equal to ten times that of the emergency relief device.

Jacketed Vessels and Heat Exchangers--

Temperature control is usually achieved via a heat transfer fluid circulating through a vessel jacket or a heat exchanger. Such equipment requires some special considerations if temperature control is to be maintained.

All vessel jackets and heat exchangers must have adequate pressure relief. Most heat exchangers and vessel jackets are designed with valves on the inlet and outlet lines. An overpressure could occur if a warm material is added to a vessel whose jacket is closed at the inlet and outlet. The warm material would heat the contents of the jacket and result in a buildup of pressure. A similar event could occur with a heat exchanger if either the shell or the tubes were valved off and warm material were circulated through the other side of the exchanger. All vessel jackets and heat exchangers must be inspected regularly to prevent a leak between the heat transfer fluid and the reaction mixture.

General design principles for jacketed systems are found in many sources. ASME has design codes for jacketed equipment and heat exchangers. As mentioned above, it is important that all jacketed vessels be supplied with adequate pressure relief. Additional general design information is discussed in Section 4 of this document.

Obtaining Chemical and Thermodynamic Data--

As mentioned above, the more chemical and thermodynamic data available, the safer the design of the temperature control system. Where hazardous materials are involved, a safe system must be based on as much data as possible.

Most plants cannot generate chemical reaction rate and thermodynamic data. This type of information and data on potential side reactions must usually be obtained from a research and development laboratory. However, a number of methods are available for estimating the chemical hazards of a system without extensive laboratory work.

Thermodynamic calculations or computer programs used to estimate thermodynamic data may be used to calculate the potential energy available in a

given molecule or system. This information can be combined with correlations
to evaluate potential hazards.

A number of empirical tests using reaction mixtures provide data that
will assist in the design of a safe temperature control system (these tests
also give information that can help in the design of all of the reaction
related control systems). Differential scanning calorimetry, accelerating
rate calorimetry, differential thermal analysis, heat flow calorimetry, and a
test method developed by DIERS will all provide information useful in the
design of a temperature control system.

2.5.3 Control Effectiveness

Measurement and Control--
    Table 2-15 lists a variety of temperature-sensing and measurement devices
and certain performance information. In addition to a temperature range, each
device will have a variety of other performance limitations, including items
such as acceptable sensitivity, reproducibility, materials of construction,
and reliability. Care must be taken when selecting an appropriate temperature
sensing device. Substituting an existing temperature measuring device with a
device more suited for the particular application may increase the system's
reliability and hence decrease the potential for an accidental release.

One author has summarized a number of considerations that apply to
thermocouple application (40). Thermocouples come in at least two grades
related to error specifications of the American National Standards Institute
(ANSI) and the Instrument Society of America (ISA). Thermocouples may change
calibration over time depending on the temperature and operating environment.
Often overlooked is that the accuracy with which a thermocouple measures the
temperature of a process fluid depends on its size, shape and location. For

TABLE 2-15.   TEMPERATURE SENSOR SELECTION GUIDE FOR NON-SEVERE SERVICE
UNDER 932°F

Sensor Type	Point Readings	Average Readings
Filled Element	Fair to good	Fair
Thermistor	Fair to excellent for some applications	Fair to excellent for some applications
Thermocouple	Fair to good	Fair for some applications
Resistance Bulb	Fair to excellent	Fair to excellent

Source: Adapted from Reference 19.

example, a thermocouple that is too small for an application or improperly
located may measure a local rather than a bulk temperature. This is especially relevant in certain equipment where poor mixing or bypassing might
occur. Leads that are too long can lead to inaccuracies, as can electrical
interference. The main point is that temperature control effectiveness
depends on accurate sensor information, which depends in turn on proper application and installation.

There may be no external indication that a temperature sensing device is
giving an incorrect reading. Some devices, such as thermocouples, require
periodic calibration. The readout from some temperature-sensing devices, such
as thermistors, may drift as the device ages. Mechanical devices may stick at
some incorrect temperature. In situations where an incorrect temperature
reading would significantly contribute to the potential for an accidental
release it may be advisable to use two different varieties of temperature
sensing equipment.

Programming the software for a typical temperature control system is often based more on empirical data than on a theoretical understanding of the process. This is because of the complexity involved in a system where a heat transfer fluid (which itself must be heated) is used to heat the wall of a process vessel which then heats a reaction melt where a heat producing or absorbing reaction is occurring. An empirical approach is adequate for handling normal operating conditions but may be inadequate for handling the abnormal event. Where hazardous materials are involved, it is therefore important to obtain as much chemical and thermodynamic data as possible and to prepare for the unexpected with temperature or pressure trip systems and adequate pressure relief.

Temperature Trip and Emergency Cooling Systems--

A temperature trip and emergency cooling system may be necessary where a toxic material is involved and where there are fairly well-defined temperature boundaries beyond which the system will be out of control. If no such temperature boundaries exist, or if the boundaries are not well defined, it may be preferable to operate the system with a pressure-activated trip system.

Careful consideration must be given to the consequences of an emergency cooling system. By its nature, an emergency cooling system will tend to temporarily shut down a portion of the process. The possibility of liquid reaction melts freezing, vapors condensing, or upstream materials accumulating must be considered to be certain that the emergency cooling system does not introduce new hazards. In addition, the thermal stress could have a drastic temperature change as a result of emergency cooling on the construction materials must be considered. A design evaluation such as a HAZOP study will help determine the safety of an emergency cooling system.

Jacketed Vessels and Heat Exchangers--

Some of the potential limitations of jacketed equipment and heat exchangers have been discussed above. There are some additional considerations

specific to vessel jackets. One additional hazard is the possibility that a
loss of temperature control would result from a jacket relief valve that
relieves and does not reseat properly. Enough heat transfer fluid could
escape to prevent adequate temperature control of the vessel. Another concern
is that there be adequate flow and mixing of the jacket fluid to avoid
temperature stratification. The location of entry and exit nozzles on the
jacket may influence this. In spite of these limitations, controlling
temperature via jacketed vessels and heat exchangers is usually preferable
because of the ability to provide controlled, uniform heating.

Obtaining Chemical and Thermodynamic Data—
    The test methods mentioned previously for use in designing a temperature
control system will provide information about what temperature ranges are
inappropriate for the system of interest. They will also give an idea about
the magnitude of the disturbance that may result if these temperatures are
exceeded. Interpretation of the test results must be made by a person with
experience in the area because the test results are qualitative as well as
quantitative and no guarantee exists that what happens in the test will happen
in the plant. Tests, however, are better than no information at all.

Reliability of Temperature Control Components—
    The reliability of temperature control depends on the reliability of the
individual components that comprise the temperature control system. The
reliability of a full multicomponent control system also depends on the
architecture of the specific system. Table 2-16 presents reliability data for
some individual components. Another factor affecting the reliability of
temperature control, which is not as significant in the control of other
variables discussed, is the relatively long response time of physical systems
and of temperature devices to temperature changes. For critical applications,
high reliability requires that delays in detecting and acting on changes in
temperature under emergency conditions have been properly accounted for.

TABLE 2-16. TYPICAL FAILURE RATES OF TEMPERATURE CONTROL COMPONENTS

Component	Failure Rate (Failures/Year)
Sensor With Thermowell	
Thermocouple (TC)	0.52
Resistance Temperature Detector (RTD)	0.41
Temperature Transducer	0.88
Temperature Controller	0.29–0.38
Control Valve	0.25–0.60
Control Loop	1.73
Cooling Water Capacity	System reliability depends on specific design and reliability of individual components.
Refrigerated Brine Capacity	System reliability depends on specific design and reliability of individual components.

Source: Adapted from References 2, 8, 19, and 37.

2.5.4   Summary of Control Technologies

Table 2-17 summarizes major hazards or hazard categories associated with
temperature, and the corresponding control technology for both new and
existing facilities. Based on the preceding discussions of this section,
numerous individual control technologies or procedural changes can be inferred
for each category.

2.5.5   Costs

The costs of components for temperature measurement and control process
modifications are presented in Table 2-18. These component types and costs
are based on a typical installation. The costs presented provide an
order-of-magnitude basis for evaluating the economic impacts of process
modifications involving temperture measurement and temperature measurement and
control systems.

2.5.6   Case Examples (42)

The explosion of a batch chlorinator caused the deaths of eight employees
and extensive damage. The reaction temperature, which was controlled
automatically by manipulating the chlorine flow, fell sharply when the
thermocouple failed. Personnel stopped the agitator and shut off the brine
cooling while the instrument was repaired, but delay in stopping the chlorine
flow led to a high temperature and decomposition reactions. The explosion
blew the reactor cover through the roof, drove the reactor into the floor, and
ruptured chlorine and ammonia lines. The released flammable gases burned and
caused the eight casualties.

In another example of loss of temperature control, localized overheating
caused the failure of a refinery reactor operating at 17,225 kPa (2500 psi),
which released a large cloud of about 250,000 lb of $>C_{10}$ hydrocarbons and H_2
that ignited and caused widespread explosions and fires. Four people were

TABLE 2-17.   MAJOR HAZARD AND CONTROL TECHNOLOGY SUMMARIES

Process Variable	Hazard	New Facility	Existing Facility
Temperature	• Runaway reaction	• Reactant feed control system	• Change reactant feed control system
		• Emergency trip to shut down reactant feeds	• Add emergency trip to shut down reactant feeds
		• Emergency cooling systems	• Add emergency cooling system
		• Emergency dump systems	• Add emergency dump system
			• Change process chemistry
	• Thermal expansion	• Design to avoid blocked-in liquid-full piping and equipment	• Change design to avoid blocked-in liquid-full piping and equipment.
		• Emergency trip to shut down heating systems	• Add emergency trip to shut down heating systems
		• Emergency cooling systems	• Add emergency cooling capacity
		• Insulation for protection from external sources of heat	• Insulate to protect from external sources of heat
	• Property changes	• Temperature control system design	• Change temperature control system design
		• Temperature control system design	• Change process chemistry

(Continued)

TABLE 2-17 (Continued)

Process Variable	Hazard	New Facility	Existing Facility
	• Equipment or material failure due to excess or deficient temperatures	• Temperature control system design • Equipment selection • Materials of construction • emergency heating system • Emergency cooling system	• Change temperature control system • Change process chemistry • Change equipment type • Change materials of construction • Add emergency heating system • Add emergency cooling system • Change process chemistry

TABLE 2-18. COSTS OF COMPONENTS ASSOCIATED WITH PROCESS MODIFICATIONS FOR TEMPERATURE MEASUREMENT AND CONTROL SYSTEMS

Basis: 4 inch diameter piping
Temperature range 0-250 degree F.

	Capital Cost Range ($)	Annual Cost Range ($/yr)	References
	(1986 Dollars)		
Sensor/with Thermowell			
Thermocouple	200 – 300	30 – 45	40
Resistance Temperature Detector (RTD)	–	–	
Thermistor	–	–	
Integrated Circuit (I.C) Sensor	–	–	
Transmitter and Indicator	990 – 1,680	150 – 250	40
Temperature switch	160 – 710	24 – 108	40
Controllers			
Simple, single loop, PID	800 – 1,600	69 – 138	26,27,28,33
Simple, interactive, PID	1,600 – 3,000	138 – 260	4,20,23,25
Programmable,PID	2,000 – 6,000	173 – 519	
Control Valve	3,000 – 6,000	450 – 900	
Control Loop			
Conventional	6,000 – 12,000	910 – 1,800	Composite
Via process control computer	3,000 – 15,000	260 – 1,298	
Additional Cooling Water Capacity (10 degree approach, 30 degree range)	30 – 75 per gpm of capacity	4.55 – 11.36 per gpm of capacity	20
Refrigerated Brine Capacity. (20 degree F evaporator)	3,000 – 8,000 per ton of capacity	450 – 1,200 per ton of capacity	20

injured and property loss was great.   Overpressures were highly directional.

2.6   QUANTITY CONTROL

The two primary objectives of quantity measurement and control are to achieve proper process material ratios and the proper level and/or weight of materials in process or storage vessels. Failure to perform either of these functions could result in the overfilling or overpressuring of process equipment, which could lead to an accidental release.

2.6.1   Quantity Measurement and Control Hazards

Hazards associated with a failure of the two functions stated above are discussed below.

Process Material Ratios--

Failure to maintain process material ratios can have both chemical and physical consequences. For most chemical reaction systems, there are two types of reactants: limiting reactants and non-limiting reactants. Often the controlled flowrate of the non-limiting feed stream will be ratioed to the flowrate of the limiting feedstream. In this situation, one feed stream controls the flowrates of all other feed streams, and a failure to measure or control the quantity of this stream may result in losing control of the quantities of all of the other process streams. Depending on the chemical reaction involved, such a loss in control could result in an overpressure caused by excess heat or material generation. For the non-limiting constituents, deviation from an acceptable quantity range may also lead to overheating or overpressure.

Level Control--

A failure to monitor and control the level in a process vessel or a storage tank may result in the overpressure of a closed vessel or in the overflow of a vented vessel.

Process problems include disturbances that lead to a change in the chemical or physical properties of a system, resulting in a loss of quantity measurement and control. Altering the physical properties of a process stream may affect quantity measurement. For flowing streams, a quantity measuring device will often be calibrated for a stream of a certain composition; a change in that composition will often alter the accuracy of a flow meter. For many types of quantity sensing and measuring equipment, the presence of solids or foams may alter the accuracy of the system.

Quantity deviations or loss of control may be caused by some other control malfunction. For example, a system failure such as a loss of temperature control results in an acceleration in the rate of reaction, then the excess reaction rate may exceed the ability of the quantity control system to respond in reducing a reactant feed and result in an overpressure. As with all sensing and control equipment, a number of equipment-related malfunctions may result in a loss of quantity control. Improper application, poor design, defective equipment, fatigue failure or corrosion may all result in a failure to measure and control process quantities. Quantity control is often a flow control problem; therefore, reliable quantity control depends on reliable flow control. A quantity control system may act as a backup to a flow control system to shut down flow when quantities are too high. Flow control considerations and their role in process hazards are discussed in Section 2.3 of this document.

Since a quantity determination and control system will only be as accurate as the standard to which the measuring device has been calibrated, improper calibration or maintenance can result in a failure in the control system. Some types of quantity determination systems, such as weighing devices, are sensitive to conditions exceeding their range, to physical abuse, ambient conditions, and other factors. Good operating practices and operating personnel who understand the limitations of the device are important to ensuring reliable service from the equipment.

2.6.2 Technology of Quantity Control

Quantity measurement and control devices can be divided into several categories according to their function:

o   Flow meters, either mass or volume;

o   Weighing devices such as load cells; and

o   Level measuring devices.

Controlling quantity with weighing devices and level measuring devices is usually accomplished via a trip system and/or alarm. A trip system will often be used to stop the feed of material once a certain weight or level of material is reached. If a level indicator is used in process equipment such as a distillation column, then a trip system may be used to alter a heating rate when a high level in a reboiler is reached. An alarm may be used in conjunction with a trip, or by itself with the quantity detection devices. Operating personnel must then decide what response to the alarm is appropriate. This type of arrangement will work only where the hazard potential associated with an overfill is low or where no appropriate controlled response to an overfill can be determined.

Flow Meters--

Probably more varieties of flow measuring devices are available than any other type of process sensing instrumentation. One author has assembled a list of well over 50 different flow-meter types (43). Flow measurement and control are discussed as a separate topic in Section 2.3 of this document.

Weighing Systems--

Weighing devices are usually used in batch operations. Weighing, a method of measuring and regulating the amounts of materials charged to a reactor, is preferred in situations where the feed material is difficult to handle. Solids, slurries, and extremely viscous materials are examples of materials that may better be quantified by weighing than by measuring flow.

Both mechanical and electrical weighing devices are available. Mechanical weighing systems are more common than electrical and include platform scales, hopper scales, tank scales and tank truck scales. Mechanical scales can be extremely accurate and are generally simple to maintain and operate.

Electrical weighing devices are often referred to as load cells. These devices have the advantage of being small and easy to install. Load cells give quick response, are not subject to the same type of wear as a mechanical scale, and provide a signal for a readout at a remote location.

It is important to design a weighing system so that the damaging effect of various environmental factors will be minimized. Methods for isolating the weighing device from vibration by strengthening support foundations or by providing vibration absorption should be considered. The design must also ensure that the system is unaffected by connecting pipe stresses. In some instances it may be necessary to add temperature compensation. Weighing devices should be housed to ensure adequate drainage and protection from moisture or chemical contamination.

Regular inspection and maintenance of weighing devices is essential. Mechanical devices in particular must be readjusted and serviced frequently to ensure accurate results.

Depending on the consequences of inaccurate weighing, a backup to the weighing system may be desirable. A level measuring device may be a suitable backup for a weighing device.

Level Measurement--

Level measuring devices have a number of different uses. A level
measuring device may be designed to sound an alarm or activate a switch only
if the level passes above or below some fixed point, or it may be used to give
a continuous level readout accurate enough to precisely track inventories.
Devices may be designed to give a local indication of process vessel levels or
to send a signal to a flow controller. Thus, level detection may be used
either as a backup for other control systems or as a primary control device.
Some devices must actually contact the liquid to sense level while some can
measure level without contacting the liquid. Some detectors function mechanically and some function electronically.

Probably the most common variety of level measuring device is the flat
glass gauge. This device is composed of a shielded external loop, portions of
which are made of glass. The liquid level is then observed in the glass
portions of the loop. Some of these devices can send a signal to another
location. Their advantage is that they provide a visual means of measuring
the level in a vessel. They are often used as a backup for a more sophisticated variety of level measuring device. Other mechanical level detection
devices, such as floats and displacers, are also available.

A common variety of level measuring device is the differential pressure
level detector, which measures level by detecting the pressure difference
between two different points in a tank. Electronic devices such as
capacitance, ultrasonic, optical, and radiation level detectors are also
available.

Generally, level devices must be inserted into process vessels and
involve additional piping or fittings. Since fittings are a potential weak
point for a process vessel, they must be designed to withstand the same
conditions that the vessel is designed to withstand. Some applicability
criteria for selected level detector devices are shown in Table 2-19.

TABLE 2-19. LIQUID LEVEL DETECTOR SELECTION GUIDE

Level Detector Type	Local Indicator	Transmitter		Switch	
		Clean Fluid	Difficult Fluid	Clean Fluid	Foaming Fluid
Float	Fair	Poor-Fair	Poor	Good	N/A
Level Gauge	Fair-Good	N/A	N/A	N/A	N/A
Capacitance Probe	Poor-Fair	Fair	Poor-Fair	Good	Poor-Fair
Conductivity Probe	N/A	N/A	N/A	Fair	Poor
Diaphragm	Poor-Fair	Poor	Poor	Fair	N/A
Differential Pressure	Fair-Good	Good	Fair	Good	N/A
Displacement	Fair-Good	Excellent	Poor-Fair	Excellent	N/A
Radiation	Good	Good	Excellent	Good	N/A
Ultrasonic	Fair-Good	Fair	Good	Good	Poor-Fair
Tape	Good-Excellent	Good	Poor-Fair	Good	N/A

Source: Adapted from Reference 19.

2.6.3 Control Effectiveness

The effectiveness of various approaches to maintaining quantity control
is evaluated in terms of performance, limitations, and reliability.

Flow Measurement and Control--
The effectiveness and secondary hazards associated with flow measurement
and control are discussed in Section 2.3.

Weighing Systems--
When functioning properly, weighing devices are an effective means of
monitoring the quantity of material added to a process vessel; however, all
weighing devices are sensitive to a variety of environmental factors that may
impair their performance.

Vibration and other related mechanical disturbances (such as impact
damage when a load is dumped onto a weighing device) can seriously affect
performance. Accuracy, stability, and repeatability can all be affected by
mechanical disturbances.

Temperature will affect the accuracy of most weighing devices. Most
mechanical scales have built-in temperature compensation; however, most
mechanical devices will not weigh accurately if subjected to a rapid temperature change. Electrical devices are also sensitive to temperature changes.
Sometimes temperature compensation will be built into an electrical device,
and sometimes it will be necessary to add temperature compensation. If
several load cells are used to weigh the contents of a reactor, any nonuni-
formity in the temperatures of the load cells is likely to result in a weighing error.

Additional environmental factors, such as moisture and chemical contamination, can damage a weighing device.

Level Measurement--

Level measurement is sometimes used directly to control the flowrates of
streams into and out of a process vessel. In some situations this may be a
simpler method for controlling flows than direct flowrate measurement and
control. In most process situations, however, vessel feed streams are not
controlled by level detection, since the failure of a level detector to
control feed flowrates could easily result in an overfill, an overpressure
and, potentially, in an accidental release. Level detection devices are
well-suited to be a backup to flow control by activating high or low level
trip systems or alarms. Where level detection is used as the primary control
device for process flows (this is often the case for filling and emptying
storage tanks and sometimes for charging to batch reactors), then two different types of level detection devices should be used in series.

The operation of some varieties of level detection equipment makes them
fairly product specific. These types of devices should be avoided where
stream compositions may vary. Many level detection devices cannot accurately
measure the level of foamy solutions.

Installation requirements for level detection equipment will influence
the suitability of a particular device for hazardous chemical service. Flat
glass gauges that stick out from the vessel must be protected from being
sheared off during a collision. Flat glass gauges are also often the first
point of failure during the overpressure of a pressure vessel. Some devices
require two points of entry into a vessel, one of which is low in the tank and
below normal liquid levels. This increases the possibility of leaks. Some
require that a probe be inserted below the liquid level. Care must be taken
to ensure that the construction materials of such a probe are compatible with
the process conditions. Additionally, care must be taken to ensure that the
additional obstruction of a probe is acceptable. Some probes can operate from
the top of a vessel without contacting the liquid. These may be preferred in
many instances; however, constructions materials are still important in

devices that operate above the liquid level. The vapor phase in a process
vessel is often more corrosive than the liquid phase.

Reliability of Quantity Control Components--

The reliability of quantity control depends on the reliability of the
individual components that comprise the quantity control system. The reliability of a full multicomponent control system also depends on the architecture of the specific system. Table 2-20 presents reliability data for some
individual components expressed as typical failure rates.

2.6.4 Summary of Control Technologies

Table 2-21 summarizes major hazards or hazard categories associated with
quantity control, and the corresponding control technology or procedural categories for both new and existing facilities. Numerous individual control
technologies or procedural changes can be inferred for each category.

2.6.5 Costs

The costs of components found in quantity measurement and control systems
are presented in Table 2-22. These component types and costs are based on a
typical installation. Since there are other types of systems, and many
variations of systems within a given type, these costs only provide an
order-of-magnitude basis for evaluating the economic effects of quantity
measurement and control system process modifications.

2.6.6 Case Examples (42)

In one example of a loss of quantity control, overfilling of a salt dome
storage well created a cloud of butane 1.25 mi. in diameter. Two explosions
occurred, one 800-1000 ft. above grade. Twenty four people were injured.

TABLE 2-20. TYPICAL FAILURE RATES OF QUANTITY CONTROL COMPONENTS

Component	Failure Rate (Failures/year)
Load Cell Weigh System	3.75
Level Detection System	
Differential Pressure Transducer	1.71
Float System	1.64
Capacitance System	0.22
Electrical Conductivity Probes	2.36
Flow Totalizer	ca. 1.0

Source: Adapted from Reference 2, 8, 19, and 37.

TABLE 2-21. MAJOR QUANTITY RELATED HAZARDS AND CONTROL TECHNOLOGY SUMMARIES

Process Variable	Hazard	New Facility	Existing Facility
Quantity	• Incorrect reactant ratio, catalyst level, or inerts concentration	• Flow control design	• Change flow control system
		• Equipment selection	• Change type of equipment
	• Incorrect level leading to overfilling or underfilling	• Level sensing and alarms	• Change or add level sensing and alarms
		• Weight sensing and alarms	• Change or add weight sensing and alarms
	• Incorrect volume or mass of material leading to overfilling or underfilling	• Emergency trip system to shut down flow	• Change or add emergency trip system to shut down flow

TABLE 2-22. COSTS OF COMPONENTS RELATED TO PROCESS MODIFICATIONS FOR QUANTITY
MEASUREMENT AND CONTROL SYSTEMS (EXCLUDING FLOW RATE MEASUREMENT.
SEE SECTION 2.3)

Basis: Load cell weigh system: 10,000 gallon batch reactor.
Level system: 10,000 gallon vessel.
Flow totalizer: 33 gpm and 667 gpm.

Component	Capital Cost Range ($)	Annual Cost Range ($/yr)	References
		(1986 dollars)	
Load cell weigh system	13,800	2,100	41
Level detection system -			
Sight gage	1,100	77	41
Float system	1,400	210	41
Capacitance system	2,400	365	41
Ultrasonic system	2,600	1,150	41
Nuclear system	14,900	2,260	41
Flow totalizer			
33 gpm	N/A	N/A	
667 gpm	N/A	N/A	

Another accident occurred in an ethylene producing plant in Holland, where failure of a level controller on a column caused cold liquid to pass out of the relief valve and into a carbon steel flare header, which cracked. The released cloud of 12,000 lb. of propylene ignited at a furnace 150 ft. away. Fourteen were killed, 104 injured and property damage was valued at almost 43 million dollars.

2.7 MIXING

Many processes can safely operate under a wide range of mixing rates. If there is an upper limit for a safe mixing rate, then it is likely to be set by foaming problems or some similar characteristic. Some processes must operate in a very specific mixing regime. Formulating processes often fall into this category. Usually there is a minimum mixing requirement below which reactants are not properly contacted or heat transfer is not sufficient and uniform.

There are three basic types of mixing systems:

●   Direct mechanical mixing;

●   Induced flow mixing; and

●   Static mixing.

Direct mechanical mixing refers to mixing by blade agitators such as turbines or propellers. Induced flow mixing is accomplished with pumps or other devices, especially where a recirculating liquid stream back to a vessel is involved. Induced flow mixing can also involve ejectors and eductors. Static mixing involves using pipeline mixers that contain stationary mixing hardware elements inserted in the piping.

2.7.1   Mixing System Hazards

One group of authors has defined three basic categories of mixing problems (44).

- Loss of agitation;

- Insufficient mixing; and

- Excess energy input from mechanical friction.

A loss of agitation is usually the primary hazard in mixing systems. Hazards associated with a loss of mixing include:

- Incomplete reactions or formation of unwanted by-products;

- Reactant accumulation in poorly mixed zones; and

- Poor heat transfer with overall overheating or overcooling of a reaction or localized hot or cold spots.

The hazards of incomplete reactions or the formation of unwanted by-products is related to the chemical and physical properties of the unreacted materials or by-products. When there is insufficient mixing to contact reactants, a potentially dangerous excess of unreacted material can accumulate in the reaction vessel. With highly reactive materials, or in exothermic reactions, such a mixture could react at an uncontrolled rate if agitation began or if additional heat were added to the system. Even without this hazard, the unintended excess of a toxic reactant in downstream processing might cause problems. Failure to react a gaseous material because of insufficient agitation might lead directly to overpressure. Unwanted byproducts could include gaseous species leading to excess pressures, or corrosive materials that could damage equipment.

A loss of mixing control can cause poor heat transfer that results in localized hot and cold spots. Runaway reaction and/or overheating and overpressure could result. Decomposition reactions might also occur with heat sensitive materials. Hazards from overcooling include freezeups, solids deposition, and heat transfer surface fouling, with all their attendant process upsets, such as plugged lines, and accumulation of unreacted material. An accidental release could occur if adequate protection were not available to handle such an event. The hazards associated with a loss of temperature or pressure control are discussed in Subsections 2.3 and 2.4 of this manual.

A potential hazard associated with too much mixing is overheating as a resulting from the energy generated by mechanical friction. A well-insulated vessel left agitated for long periods of time may experience a temperature rise from mixing friction. Such a temperature rise could begin the chain of events that leads to an accidental release. For example, the additional heat input could start a decomposition reaction which could lead to gas evolution and overpressure.

A hazard associated with mixing is the generation of a static charge between the process stream and the process vessel. This is a particular problem where process vessels and piping are coated with glass or plastic materials that act as electrical insulators. Static charge can arc and result in fire and explosion if a flammable atmosphere is present.

Any event that leads to a dramatic change in the physical characteristics of the process materials being mixed may result in a decrease or loss of mixing. For example, a loss of heating resulting in frozen process materials would result in a loss of agitation. An increase in liquid viscosity could be caused by a decrease in temperature or by other unexpected conditions (e.g., improper reactant ratios). The viscosity of some liquids can increase with increased temperature, which is particularly hazardous if the increase in temperature leads to an increase in reaction rate and impaired mixing.

The hazard potential associated with freezing or viscosity changes is most severe where high speed, high shear mixing is involved.

Probably the most frequent cause of a loss of mixing control is some type of mechanical failure, which includes:

- Electric motor failure (either power outage or mechanical failure);

- Shaft seizure or breakage in the mixing equipment; and

- Breakage or detachment of a mixing impeller.

A variety of mechanical agitation devices are available. The possible failure modes of these devices depends on the specific type of equipment involved. Most systems involve rotating equipment subject to common failures such as bearing failure, shear pin failure, belt slippage or breakage, electrical malfunction, and others. Corrosion or wear may also cause a failure in the portion of the agitator system that actually contacts the process fluid. Agitators tend to experience more severe erosion and corrosion than do process vessels. Induced flow agitation systems may be subject to more potential failure modes than a mechanical agitation system. An induced flow system may fail because of a pump failure, valve failure, or piping failure.

Static mixers do not have rotating or moving components and are therefore not directly subject to all of the failure modes associated with mechanical agitators; however, a static mixer requires sufficient flow to induce adequate mixing. Therefore, a reduced flow through the static mixer may result in insufficient mixing. As with mechanical mixing systems, the static mixer is subject to corrosion or erosion from the process fluid.

Where mechanical agitation is involved in a batch process, it is possible for operators to forget to begin agitation at the proper time. Beginning agitation after materials have been charged may result in a runaway reaction, overpressure, and in an accidental release.

2.7.2   <u>Technology of Mixing Control</u>

Before an adequate mixing system can be designed it is important to
determine what the effect of various mixing rates, including no mixing, may
have on the process. One author has suggested that the effect of mixing on a
chemical reaction system may be assessed using a heat flow calorimeter, which
is capable of measuring instantaneous heat generation rates, heats of reaction, reactant heat accumulation, specific heats, and heat transfer data under
simulated industrial process conditions (44).

Once the effect of mixing on a chemical process is understood, the
effects of a mixing failure can be evaluated. It is then possible to design a
system that minimizes the adverse effects of such a failure. If a mixing
failure is potentially hazardous, it may be appropriate to provide an agitation detection system that is tied into flow and/or temperature control
systems so that protective measures may be taken if agitation stops. An
example would be to stop or reduce the flows of reactor feeds when agitation
ceases. An agitation detection system does not necessarily have to directly
measure agitation; it may be more appropriate to place a temperature probe or
flow meter at a location that would be sensitive to changes occurring as a
result of a loss of agitation. Where mechanical agitation is involved it is
possible to monitor the mechanical equipment as an indicator of agitation.

As discussed above, in some situations it will be hazardous for an
operator to begin a reaction addition sequence without agitation. For these
situations it may be appropriate to provide an interlock that prevents the
addition of a reactant when agitation is not present.

Proper maintenance of a mixing system is important. Mechanical agitators
are used for batch or semibatch reactors of moderate size. They are appropriate when thorough and continuous mixing of the bulk liquid is required. An
induced flow system may be used in conjunction with or independent of mechanical agitation. When applied by itself, an induced flow system is used when

less agitation is required. A storage tank is often mixed using an induced flow mixing system. An induced flow system is also used when it is desirable to inject a reactant into a flowing stream before it is blended with the bulk solution. With this type of system there will be high shear and good mixing at the point of injection without going to the expense of more thorough agitation of the entire bulk solution. For all but small vessels, an induced flow system would be less expensive than a mechanical agitation system. An in-line static mixer is used in continuous processes where short duration mixing is satisfactory.

2.7.3 Control Effectiveness

People with experience in the area must determine the potential hazards associated with a loss of mixing control. The data obtained from a device such as a heat flow calorimeter may not be scaled up for a full-size process without interpretation or pilot plant experimentation.

When selecting an indirect method for agitation detection, it is important to evaluate whether the device selected will reliably detect a problem with agitation for all potential failure modes.

Mechanical agitation is generally very effective in providing good bulk mixing. Because mechanical agitation involves a large rotating piece of equipment, it is possible that a severe failure in the agitator will result in additional equipment damage. Agitators may become off balance, a situation that tends to become more severe as the device continues to rotate, and may destroy other portions of the system. If jammed, an agitator may shear and puncture a vessel. Most agitators have shear pins designed to prevent excessive torque on the agitator; however, with a shear pin an agitator system could stop at an inappropriate time because of excessive liquid viscosity and premature shear pin failure.

While providing a high rate of local mixing to the material in the
pump-around loop, an induced-flow mixing system will usually provide less
mixing to the bulk liquid than will a mechanical mixing system. This type of
mixing system requires additional piping and therefore creates additional
potential for a piping leak. It is possible to run the mixing loop pump with
a valve closed or with a line plugged and be unaware that no mixing is
occurring.

An in-line mixing device is generally physically smaller and less expensive than an agitator-type mixing system. Such a device is appropriate where
single pass, short duration mixing is acceptable and where heat transfer
performance is satisfactory and where gas or vapor evolution are not of
concern. The extent of mixing for these devices depends on the length of the
mixing section, the design of the internals and the flow rate through the
section. Higher flow rates create a higher shear and better mixing, but they
also require more energy input to overcome pressure drop.

Since mixing involves high shear, a mixing device is often more prone to
corrosion and erosion than are other equipment components of a process system.

The reliability of mixing system depends on the reliability of individual
components that comprise the system. The reliability of a full multicomponent
control system also depends on the specific equipment and detailed design of
the system. Table 2-23 shows reliability data, expressed as typical failure
rates, for some individual components.

2.7.4 Summary of Control Technologies

Table 2-24 summarizes, from the point of view of both new and existing
facilities, major hazards or hazard categories associated with mixing control,
and the corresponding control technology. Numerous individual control
technologies or procedural changes can be inferred for each category.

TABLE 2-23. TYPICAL FAILURE RATES OF MIXING SYSTEM COMPONENTS

Components	Failure Rate (Failures/year)
Mechanical Agitator (agitator motor only)	
- "Normal" Service	0.088
- "Severe" Service	8.8
Induced Flow Pump System (pump and motor only)	
- "Normal" Service	0.26
- "Severe" Service	8.8
Static Mixer	[a]
Flow Switch	1.12
Pressure Switch	0.34
Tachometer	0.044

[a]Not Available

Source: Adapted from References 8, 19, 37.

TABLE 2-24. MAJOR MIXING-RELATED HAZARDS AND CONTROL TECHNOLOGY SUMMARIES

Process Variable	Hazard	Control Technology or Procedural Category	
		New Facility	Existing Facility
Mixing	• Loss of cooling	• Mixing detection	• Add mixing detection
	• Reactant accumulation	• Backup power supply	• Add backup power supply
	• Loss of heating	• Materials of construction	• Change materials of construction
		• Equipment selection	• Change type of mixing system

2.7.5 Costs

Table 2-25 presents costs for components related to mixing system process modifications. These component types and costs are based on a typical installation. Since there are other types of systems and many variations of systems within a given type, these costs provide an order-of-magnitude basis for evaluating the economic effects of possible process modifications involving mixing systems.

2.7.6 Case Example (1)

In one instance of a loss of mixing control, the agitator stopped in a batch nitration reactor, but the process operator was unaware of this since instrumentation that would have stopped the acid feed to the reactor and given an alarm signal of agitator stoppage failed to work. When the agitator started up again, the reactor exploded.

2.8 COMPOSITION CONTROL

Varying a stream's composition affects its chemical and physical properties. Stream composition must usually fall within a fairly narrow range if a process system is to operate within design specifications. Proper design and operation of composition control systems may be important in preventing accidental releases.

2.8.1 Hazards Associated With the Loss of Composition Control

As a stream's composition varies from the design specifications, its chemical and physical properties also vary. The consequences are process-specific and may range from a lower quality product to an explosion and massive chemical release.

TABLE 2-25. COSTS OF COMPONENTS ASSOCIATED WITH PROCESS MODIFICATIONS FOR
MIXING SYSTEMS

Basis: Mixing vessel volume: 10,000 gallon
Agitator power: 2 hp/1,000 gallons (20 hp)
Mixing rate: 10,000 gallons/15 minutes (667 gpm)
Static mixer diameter: 50 inches (200 gpm)
Induce flow pump: 667 gpm, 50 psig

Component	Capital Cost Range ($)	Annual Cost Range ($/yr)	References
	(1986 Dollars)		
Mechanical agitator (turbine), single impeller, impeller speed to 45 rpm) carbon steel	31,600	4,800	20
316 stainless steel	50,600	11,800	20
Induced flow pump system	8,300	1,250	20
Static mixer	N/A	N/A	N/A
Flow switch	530	80	22
Pressure switch	530	80	38

Physical properties such as vapor pressures and boiling and freezing temperatures will vary as composition varies. Some possible consequences include an overpressure caused by increased vapor pressure or a ruptured line caused by frozen process materials.

The chemical properties of a process stream will vary as the composition varies. Potential reactions and their corresponding rates will be affected by variations in the composition that could result in a variety of consequences. An altered composition could lead to a much more rapid reaction rate than desired, which could result in an overpressure, which could result in unwanted side reactions, or in an altered composition that could stop the desired reaction completely.

Primary composition considerations involve reaction rates. Chemical reaction rates are usually highly temperature dependent and are sometimes pressure dependent. For a system where chemical reactions are involved, anything that affects pressures or temperatures may affect stream compositions. Additionally, any alternate event (such as catalyst decay) that alters the nature of the chemical processes involved will affect stream composition.

Any number of mechanical or electrical malfunctions could alter stream compositions. A loss of flow control resulting from equipment malfunction may lead to a loss of composition control. No flow, too much flow or reverse flow of feed streams will all affect the composition of the process unit into which they feed. Contamination of the stream via the rupture of a heat exchanger tube is another example of a mechanical failure resulting in a loss of stream composition control.

The instruments used to monitor composition are often sensitive, and in some cases, prone to failure when exposed to adverse conditions. For some varieties there is a time lag from when the sample is taken until the results are known. In this situation it will be difficult to determine the actual composition at any given time if the composition is changing rapidly. The consequences of this time delay will depend on the nature of the reaction system involved.

A potential shortcoming of many composition monitors is that they often have a fairly specific range of materials that they can detect. An ideal detector would be able to monitor the products and reactant in a given reaction and all of the unwanted by-products and contaminants. Not selecting the best detector for the given application is another potential cause of loss of composition control.

Operational considerations also influence composition control. A lack of proper maintenance and training will contribute to a loss of composition determination and control. Composition determination systems require regular maintenance to function properly. Since most systems will monitor only a few components in the system, an operator must be trained to interpret what may actually be happening to the entire system.

2.8.2 Technology of Composition Control

A composition analyzer is an excellent device for monitoring the overall condition of the process, but it is rarely used directly to control a single variable such as flow because many process variables may affect composition, and rarely can composition be entirely controlled by one variable. An exception might be controlling pH by metering an alkaline or acid solution into a process stream. In most cases, a composition analyzer is usually used to warn operators when the composition of a process stream has deviated from normal conditions.

Composition analysis may be achieved by manually taking samples from a process stream and analyzing them in the laboratory or by installing an automatic on-line analyzer. An industrial equivalent exists for most common instruments used in laboratory analysis. Laboratory analysis of a process sample is usually much slower than on-line analysis, though it may be more accurate than on-line analysis. Where a process involves hazardous materials, some type of on-line composition analysis that provides an early warning of significant deviations may be desirable for critical streams.

Composition analysis equipment may be divided into two broad categories: devices that analyze for chemical species in stream, and devices that measure a chemical or physical property of the stream. Devices for measuring the chemical and physical properties of a stream are typically used where the cost of individual chemical species determination is excessive, or where the information obtained from a chemical or physical property measurement is adequate.

Common instruments used for on-line chemical species analysis include oxygen and moisture analyzers, chromatographs, ultraviolet analyzers, flame ionization analyzers, mass spectrometers, and infrared analyzers. Common chemical and physical properties measured on line include density, molecular weight, pH, viscosity, and oxidation-reduction potential. Some of these devices take a continuous reading from an in-line probe, some take continuous readings from a slip stream, while others automatically withdraw samples periodically.

The output from composition analyzers varies according to the type and sophistication of the instrument. Many of the physical and chemical property probes generate a signal that may be read directly from a scaled meter. These devices usually require periodic calibration. Most analyzers generate some type of graphical spectrum. The presence of various chemical species is then interpreted either by the operator or by the computer software available with some modern systems.

Two important considerations in the design of a composition analyzer systems are: 1) the location of the sensing device, and 2) the composition data required. These considerations might be addressed by a formal evaluation of the process using a hazard and operability study or some similar method that would help identify which stream compositions are most important from the perspective of safety and which stream compositions best indicate the status of the process.

Once the location and chemical species to be monitored have been selected, the analyzer is selected. It may be impossible or prohibitively expensive to monitor all the species desired. In such cases, the potential hazard associated with a change in composition and the amount of useful information that may be obtained from monitoring the composition must be weighed against the expense.

2.8.3 Control Effectiveness

Composition deviations indicate that something is potentially wrong with the process and provide information for diagnosing the problem's source. However, an incorrect result on a composition analyzer may cause a process problem to be missed and a good reading from a composition analyzer is no guarantee that everything is in order.

The information obtained from all varieties of composition instrumentation is limited. There are no perfect analyzers, and it is likely that no single analyzer is available to monitor every potential species within a given process stream. Since cost and complexity prevent every process stream from being monitored, it is important to carefully choose which streams should be monitored and which type of composition analyzer will give the most meaningful cost-effective information. As with all instrumentation systems accuracy, precision, sensitivity, and tolerance to overrange are important considerations in analyzer selection.

Interpreting the results from a composition analyzer is not necessarily straightforward. The problem with interpretation is two-fold: 1) the output from the device may be misread or misunderstood or 2) a correct instrument reading may be made and a wrong conclusion drawn. Instruments that provide a graphical trace are most easily misread or misunderstood. In chemical analyses impurities may be obscured by the signal of other species, the chemical character of a compound may make it invisible to the detector or it may respond more intensely than other compounds and appear to be present in a much higher concentration than it actually is. Any number of process deviations that may result in composition deviations, and correct interpretation of a composition analysis does not mean that a correct assessment of the state of the process will be made.

The maintenance requirements of a composition analyzer depend on the device. The sensing element in most devices must be cleaned and usually recalibrated regularly.

The reliability of composition control depends on the reliability of the individual components that comprise the composition control system. The reliability of a multicomponent system also depends on the architecture of the specific system. Table 2-26 presents reliability data expressed as typical failure rates for some individual components.

2.8.4 Summary of Control Technologies

Table 2-27 summarizes, from the point of view of both new and existing facilities, major hazards or hazard categories associated with composition and corresponding control technology. Numerous individual control technologies or procedural changes can be inferred for each category.

2.8.5 Costs

Costs of components related to composition measurement and control system process modifications are presented in Table 2-28. These component types and costs are based on typical installations. Since there are other types of systems, and many variations of systems within a given type, these costs

TABLE 2-26. TYPICAL FAILURE RATES OF COMPOSITION SYSTEM COMPONENTS

Component	Failure Rate (Failures/Year)
Composition Determination Equipment	
Density Sensor	
- indicator	a
- transmitter	—a
pH Meter	5.88
Viscosity Sensor	
- indicator	a
- transmitter	—a
Chemical Species Analyzers	
Chromatograph	30.6
Infrared Analyzer	1.40
Oxygen Analyzer	2.5-5.65
Moisture Analyzer (gases)	8.0
Conductivity Sensor	14.2-16.7

[a] Not available

Source: Adapted from References 1, 8, 19, and 37.

TABLE 2-27. MAJOR COMPOSITION-RELATED HAZARDS AND CONTROL TECHNOLOGY SUMMARIES

Process Variable	Hazard	Control Technology or Procedural Category	
		New Facility	Existing Facility
Composition	• Excess or deficient reactant	• Initial design and selection of composition analysis system	• Change type of composition analysis
	• Excess or deficient catalyst		• Add composition analysis to process stream
	• Excess corrodant		
	• Excess toxic material in wrong stream		

TABLE 2-28.   COSTS OF COMPONENTS ASSOCIATED WITH PROCESS MODIFICATIONS
FOR COMPOSITION DETERMINATION AND CONTROL

	Capital Cost Range ($)	Annual Cost Range ($/Yr) (1986 Dollars)	References
Composition Determination Equipment			
Density Sensor			
- indicator	500-1,500	92-275	20,41
- transmitter	1,500-5,000	275-917	20,41
Molecular Weight Sensor	5,000-10,000	917-8,834	20,41
pH Detector	4,000-5,000	734-917	20,41
Viscosity Sensor			
- indicator	2,000-5,000	367-917	20,41
- transmitter	5,000-12,000	917-2,200	20,41
Chemical Species Analyzers			
Chromatograph	10,000-40,000	1,834-7,340	20,41
Infrared Analyzer	3,000-10,000	550-1,834	20,41
Refractometer	4,000-10,000	734-1,834	20,41
Oxygen Analyzer	3,000-8,000	550-1,467	20,41
Moisture Analyzer	4,000-15,000	734-2,751	20,41
Spectrometer	10,000-40,000	1,834-7,340	20,41
Conductivity Sensor	700-2,000	129-367	20,41

provide an order-of-magnitude basis for evaluating the economic effects of possible process modifications involving composition measurement and control systems.

2.8.6 Case Example (42)

A day tank containing 6500 gallons of ethylene oxide became contaminated with ammonia. The tank ruptured, dispersing ethylene oxide into the atmosphere, where the cloud ignited and created an explosive force equal to 18 tons of TNT. One person was killed, nine were injured and property loss was valued at approximately 16 million dollars.

2.9 REFERENCES

1. Lees, F.P. Loss Prevention in the Process Industries, Volume 1 and 2.
    Butterworths, London, England, 1983.

2. Hix, A.H. Safety and Instrumentation Systems. Loss Prevention, Volume
    6. American Institute of Chemical Engineers, New York, NY, 1972.

3. Considine, D.M. (ed.). Process Instruments and Controls Handbook,
    McGraw-Hill Book Company, New York, N.Y., 1985.

4. Shinsky, F.G. Process Control Systems. McGraw-Hill Book Company, New
    York, NY, 1979.

5. Green, D.W. (ed.). Perry's Chemical Engineers' Handbook, (Sixth
    Edition). McGraw-Hill Book Company, New York, NY, 1984.

6. Control Engineering. Any issue. Technical Publishing, Inc., Barrington,
    Illinois.

7. Cost indices obtained from Chemical Engineering. McGraw-Hill Publishing
    Company, New York, NY. November 1972, June 1974, December 1985, and
    August 1986.

8. U.S Nuclear Regulatory Commission. Reactor Safety Study. National
    Technical Information Service, WASH-1400 (NUREG 75/014), October 1975.

9. Process Safety Management. Chemical Manufacturer's Association.
    Washington, D.C., May 1985.

10. Guidelines for Hazard Evaluation Procedures. The Center for Chemical
    Plant Safety. American Institute of Chemical Engineers, 1985.

11. Kletz, T.A. Make Plants Inherently Safe. Hydrocarbon Processing.
    September 1985.

12. Hazard Survey of the Chemical and Allied Industries. American Insurance
    Association. Engineering and Safety Service, AIA, New York, NY, 1979.

13. Groggins, P.H. (ed.). Unit Processes in Organic Synthesis, Fourth
    Edition. McGraw-Hill Book Company, New York, NY, 1952.

14. Herrick, E.C., J.A.King, R.P. Ouellette and P.N. Cheremisinoff. Unit
    Process Guide to Organic Chemical Industries. Ann Arbor Publishers, Ann
    Arbor, MI, 1979.

15. Courty, P.H., J.P. Artie, A. Converse, P. Mikitinko, and A. Sugier.
    C_1-C_6 Alcohols From Syngas. Hydrocarbon Processing. November 1984. pp.
    105ff.

16. Kremers, J. Avoid Water Hammer. Hydrocarbon Processing. March 1983.

17. Rasmussen, E.J. Alarm and Shutdown Devices Protect Process Equipment.
    Chemical Engineering. May 12, 1975.

18. Rase, H.F. Piping Design for Process Plants. John Wiley and Sons, Inc.,
    New York, NY, 1963.

19. Liptak, B.G. and V. Christa (ed.). Instrument Engineers' Handbook,
    Revised Edition. Chilton Book Company, Radnor, PA, 1982.

20. Peters, M.S. and K.D. Timmerhaus Plant Design and Economics for Chemical
    Engineers, Third Edition. McGraw-Hill Book Company, New York, NY, 1980.

21. Liptak, B.G. Costs of Process Instruments. Chemical Engineering.
    September 7, 1970.

22. Flow Measurement Handbook. Omega Engineering Corporation. 1985.

23. The Richardsen Rapid Construction Cost Estimating System, Volumes 1-4.
    Richardsen Engineering Services, Incorporated, San Marcos, CA, 1986.

24. Floar, P.C. Programmable Controllers Directory, First Edition.
    Technical Data Base Corporation, Conroe, Texas, 1985.

25. Telephone conversations between J.D. Quass of Radian Corporation and a
    representative of Foxboro Corporation. Corpus Christi, Texas, 1986.

26. Telephone conversation between J.D. Quass of Radian Corporation and a
    representative of Fisher Controls. Stafford, TX, 1986.

27. Liptak, B.G. Safety Instruments and Control Value Costs. Chemical
    Engineering. November 2, 1970.

28. Fauske, Haus K. Emergency Relief System Design. Chemical Engineering
    Progress. August, 1985.

29. Recommended Practice for the Design and Installation of Pressure
    Relieving Systems in Refineries: Part I - Design; Part II -
    Installation. American Petroleum Institute. API RP 520, 1977.

30. Guide for Explosion Venting. National Fire Protection Association. Vol.
    68, 1978.

31. American Society of Mechanical Engineers. Code for Pressure Piping B31.
    Chemical Plant and Petroleum Refinery Piping. New York, NY.

32. Scully, William A. Safety-Relief-Valve Manfunctions: Symptoms, Causes
    and Cures. Chemical Engineering. August 10, 1981.

33. Papa, D.M. How Back Pressure Affects Safety Relief Valve. Hydrocarbon Processing. May 1983.

34   Bradford, Mike and David G Durrett. Avoiding Common Mistakes in Sizing Distillation Safety Valves. Chemical Engineering. July 9, 1984.

35. Fitt, J.S. The Process Engineering of Pressure Relief and Blowdown Systems, Loss Prevention and Safety Promotion in the Process Industries. First International Loss Prevention Symposium, Amsterdam, Holland. 1974.

36. Lawley, Herbert G. and Trevor A. Kletz. High-Pressure-Trip Systems for Vessel Protection. Chemical Engineering. May 12, 1975.

37. Anyakora, S.N. G.F.M. Engel, and F.P. Lees. Some Data on the Reliability of Instruments in the Chemical Plant Environment. The Chemical Engineer. Number 255, 1971.

38. Pressure and Strain Measurement Handbook and Encyclopedia. Omega Engineering Corporation, 1985.

39. Temperature Measurement Handbook and Encyclopedia. Omega Engineering Corporation, 1985.

40. Bartosiak, G. Guide to Thermocouples. Instruments and Control Systems. November, 1978.

41. Liptak, B.G. Costs of Viscosity, Weight, and Analytical Instruments. Chemical Engineering. September 21, 1970.

42. Davenport, J.A. A Survey of Vapor Cloud Incidents. Loss Prevention. American Institute of Chemical Engineers. Volume 17, September 1977.

43. Weir, E.D, G.W. Gravenstine, and T.F. Hoppe. Thermal Runaways: Problems with Agitation. Plant/Operations Progress. July 1986.

44. Wilmot, D.A., and A.P. Leong. Another Way to Detect Agitation. Loss Prevention. American Institute of Chemical Engineers. Volume II, New York, NY, 1977.

3. Physical Plant Design Considerations

The preceding section of this manual addressed process design considerations associated with basic process variables, the chemistry of a process, and process control. This section addresses the physical plant design and hardware of a process facility. Process and physical plant design taken together, whether of a new facility or of the modification of an existing facility, are the basis for preventing accidental chemical releases. In the words of one author: "The safety of the plant is determined primarily by the quality of the basic design rather than the addition of special safety features. It is difficult to overemphasize this point" (1). Safety features are part of that basic design, but in some facilities there may have been oversights in design and construction that warrant the addition of safety features to an existing facility.

Physical plant design considerations address the specific hazards caused by hardware failure, the proper design and construction of equipment to reduce those hazards, and the siting and layout of the equipment within the process facility. Specific hardware-related prevention measures can be identified to reduce the probability of accidental chemical releases.

General and detailed principles for the design of chemical process facilities comprise a vast technical literature. It is not within the scope of this manual to review all aspects of design, nor the intent of this manual to be used for design. The reader is referred to other technical literature for these purposes. It is the purpose here to highlight significant general considerations related to accidental release prevention.

Since the fundamental purpose of the physical plant and its equipment is to contain the chemicals under normal process operating conditions and within

limited ranges of deviation, sound design practices that are thoroughly
understood and faithfully followed are essential to prevent releases. Design
principles related specifically to release prevention can be grouped into four
broad categories:

- Standards, codes, recommended practices, and guidelines;

- Siting and layout;

- Miscellaneous considerations; and

- Equipment.

Each of these categories is discussed in the subsections that follow.
Some important potential hazards and their control, related to the equipment
and its layout in a process facility, are examined.

3.1   STANDARDS, CODES, AND RECOMMENDED PRACTICES

Equipment used in a process facility must be selected to function under
the specified process conditions and under upset conditions; it must ensure
containment of the chemicals being processed and must be resistant to fire and
explosion. To help industry meet these requirements, numerous codes, standards of practice, recommended practices, and guidelines (referred to here-
after as standard design criteria) exist for categories and specific kinds of
equipment. Different organizations developed standard design criteria over
the years in response to knowledge gained from actual accidents or in antici-
pation of operating and safety problems based on technical analysis. Many of
the standard design criteria were developed to meet the needs of property
insurer's and worker health and safety concerns. It is beyond the scope of
this manual to even present a detailed summary of all the standard design
criteria applicable in the chemical process industries. Table 3-1, however,
lists some major organizations involved in developing such standard design

criteria. Table 3-2 summarizes some of the areas applicable to chemical process facilities (2).

Two major equipment categories addressed by codes are vessels and piping. Table 3-3 shows the design and physical components of vessels covered by the American Society of Mechanical Engineer's (ASME) code for pressure vessels (3). This is shown as an example of how much detail must be addressed by fabricators and constructors of process equipment and facilities. This code has been adopted as a legal standard in many areas. For piping, an important code was developed by the American Petroleum Institute (API); it applies to petroleum refineries and petrochemical plants, and may be applicable to other chemical process facilities as well (4). Many individual companies have their own standard design criteria, especially major corporations in the chemical process industries. Design criteria must always be carefully scrutinized when toxic chemicals are involved, especially in new situations, because the such criteria have often been developed for known sets of conditions, and it is possible that adherence to a standard criterion such as a code without a full appreciation of its basis or limitations could lead to unforeseen secondary hazards.

In the context of accidental release prevention, any standard design criteria, even codes, should be viewed as a minimum basis or starting point for equipment and plant design. Many specific situations may suggest more stringent specifications than the standard design criteria require. This if often an overlooked aspect of code and standards compliance that can lead to problems.

3.2 SITING AND LAYOUT CONSIDERATIONS

Siting and layout considerations are an important aspect of accidental release prevention. Siting refers to the location of the process facility within a community, while layout refers to the positioning of equipment within the process facility.

TABLE 3-1.   SOME OF THE MAJOR ORGANIZATIONS PROVIDING CODES, STANDARDS, RECOMMENDED PRACTICES, AND GUIDELINES FOR EQUIPMENT FOR CHEMICAL AND ALLIED INDUSTRY PROCESS PLANTS

Name	Abbreviation Symbol
Technical and Trade Groups	
American Water Works Association	AWWA
Air Conditioning & Refrigeration Institute	ARI
Air Moving and Conditioning Association	AMCA
American Association of Railroads	AAR
American Gas Association	AGA
American Petroleum Institute	API
Chlorine Institute	CI
Compressed Gas Association	CGA
Cooling Tower Institute	CTI
Chemical Manufacture's Association	CMA
Manufacturers Standardization Society	MSS
National Electrical Manufacturers Association	NEMA
Pipe Fabrication Institute	PFI
Scientific Apparatus Makers Association	SAMA
Society of Plastics Industry	SPI
Steel Structures Painting Council	SSPC
Tubular Exchanger Manufacturers Association	TEMA
U.S. Government Agencies	
Bureau of Mines	BM
Department of Transportation	DOT
U.S. Coast Guard	SCG
Hazardous Materials Regulation Board	HMRB
Federal Aviation Administration	FAA
Environmental Protection Agency	EPA
National Bureau of Standards	NBS
Occupational Safety and Health Administration	OSHA
Testing Standards and Safety Groups	
American National Standards Institute	ANSI
American Society for Testing and Materials	ASTM
National Fire Protection Association	NFPA
Underwriters Laboratories, Inc.	UL
National Safety Council	NSC
Insuring Associations	
American Insurance Association	AIA
Factory Insurance Association	FIA
Factory Mutual System	FM
Oil Insurance Association	OIA

(Continued)

TABLE 3-1 (Continued)

Name	Abbreviation Symbol
Professional Societies	
American Conference of Governmental Industrial Hygienists	ACGIH
American Industrial Hygiene Association	AIHA
American Institute of Chemical Engineers	AIChE
American Society of Mechanical Engineers	ASME
Amer. Soc. of Htg., Refrig. & Air-Cond. Engs.	ASHRAE
Illumination Engineers Society	IES
Institute of Electrical and Electronic Engineers	IEEE
Instrument Society of America	ISA

Source: Adapted from Reference 2.

TABLE 3-2.   SOME OF THE AREAS COVERED BY CODES, STANDARDS, GUIDELINES,
AND RECOMMENDED PRACTICES OF DESIGNATED ORGANIZATIONS
(SEE TABLE 3-1 FOR SYMBOLS DEFINITIONS)

Accident Case History - NFPA, NSC, AIA, FIA, FM, OIA, AIChE, AGA, API, CMA, USCG, OSHA

Plant & Equipment Layout - NFPA, NSC, AIA, FIA, FM, OIA, AWWA, AAR, API, CGA, CMA, USCG, HMRB

Electrical Area Classification - ANSI, NFPA, NSC, AIA, FIA, FM, OIA, API, CMA, USCG, OSHA

Electrical Control and Enclosures - ANSI, NFPA, UL, NSC, AIA, FIA, FM, OIA, IEEE, ISA, ARI, MCA, NEMA, USCG, OSHA

Grounding and Static Electrical - ANSI, NFPA, UL, NSC, AIA, FIA, FM, OIA, IEEE, API, NEMA, USCG, OSHA

Power Wiring - ANSI, NFPA, UL, FIA, FM, OIA, IEEE, API, NEMA, USCG, OSHA

Lighting - ANSI, NFPA, UL, NSC, FM, IEEE, IES, NEMA, USCG

Emergency Electrical Systems - NFPA, AIA, FM, IEEE, AGA, NEMA, USCG

Instrumentation - ANSI, ASTM, NFPA, UL, AIA, FIA, FM, OIA, IEEE, ISA, AWWA, ARI, API, CGA, SAMA, USCG, HMRB, NBS

Shutdown Systems - NFPA, UL, AIA, FIA, OIA, API, USCG

Pressure Relief Equipment Systems - NFPA, AIA, FIA, FM, OIA, ASME, API, CI, CGA, USCG, HMRB, OSHA

Venting Requirements - NFPA, FIA, FM, API, USCG, HMRB

Product Storage and Handling - ANSI, NFPA, AIA, FIA, FM, OIA, AIChE, AAR, API, CI, CGA, MCA, USCG, OSHA

Piping Materials and Systems - ANSI, ASTM, NFPA, UL, NSC, AIA, FIA, FM, NBS ASHRAE, IES, AWWA, ARI, AGA, API, CI, CGA, MSS, NFPA, PFI, SPI, USCG, HMRB,

Materials of Construction - ASTM, ANSI, NFPA, UL, NSC, AIA, FM, OIA, ISA, AWWA, CI, CGA, CTI, MCA, TEMA, USCG, HMRB, NBS

Insulation and Fireproofing - ANSI, ASTM, UL, AIA, FM, OIA, ASHRAE, USCG

Painting and Coating - ANSI, ASTM, UL, AIChE, AWWA, SSPC, HMRB, NBS, OSHA

Ventilation - ANSI, NFPA, UL, NSC, FIA, FM, ACGIH, AIHA, BM, USCG

(continued)

TABLE 3-2 (continued)

Dust Hazards - ANSI, NFPA, UL, NSC, FIA, FM, ACGIH, AIHA, BM, USCG

Noise and Vibration - ANSI, ASTM, NFPA, UL, NSC, AIHA, AIChE, ASHRAE, ISA, ARI, AMCA, AGA, API, NFPA, EPA, OSHA

Lubrication - ANSI, NFPA, ASME, AMCA

Fire Protection Equipment - ANSI, NFPA, UL, NSC, AIA, FIA, OIA, AWWA, API, CGA, CMA, NEMA, BM, USCG, OSHA

Safety Equipment - ANSI, UL, NSC, FM, ACGIH, AIHA, CI, CGA, MCA, BM, USCG, OSHA

Pumps - ANSI, UL, OIA, AIChE, AWWA, HI NFPA, USCG

Fire Pumps - ANSI, NFPA, UL, FM, IEEE, HI, USCG

Fans and Blowers - FM, ACGIH, AIHA, ASME, ARI, AMCA, USCG

Compressors - AIA, FM, OIA, ASME, ASHRAE, ARI, USCG

Air Compressors - ANSI, AIA, FM, USCG

Steam Turbines - AIA, FM, OIA, IEEE, USCG

Gas Turbines - NFPA, FIA, FM, OIA, AGA, USCG

Gas Engines - NFPA, FM, OIA, USCG

Electric Motors - ANSI, NFPA, UL, IEEE, CMA, USCG

Shell & Tube Exchangers - AIChE, ASME, ASHRAE, ARI, AGA, CGA, PFI, USCG

Air-Fin Coolers - OIA, ASHRAE, ARI, USCG

Cooling Towers - NFPA, FM, OIA, CTI

Boilers - ANSI, NFPA, UL, NSC

Fired Heaters - ANSI, NFPA, UL, FIA, FM, OIA, ASME, USCG

Combustion Equipment & Controls - ANSI, NFPA, UL, NSC, FIA, FM, OIA, USCG

Refrigeration Equipment - ANSI, NFPA, UL, FM, ASHRAE, ARI, USCG

(continued)

TABLE 3-2 (continued)

Dust Collection Equipment - NFPA, FIA, FM, USCG

Pneumatic Conveying - ANSI, NFPA, FIA, USCG

Solids Conveyors - CMA

Storage Tanks - NFPA, UL, OIA, AWWA, CI, USCG, NBS, OSHA

Pressure Vessels - NFPA, NSC, AIA, ASME, ARI, CGA, HEI, DOT, USCG, OSHA

Material Handling - NFPA, NSC, CMA, OSHA

Jets and Ejectors - HEI, USCG

Gear Drives Power Transmission - ANSI, NSC, AIA, AGMA, USCG

Stacks and Flares - OIA, USCG, FAA

Drain and Waste Systems - AIChE, AWWA, CMA, USCG

Inspection and Testing - ASTM, NFPA, NSC, AIChE, ASHRAE, IEEE, AMCA, ABMA, API, AGMA, AWWA, CGA, CTI, HEI, HI, MSS, NFPA, PFI, DOT, USCG

Source: Adapted from Reference 2.

TABLE 3-3. PHYSICAL COMPONENTS, DESIGN, AND FABRICATION FEATURES
OF VESSELS COVERED BY ASME CODES AS INDICATED

Component	ASME Code Section

Component Codes

Component	ASME Code Section
Full face gasket	UA-6, UA-45
Welded connection	UW-15, UW-16, Fig. UW-16.1
Reinforcement pad	G-37, UG-40, UG-41, UG-82, UW-15, UA-280
Code termination of vessel	U-1 (e)
Lap Joint stub end	UG-11, G-44, UG-45
Loose type flange	UG-44, UA-45 to 52, Fig. UA-48
Ellipoidal head, pressures, int.	UG-32, Ext.UG-33, UA-4, UA-275
Head skirt	UG-32, Fig. UW-13.1, UW-13
Optional type flanges	UG-14, UG-44, UW-13, Fig. UW-13.2, UA-45 to 52, UA-55, Fig. UA-48, Appendix S
Nuts and washers	UG-13, UCS-11, UNF-13
Studs and bolts	UG-12, CS-10, UNF-12
Applied linings	Part UCL., UG-26, Appendix F
Integrally clad plate	Part UCL, Appendix F
Corrosion	G-25, CS-25, UCL-25, UA-155 to UA-160
Stiffener plate	UG-6, UG-22, UG-54, UG-82
Support lugs	G-6, UG-54, UG-82, Appendix G
Longitudinal joints	UW-33, UW-3, UW-35, UW-9
Tell tale holes	UG-25, UCL-25
Attachment of jacket	Fig. UA-104, Fig. UA-105
Jacket vessels	UG-28, UG-47(c) Appendix IX
Plug welds	UW-17, UW-37
Bars and structural shapes	UG-14, UW-19, Fig. W-19.2 Stayed surfaces, UG-47
Stay bolts	UG-14, UG-27f, UG-47 toUG-50, UW-19, Fig. UW-19.1
1/2 Apex angle	G-32
Support skirt	UG-6, UG-22, UG-54, UA-185 to UA-189, Toriconical head pressures, Int. UG-32, UG-36, Fig. UG-36, Ext. UG-33, A-275
Studded connections	UG-43, UG-44, W-16, Fig. UW-16.1, UW-15
Optional type flange	UG-14, UG-44, UW-13, Fig. W-13.2, UA-45 to 52, UA-55, Fig. UA-48, Appendix S
Bolted flange, spherical cover	UA-6
Manhole cover plate	UG-11, UG-46
Flued openings	UG-32, UG-38, Fig. UG-38

(Continued)

TABLE 3-3 (Continued)

Component	ASME Code Section

Component Codes

Yoke	UG-11
Studs, nuts, washers	UG-12, UG-13, UCS-10, UCS-11, UNF-12, UNF-13
Spherically dished covers	UA-6, Fig. UA-6
Flat face flange, Appendix Y	Fig. UA-1110
Welded connection	UW-15, UW-16, Fig. UW-16.1
Opening	UG-36 to UG-42, UA-7, UA-280
Multiple openings	UG-42
Non-pressure parts	G-6, UG-22, UG-55, UG-82
Hemispherical head, pressures, int.	UG-32, UA-4, UA-3, Ext. UG-32, UA-275
Unequal thickness	W-9, Fig. UW-9, UW-13, Fig. UW-13.1
Shell thickness	UG-16, UHA-20, Pressures, Int. UG-27, UA-1, UA-2, UA-274, Ext. UG-28, UA-270 to UA-272
Stiffening rings	UG-29, UG-30, UA-272
Welded connection	W-15, UW-16, Fig. UW-16.1
Flat head	UG-34, Fig. UG-34, UW-13, UG-93(d)(3), Fig's. UW-13.2 & UW-13.3
Openings, flat heads	UG-39
Backing strip	Table UW-12, UW-35
Circumferential joints	UW-3, UW-33, UW-35
Flat head	UG-34, Fig. UG-34, UG-39
Tube sheet, no code, TEMA acceptable.	U2(g)
Tubes	UG-9, Pressure, Int. UG-31, Ext. UG-28, UG-31
Baffle	G-6
Channel section, cast steel	UG-24 part UCS, UHA, Cast Iron, UC1
Integral type flange	UG-44, UA-45 to UA-52, Fig. UA-48, UA-55, Appendix S
Reinforcement pad	UG-22, UG-37, UG-40, UG-41, UG-82, UW-16, UA-280
Compression ring	UA-5
1/2 Apex angle	UG-32
Conical heads, pressures	Pressures, Ext. UG-33, UA-275, Int. UG-32, UG-36, Fig. UG-36, A-4,
Small welded fittings	UG-11, UG-43, UW-15, UW-16, Fig. UW-16.1, Fig. UW-16.2

(Continued)

TABLE 3-3 (Continued)

Acitivity	ASME Code Section

Component Codes

Threaded openings	UG-43(e)
Heat attachment	UW-13, Fig. UW-13.1
Fillet welds	UW-18, UW-36 Table UW-12
Knuckle radius.	UG-32, UCS-79
Torispherical head, pressures	Int. UG-32, UA-4, Ext. UG-33, A-275

General Design and Fabrication Codes

Heat Treatment	UG-85, UW-10, UW-40, UCS-56, Table UCS-56, UCS-79(d), UCS-85, UNF-56, UHA-32, UHA-105, & UCL-34
Inspection	UG-90 thru UG-97, U-1 (j)
Joint Efficiency	UW-12, & Table UW-12
Lethal Service	UW-2(a), UCD-2, & UCl-2
Loadings	UG-22
Low Temperature	UG-84, UW-2(b), UCS-65, UCS-66, UCS-67, UNF-65, & UCL-27
Materials	UG-5 thru UG-15, UG-18, UG-77, UCL-11 & UW-5, Tables NF-1 & NF-2
Pressure, Design	UG-19, & UG-21, Max. Allowable Working UG-98
Temperature, Design	UG-19, UG-20
Pressure Vessels Subject to Direct Firing	UW-2(d), U-1(h)
Radiographic Exam	UW-11, UW-51, UW-52, UCS-57, UNF-57, UHA-33, & UCLK-35, Spot Exam of Welded Joint UW-52, No Radiograph "W-11(c)
Relief Devices	UG-125 through UG-136, App. XI
Repairs	UG-78, UW-38, UW-40(d)
Stress	Max, Allow., Value UG-23, W-12(c), UNF-23, UHA-23, UCL-23
Test, Hydrostatic	G-99, UC1-99, UCL-52, & UA-60, Pneumatic UW-50 & UG-100; Proof, UG-101; Non-Destructive, UG-103, UNF-58, & UHA-34; Mag. Part, UA-70 thru UA-73; Liq. Pene, UA-91 thru UA-95; Ultrasonic, UA-901 thru UA-904; Impact, UG-84, UCS-66, UHA-51,NF-6
Stamping and Data	UG-115 thru UG-120
Unfired Steam Boilers	UW-2(c), U-1(g)

Source: Adapted from Reference 3.

3.2.1 Siting

Siting, as related to accidental chemical releases, is usually thought of in terms of mitigating the consequences of a release rather than in preventing it. Siting from the perspective of mitigation, the consequences of a release is dealt with in a companion manual on mitigation in this series. Certain conditions of siting, however, can influence the probability of a release. These conditions primarily are related to natural disasters and climate, but also to circumstances at neighboring facilities, or transportation accidents, that could play a role in causing releases.

Examples of natural phenomena and disasters include lightening, windstorms, floods, earthquakes, subsidence, and landslides. Climatic factors include rainfall patterns, temperature extremes, and temperature variability.

These factors directly influence physical plant design, and if not properly accounted for they can result in hazardous design deficiencies. For example, a facility in a flood-prone area may require different types of foundations for storage tanks or process equipment. Ambient temperatures might be important if a line freezeup in a particular process could lead to a failure that causes an accidental release.

Lees discusses siting from a process hazard perspective and cites a number of literature references on siting (1).

Materials of construction are influenced by the temperature extremes the facility is subjected to and by any corrosive characteristics of the ambient air. For example, areas of high humidity, or the presence of nearby salt water or of a facility that emits a corrosive gas should influence the selection of construction materials. The level of dust in the air and the

potential for severe dust storms may require special protection for certain
types of equipment. Special precautions should be taken wherever subfreezing
temperatures are common. Such precautions may include the addition of extra
insulation or heat tracing and may dictate the types of seals and gaskets used
on equipment.

Material strength requirements are influenced by the frequency of severe
wind and weather conditions. Process equipment and their supporting structures may need extra reinforcement or protection where high winds or severe
hail occur.

As a final example, the presence of a neighboring facility where a fire,
explosion, or accidental release could trigger an incident in one's own plant
may require consideration.

3.2.2 Layout

Layout refers to the placement and spacing of the various parts of a
process facility, including the individual equipment in the various parts. A
properly designed layout reduces the potential for and the consequences of an
accidental release by enhancing process operability and by segregating hazardous areas within the facility. A great deal has been written on plant layout
in the technical literature. Both the Chemical Manufacturers Association
(CMA) and the National Fire Protection Association (NFPA) have issued standards or guidelines for plant layout (5).

Increased distances between process units tend to reduce the potential
for, and the impact of an accidental release. The value of distance depends
on the nature of the hazard. An explosion in one process unit can result in
an accidental release in another process unit because of damage from the blast
wave. The blast wave pressure front travels rapidly but its intensity decreases rapidly with distance. A vapor cloud associated with an accidental
release travels much more slowly than a blast wave but usually affects a

larger area before it is diluted enough to present no danger. A fundamental
layout consideration where flammable or explosive vapor clouds may occur is to
locate away from obvious ignition sources. While unexpected ignition sources
may still ignite a flammable cloud, there is no reason to make the process
easier.

Whenever possible, a quantitative method should be used to aid in
choosing the spacing between process units. This can be done by first developing a relative hazard ranking of individual process units. One method of
quantifying the hazard of a given process unit assigns a hazard number to
various features of a substance or process. These numbers are based on an
arbitrary scale. The numbers for various portions of the system are combined
by multiplying by weighting factors and summing the results to create an
overall hazard rating. The Dow Index is an example of this type of approach
for ranking processes in terms of fire and explosion potential (6). An
extension of the DOW Index is the Mond Index, which extends the ranking to
include toxicity. Other methods for quantifying relative hazards and developing spacing requirements have been developed (7,8,9).

A list of basic concepts for achieving optimal layout has been assembled
by Lewis (8):

●   Roads need to allow entry to the plant from at least two
     points on the site perimeter, preferably from opposite
     sides.

●   All units in the area with a moderate or high fire risk
     should have access for emergency vehicles from at least
     two directions.

●   Control rooms, amenity buildings, workshops, laboratories
     and offices should be sited close to the site perimeter.

- Pipebridges should be laid out to minimize easy transfer of incidents from one unit to another. Key pipebridges should be assessed independently of adjacent units for hazard potential.

- Wherever possible, units of high risk should be separated from each other by units of mild, low or medium risk.

- Appropriate distances should separate units from identified ignition sources, such as furnaces, electrical switchgear, flarestacks, etc.

- Control rooms, amenity buildings, workshops, laboratories and offices should be adjacent to units of mild or low risk, which act as a barrier from higher risk units. Medium risk units are only acceptable adjacent to populated buildings (a) if lower risk units are not available for separation purposes, and (b) if the risk level is only just inside the "medium" band of Overall Risk Rating as assigned by the Mond Index method.

- Units with the highest values of the Major Toxicity Incident Index, assigned by the Mond Index method, should be suitably distant from all facility buildings containing many people, and also from activities outside the works boundary. (This particularly applies to the locations of schools, hospitals, places of entertainment etc.)

- Units having the highest values of the Aerial Explosion Index, assigned by the Mond Index method, should not be located close to a plant or facilities boundary; they should be separated by areas occupied by low risk activities with low population densities (up to 25 people per acre).

- Major pipebridges with a medium to high "Overall Risk Rating" (assigned by the Mond Index method) should be located so that they are at minimal risk from incidents on tall process units and from transport accidents arising from the regular movement of vehicles taking materials to or away from the plant site.

- Units previously and separately assessed for hazard ratings can be combined into a larger single unit providing (a) the risks are similar, (b) the potential direct and consequential losses do not become excessive and (c) a reassessment of the rating of the combined unit is acceptable.

- As far as possible, units should be laid out for a logical process flow to minimize the pipebridge requirement.

- Pipebridge routes should be chosen so that they are not likely to contribute to the spread of an accident. Alternative route selection should allow as much process control as possible.

- Storage units should be adequately separated from operational areas and, as far as possible, located away from road and rail traffic routes within the plant site.

In general, optimum spacing arrangements provide straight, unobstructed access ways that are continuous from one end of a processing unit to the other and are connected to roads that surround the unit. Piping should be laid out in a way that minimizes the amount of piping. Piping associated with various

utilities and process units should be segregated, since this will help to
avoid confusion during operations. Adequate spacing must be provided for
maintenance and construction. More than one escape route should be available
for any location within the facility. In the initial facility design, extra
room should be provided for future expansion to avoid creating future congestion hazards.

The layout for indoor process units requires special considerations.
Hazardous or flammable vapors can accumulate in an indoor facility. Proper
ventilation is a primary consideration for indoor facilities that handle
hazardous materials. Many air changes per hour with little recirculation may
be in order where a hazardous material is involved. Any potential leak source
should be protected. Vents and pressure relief discharge lines which may
release toxic or flammable materials should be piped to a safe area. The
holdup of hazardous materials inside of process buildings should be minimized
and generally limited to material in process equipment and piping. Some
additional considerations for indoor facilities are listed in this manual in
Section 3.3.4.

A safe control room is essential if a plant is to respond to an accidental release once it has occurred. As far as possible, a control room should
be located in an area where exposure to fires, explosions or toxic releases is
minimum. Wherever a control room is located, it should be designed to minimize the potential for injury to the employees in the event of an accident.
Examples of precautions include, explosion shielding, shatterproof windows, a
ventilation system that provides clean air while keeping the control room
under slight positive pressure, and extra fire protection. If a control room
is located within a process unit, then other buildings such as laboratories,
lunchrooms or offices should not be added onto the control room. Support
structures where personnel tend to congregate should be located outside of the
processing area.

Many of the layout precautions listed above are necessary for the sake of fire prevention. Additional layout considerations specifically for fire protection are discussed in Subsection 3.3.4.

3.2.3   Storage Layout Considerations

The arrangement of storage facilities follows the same general principles just discussed for general siting and layout. Storage facilities may merit special consideration, however, because they contain large inventories of chemicals.

Storage near high-hazard process areas and loading and unloading terminals should be the minimum required for the needs of the process or shipping requirements. Storage facilities should be located on, at most, two sides of a processing area so that the area is open on at least two sides. Inventory should be kept to a minimum. Tank storage areas should be located away from offices and other locations where people congregate. A tank storage area should be located as far from a community population center as possible. Prevailing wind direction should be considered so that the effects of a large release from storage facilities are minimized. The natural ground contour should be considered so that runoff from a spill would not contaminate a large area or would not find its way to a source of ignition.

One basis for selecting tank spacings are NFPA standards (5). These are based on flammable materials but are a useful guide for separating flammable and toxic materials. Two materials that may react and generate heat upon contact should not be stored within the same diked area. A chemical should not be stored in a diked area with a tank constructed of material that it would react with. For example, concentrated hydrochloric acid should not be stored in a diked area with a carbon steel tank that contains another hazardous chemical.

Adequate spacing for fire control and maintenance should be available for all tanks. Access roads should surround groups of tanks. As far as possible, tanks should be located away from heavily traveled roadways to minimize the potential for vehicle collisions with tanks. Fittings and pipe runs should be protected from vehicle collision. This applies especially to pipe bridges with pipes carrying toxic chemicals. Appropriate fire protection in the form of deluge systems, and/or foam systems should be permanently in place.

3.2.4 Miscellaneous Design Considerations

A number of miscellaneous design considerations are an inherent part of physical plant design. These include:

- Availability and dependability of utilities;

- Operability;

- Fail-Safe Design;

- Fire prevention and protection;

- Electrical system requirements;

- In-process inventories; and

- Special safety requirements.

Availability and Dependability of Utilities--
The loss of a utility can result in an accidental release if control of the process is consequently lost. The potential for an accidental release may be reduced if two principles are followed in designing a utility system.

The first principle of safe utility design is, if the failure of a utility to a portion of a facility can result in a loss of control in that section of the facility, then that utility should be supplied with a backup in at least that portion of the facility. A facility electrical system backup could be a diesel generator system. An instrument air system could be backed up with a nitrogen system. Backup for a utility does not mean that there must be a duplicate of the entire system. Backup for a cooling water system could be a set of pumps that deliver municipal water to a few crucial portions of the cooling system in the event of a cooling system breakdown. It may not always be possible to provide backup to a utility, but when a failure could lead to an accidental release a backup should be considered.

The second principle of safe utility design is that the loss of a utility in one section of a plant should not result in the loss of that utility to other sections of the plant. Breaking or plugging a utility line in one process unit should not result in the loss of that utility to another area of the plant. This implies parallel rather than series distribution for critical utility systems. Loop piping systems are more effective in meeting this criterion than are branch systems. Complete adherence to this principle will sometimes be difficult to achieve, but it should be attempted whenever possible.

Operability--

Considering the human aspect of process operation and maintenance during the design phase of a project, may help to reduce the potential for an accidental release by enhancing operability of a plant. A plant should be capable of running under manual control long enough to complete a safe shutdown. Automatic process control systems can be expected to fail from time to time. Valves, gages, and other instruments and controls should be located where they are easy to use. Important instruments and controls should be readily distinguishable from less important instruments. Unnecessary complexity should be avoided. An overcontrolled plant may be as difficult to operate as an uncontrolled plant. Operators may bypass control systems that are difficult to maintain or that result in frequent process upsets or interruptions.

Any actions that are essential for safe operations should be set up in a way that minimizes effort to the operator in order to facilitate their implementation. An important task that is difficult to perform may be overlooked or implemented too late. Preventing this kind of error is the responsibility of both management and the design engineer. For example, if water must be regularly drained from a vessel or if samples must be regularly withdrawn, then the drain or sample loop should not be located at the top of a ladder. Important valves should be readily accessible. Movement of personnel from one task to the next should be taken into consideration. For example, people may take the easiest route between two points even if it means inappropriately climbing over equipment.

Unnecessary clutter such as extra piping and equipment can make a process difficult to operate and may increase the potential for an accidental release. Facilities often evolve as modifications and additions are made. Care should be taken to streamline and simplify by removing unnecessary piping and equipment as modifications are made.

Fail-safe Design--

A fail-safe control device is one designed to move to a preset position upon failure of the instrument air or electricity. For example, fail-safe pneumatic control valves are designed to go to the fully open or fully closed position on air failure. The purpose of fail-safe equipment is to reduce the potential for a process upset by preassigning the failure modes for process controls. A correct decision as to whether a device should close or open on failure is essential for safe plant design. For most processes, some type of formal design evaluation procedure, such as a Hazardous and Operability Study (HAZOP) study, may be useful for evaluating what failure mode should be selected.

Fire Prevention and Protection--

A properly designed plant will have fire protection systems external to the process unit, such as sprinklers, as well as inherent fire protection design features. From an accidental release perspective, fire prevention and protection is important because fire damage could lead to an accidental release and because many design considerations associated with fire prevention

are actually accidental release prevention measures. Much has been written on
fire prevention and protection.

The National Fire Protection Association has issued guidelines in the
form of "Flammable and Combustible Liquids Code" (5). Generally, local
ordinances will have specific guidelines for fire protection systems. A
detailed discussion of the design of fire protection systems such as sprinklers, foams, etc. is outside the scope of this text; however, this section
provides an overview of some fire protection design measures.

The supply of fire fighting water and the availability of utilities in
the event of a fire is important. In addition to controlling fires, firefighting water (and steam) are often used to control the spread of vapors from
an accidental release. The loss of utilities to the entire facility because
of a fire in one process unit may result in a loss of process control that
could end in an accidental release.

The fire fighting water distribution system is important. A loop design
rather than a branch layout design may be the best arrangement for water lines
in high hazard areas because a loop design ensures water availability even if
lines have been cut off in one area of the plant. Valves should be placed
throughout the distribution system so that a damaged line can be isolated.
These valves may need to be remotely operable.

The primary water supply is usually designed to supply water for fire
fighting for at least four hours. All sources combined are usually designed
to be able to supply water for an additional two hours (5). Many other
features of a fire fighting water system, including specifications on water
requirements, pumps and pumping, sprinkler systems, fire extinguishers and
other aspects of fire protection, are presented in additional NFPA publications. Specialized fire fighting equipment, such as foams or extra sprinklers
or water curtains, should also be considered in highly hazardous areas.

Fireproofing of structural steel, piping, and machinery is often necessary. Fireproofing may include special insulations or materials of construction.

Adequate drainage is important for fire prevention, especially for flammable liquid spills. Flammable solvents may be carried to other locations within a plant on the surface of water. The contents of a drainage trench may ignite and spread fire throughout a plant. Submerged equipment is likely to malfunction and contribute to the hazard. Drains should be sized to handle the maximum anticipated load. The presence of debris such as insulation and tags that may clog drains should be accounted for in the design. To avoid this, trenches sometimes have a cover two thirds closed and one third grate (5). This dampens flames within the trench if a fire occurs. Flame traps are usually spaced periodically throughout a trench to prevent the spread of a fire.

Indoor facilities require additional fire protection. Where flammable dust or vapors may be present, buildings should be constructed with nonload-bearing walls so that walls lost because of an explosion will not result in collapse of the building. In addition to area sprinklers, it may be necessary for each piece of equipment to be protected by a local device. Local flammable gas detectors and/or area monitors are often used and placed near likely release points. Air is handled through ducts in indoor facilities. Flammable dusts, vapors and vapor condensate can collect in these ducts. For this reason, vent and duct systems are specifically designed with safety features such as those suggested in the literature (8).

Fugitive emissions of flammable materials are sometimes controlled by constructing vented enclosures around equipment. Process fittings such as valves or flanges, or process machinery such as pumps or compressors are potential candidates for this type of control. The vent lines from these enclosures are often piped into a common manifold and sent to a treatment device. Air in the vent manifold can lead to fire and explosion. Air leaks into the vent system can be prevented by creating a slight positive pressure with an inert gas. Diluting with an excess of air is another approach to keep the concentration below the flammable or explosive limits. A probe could be installed to monitor oxygen levels in the vent.

Electrical Systems--

Fire hazards are reduced by selecting the appropriate class of electrical equipment. Electrical equipment is given a classification and a division within that classification by the manufacturer which indicates what type of environment the equipment can safely operate in without becoming an ignition source.

Class I equipment is often referred to as "explosion proof" and is the class required where flammable atmospheres may be present. A summary of some locations encountered in a chemical plant and the equipment classification that would apply to each location is specified by the National Fire Protection Association (5). The following definitions of Class I, Divisions 1 and 2, are provided by the National Fire Protection Association:

Class I, Division 1. A Class I, Division 1 location is a location: 1) in which ignitible concentrations of flammable gases or vapors can exist under normal operating conditions; or 2) in which ignitible concentrations of such gases or vapors may exist frequently because of repair or maintenance operations or because of leakage; or 3) in which breakdown or faulty operation of equipment or processes might release ignitible concentrations of flammable gases or vapors, and might also cause simultaneous failure of electric equipment.

Class I, Division 2. A Class I, Division 2 location is a location: 1) in which volatile flammable liquids or flammable gases are handled, processed, or used, but in which the liquids, vapors, or gases will normally be confined within closed containers or closed systems from which they can escape only in case of accidental rupture or breakdown of such containers or systems, or in case of abnormal operation of equipment; or 2) in which ignitible concentrations of gases or vapors are normally prevented by positive mechanical ventilation, and which might become hazardous through failure or abnormal operation of the ventilating equipment; or 3) that is adjacent to a Class I, Division 1 location, and to which ignitible concentrations of gases or vapors might occasionally be communicated unless such communication is prevented by adequate positive-pressure ventilation from a source of clean air, and effective safeguards against ventilation failure are provided.

It may be appropriate to use Class I equipment exclusively throughout a plant that handles highly toxic chemicals. Ordinary electrical equipment is usually considered acceptable in these locations if it is installed in a room or enclosure maintained under positive pressure by ventilation air that is not contaminated by flammable vapors. If such enclosures are used where toxic chemicals are involved, then it would be appropriate to install an alarm system to sound when the enclosure loses pressure. Additional details may be found in the National Electrical Code, 1984, put out by the National Fire Protection Association (10).

Storage and In-Process Inventories--

The potential effects of an accidental release will be reduced as both storage and in-process inventories of toxic and flammable materials are reduced. No generalized formal methods appear to have been developed for reducing inventories. Each process is unique and potential reductions will be specific for that process.

Reducing the volume of process vessels is probably the most obvious method available for in-process inventory reduction. While it is sometimes possible to reduce the volume of intermediate hold and feed vessels or of process piping in an existing plant, this is best accomplished in the initial design of the plant. A more practical approach for an existing plant is to reduce the volume contained in process vessels rather than to reduce the actual volume of the vessel. Improved process control would assist in this type of volume reduction. Better process control would reduce fluctuations in upstream and downstream processes and allow for a smaller intermediate storage volumes.

Substitution of chemicals or equipment may be used to reduce both storage in-process inventories. Some of these substitutions are practical modifications for an existing plant and some are best considered in the initial plant design. Heat transfer systems are good candidates for substitutions. A flammable heating fluid may be replaced with a nonflammable heating fluid. High temperature unit operations such as distillation may possibly be replaced with lower temperature unit operations such as adsorption, crystallization or liquid-liquid extraction. An example of an equipment substitution is suggested by Lees (1): replace a kettle reboiler with a thermosiphon reboiler. Thermosiphon reboilers hold a smaller inventory than do kettle reboilers.

Altering the process chemistry may reduce inventories. For example, the necessity of storing an intermediate could be eliminated by combining two reactions and running them simultaneously in the same reactor. A more efficient catalyst could reduce reactor residence time and inventory at an equivalent production rate. A non-toxic or nonflammable solvent could be substituted for a flammable or toxic solvent, or it might be possible to eliminate the solvent altogether and run the reaction "neat." It may be possible to substitute a nonflammable or non-toxic raw material for toxic or flammable material. As with all potential inventory reduction methods, some of these alternatives would best be explored in the process development stage and are more easily applied to new rather than existing facilities, but certain changes still might be developed for existing facilities.

Special Safety Requirements--
A hazardous process, or a process that handles toxic or flammable materials may require special safety requirements that were not called for in the previous principles of design. Such features are generally mitigative in nature but may also include the addition of extra preventive or protection measures. Some of these, discussed later in this manual, are scrubbers and flares. Blast shields, water or steam curtains, specialized spill cleanup equipment and specialized release detection equipment and alarm systems are

examples of mitigation measures that may be required. Mitigation measures are discussed in the companion manual on mitigation, in this series.

3.3   EQUIPMENT DESIGN CONSIDERATIONS

The equipment in a chemical process may be broadly divided into four categories:

- Vessels;

- Piping (including valves);

- Process Machinery; and

- Process Instrumentation.

This section focuses on some fundamental design considerations, typical failure modes, and control measures for these process equipment categories that are necessary to prevent failures from causing accidental releases. Certain general topics applicable to all equipment are discussed first. These include materials of construction and general equipment failure modes.

3.3.1   Materials of Construction

Selecting the wrong material of construction can lead to equipment failure by corrosion, erosion, or mechanical wear. A variety of process characteristics determine an appropriate material of construction. Process fluid characteristics such as alkalinity, acidity, abrasiveness, and re-activity must be considered in combination with pressure and temperature extremes and the frequency and magnitude of temperature and pressure fluctuations. General procedures for selecting appropriate materials of construction are well defined in the technical literature (11). Even if an appropriate material is specified in the original design, a contractor or vendor may substitute one material for another or maintenance personnel may replace a part with a part made of an inappropriate material. It is valves, flanges,

gaskets, bolts or other small parts of a process unit that are likely to be
made of an incompatible material. These small substitutions are potentially
dangerous because their significance as a potential cause of system failure
leading to a release may not be fully appreciated.

An example qualitative guideline for evaluating the appropriateness of a
material of construction under acid conditions is presented in Figure 3-1
(11). The graph presents qualitative information on the range of conditions
for different materials. The variables include oxidizing environment, reducing environment and chloride content. This chart does not show temperature
dependency, which is also an important consideration.

The material strength requirements depend on the temperature and pressure
to which a piece of equipment is subjected. Calculation of strength requirements should be based on the anticipated normal operating conditions, extremes, and on the frequency of heating-cooling, pressurization-depressuriza-
tion cycles that the system will undergo. A detailed discussion of the design
specifications that would be required for different conditions is outside the
scope of this text, but is fundamental in the field of design and fabrication
for vessels and other equipment.

For vessels, piping, pipe fittings and other hardware, design codes or
standards are available that define in detail the steps required to properly
account for stresses that equipment will experience. These procedures first
define specific criteria for establishing the design pressure and temperature.
The design conditions are established by taking into account the maximum
temperature and pressure that will be experienced under normal operations, and
the frequency and severity of occasional excursions from the normal maximum
conditions. After the design conditions are set, the standards outline
methods for determining wall thickness based on the design temperature and
pressure and on forces other than pressure forces that result in torsion,
axial compression or tension, and bending. Examples of additional forces are
stresses caused by repeated temperature and pressure cycles or stresses caused
by the thermal expansion and contraction of piping. An allowance for

MATERIAL OF CONSTRUCTION	REDUCING ENVIRONMENT	OXIDIZING ENVIRONMENT
TANTALUM	XXXXXXXXXXXXXXXXXXXXXX	XXXXXXXXXXXXXXXXXXXXXX
ZIRCONIUM	XXXXXXXXXXXXXXXXXXX	XXXXXXXXXXXXXXXXX
HASTELLOY B-2	XXXXXXXXXXXXX	
Ti-Pd	XXXXXXX	XXXXXXXXXXXXXX
TICODE-12	XXXXX	XXXXXXXXXXXXXXXXXXXXX
Ti		XXXXXXXXXXXXXXXXXXXXX
HASTELLOY C-4	XXXX	XXXXXXXXXXXXX
MONEL 400	XXX	
HASTELLOY G	X	XXXXXXXX
		CHLORIDES

MATERIAL OF CONSTRUCTION	REDUCING ENVIRONMENT	OXIDIZING ENVIRONMENT
COPPER ALLOYS	XXXXXXXXXXXXXXXX	
ALUMINUM	XXXXXXXXXXXXX	XXXXXXXXXXXXXXXXXXXXXX
MONEL 400	XXXXXXXXXXXXXXXX	
NICKEL 200	XXXXXXXXXXXXXXXXXXXX	XX
INCONEL 600	XXXXXXXXXXXXXXXXXXXX	XXXXXXXXX
430 SS		XXXXXXXXXXXX
304 SS		XXXXXXXXXXXXXXXXX
316 SS		XXXXXXXXXXXXXXXXXXXXXXX
CARPENTER 20 Cb-3	X	XXXXXXXXXXXXXXXXXXXXXXXX
INCOLOY 825	XXX	XXXXXXXXXXXXX
HASTELLOY B-2	XXXXXXXXXXXXXXXXX	XXXXXXXXXXXXXXXXX
HASTELLOY C-4	XXXXXXXXXXX	XXXXXXXXXXXX
HASTELLOY G	XXXXXXXXXXX	XXXXXXXXXXX
Ti		XXXXXXXXXXXXXXXXXXXXXXX
Ti-Pd	XXXXXXXXXXX	XXXXXXXXXXXXXXXXXXXXXX
CAST IRON, STEEL	XXXXXXXXXXXXXXXXXXXXXX	
ZIRCONIUM	XXXXXXXXXXXXXXXXXX	XXXXXXXXXXXXXXXX
		NO CHLORIDES

Figure 3-1.   Example guide for material selection under acidic conditions.

corrosion, erosion, or the depth of threads or other grooves must be made by the designer. Such an allowance is necessary since these factors weaken the basic material. An additional allowance is usually provided where toxic chemicals are involved.

The published standards for determining the strength requirements for process equipment should be considered minimum standards where toxic chemicals are involved. These standards are general and cannot cover all possible design, construction, and operating conditions. For example, standards cannot account for unrelieved stresses created during fabrication or installation. The equipment designer of a facility that handles toxic chemicals should try to account for all stresses and allow a safety factor consistent with the hazards involved.

At high pressures, over 3000 psig, special designs not necessarily covered by design standards or codes are sometimes used. For example, at such pressures, a vessel constructed from ordinary low-carbon-steel plate would become too thick for practical fabrication by ordinary methods. The alternatives are to make the vessel of high-strength plate, use a solid forging, or use multilayer construction. Such specialized fabrication methods may not be subject to any particular standards or codes.

Very low or high temperature operations also require special consideration. These extremes are more a problem of material selection than they are of material strength requirements (though one is closely related to the other). Brittle fracture is possible wherever very low temperatures are involved. The materials used in plants handling low temperature fluids should have a ductile/brittle transition temperature below not only the normal operating temperature but also below the minimum temperature that may be expected to occur under abnormal conditions. Brittle fractures are serious since the fracture can propagate at a velocity close to that of sound. For high temperatures a lining is often used to protect the outer equipment from the high temperatures. A specialized lining must be placed inside a vessel of special construction when high temperatures and pressures are involved.

3.3.2 General Equipment Failure Modes

A number of failure modes are common to all categories of equipment, and each of the three fundamental equipment categories of vessels, piping, and process machinery have failure modes to which they are especially susceptible. This subsection presents a general discussion of potential failure modes and is a basis for the discussions specific for each of the equipment categories presented in later subsections.

Categories of failure modes that may lead to an accidental release are:

- Process upset causing pressure or temperature to exceed design limits of the equipment;

- Faulty fabrication;

- Faulty repair or installation;

- Corrosion;

- Erosion; and

- Mechanical failure.

Process Upsets--

A process upset, as meant here, is any event that leads to a loss of process control that could result in an accidental release. A runaway reaction and overpressure, an internal explosion, or even failure caused by an external explosion in an adjacent process or storage unit would all be examples of process upset failures. Thus a failure occurring as a result of a fire would also be a process upset failure. Numerous events, including instrument failure, operator failure, or mechanical failure, can result in a loss of process control. Process upsets, which can lead to equipment failure, were discussed in Section 2 of this manual on Process Design Considerations.

Faulty Fabrication--

An equipment failure could result from faulty fabrication. A poor weld would be an example of a fabrication fault. Incomplete penetration, lack of fusion, or a porous weld could result from improper or careless welding techniques. Improper heat treatment of a metal component can result in unrelieved stresses. As will be discussed latter, these stresses can contribute to several forms of structural failure. Poor dimensional tolerances can result in excess mechanical stress or poor fit at gasketed joints. The use of an improper material of construction can result in premature corrosion and failure. Additionally, attempted cost saving steps such as substituting one grade of equipment for a lower grade can result in premature equipment failure. Detailed consideration of fabrication faults is beyond the scope of this manual, but there is a voluminous literature on the subject (e.g., 13).

Faulty Repair or Installation--

Another contributor to equipment failure can be faulty repair or installation. One author believes that most initial process plant designs are relatively safe but that problems arise because changes, substitutions or mistakes are made during the initial construction of the plant (14).

Corrosion--

The materials of construction in a chemical plant are subject to a number of types of corrosion. Corrosion can structurally weaken equipment to the point of failure under either normal process operating conditions or process upset conditions.

There are numerous categories of corrosion. Some of these are (1):

1)   General corrosion,
2)   Scaling,
3)   Exfoliation,

4)   Intergranular corrosion,

5)   Stress-related corrosion,

    a) stress corrosion cracking,

    b) corrosion fatigue,

    c) stress-enhanced corrosion,

6)   Galvanic corrosion,

7)   Corrosion pitting,

8)   Knife-line corrosion,

9)   Crevice corrosion,

10) External corrosion.

These are all discussed in the cited reference.   Some categories are discussed further below.

Stress corrosion cracking is the brittle failure of normally ductile metals that occurs under the combination of corrosion and tensile stress. The stress may be internal or external.   The American Petroleum Institute states that almost any alloy can be made to fail by a stress corrosion cracking mechanism (15).   Caustic embrittlement is a common form of stress corrosion cracking, as is the well known chloride stress corrosion cracking of stainless steels.

Intergranular corrosion occurs when austenitic steels are heated in the temperature range of 750 to 1650°F, or cooled through this range.   A complex carbide precipitates out and collects along grain boundaries.   These pockets of carbide are susceptible to corrosion by relatively mild aqueous corrodants. The end result of this type of corrosion is the formation of cracks.

Galvanic corrosion is used to describe an accelerated electrochemical type of corrosion that occurs when two different metals are in contact with each other in the presence of an electrically conductive solution such as an aqueous salt or acid solution.   An electrical current flows between the two metals and rapidly corrodes the metal that acts as the anode.

Other general categories of corrosion include graphitic corrosion, dezincification and biological corrosion. Many additional chemical-specific varieties of corrosion are possible. Discussions of corrosion are numerous in the technical literature (16,17,18).

Erosion--

Erosion is the physical wearing away of process equipment materials by the abrasive action of rapidly moving liquids, gases, or solids. Erosion can result in an equipment failure because of physical weakening. The rate of erosion is a function of the velocity of the process stream, the angle of impingement upon the surface of concern, the solids concentration of the stream and the temperature of the stream. Erosion and corrosion may sometimes occur together. Some materials form a layer of corrosion that acts as a barrier to further corrosion. These protective layers usually have less physical strength than the material on which they are formed. If erosion destroys this layer, then the surface of the material will be eroded away as new corrosion layers are formed and eroded off. A common site of erosion is in the bends of piping.

Mechanical Failure--

Mechanical failure of equipment results from overstress, whether internal or external. Extremes or cycles of temperature and pressure can result in mechanical failure of process equipment. Externally applied stresses such as vibration and supported loads can also result in the mechanical failure of process equipment.

Creep is the flow of metals held for long periods of time at stresses lower than the normal yield strength. Stress rupture is the failure of a metal at stresses lower than the normal yield strength after it has reached a point beyond which it cannot creep.

Repeated heating and cooling cycles may sometimes result in thermal fatigue. Thermal fatigue produces cracks and may ultimately result in a

structural failure.   A sudden change in temperature can sometimes result in
thermal shock.   Thermal shock which causes the sudden unequal expansion or
contraction of different parts of a piece of equipment, can result in cracking
or separation.

Vibration can be very destructive.   Metals can become fatigued and fail
in a manner similar to that of thermal fatigue. Concrete foundations can be
destroyed by vibrating equipment.   The weight of insufficiently supported
equipment or the stress caused by settling foundations or supports is also
destructive.

The ensuing subsections discuss some of the design considerations,
failure modes, and control technologies specific to each of the previously
named equipment categories: vessels, piping, and process machinery.

3.3.3   Vessels

As used here, vessels include all major items of equipment containing
significant inventories of liquids or gases; this includes both process and
storage equipment.   Vessels include tanks, reactors, heat exchanger shells,
receivers, columns, and similar equipment.   Vessels may be atmospheric,
pressurized, or operate under vacuum.

Vessel equipment failures that could result in an accidental release
could occur in one of the following locations:

- Piping joints at vessel nozzles;

- Vessel seams, including welded attachments such as nozzles
  and support lugs;

- Vessel shell wall failure; and

- Vessel flanged joints such as where a head joins the shell.

Vessel Pipe Joint Failure--

Joints of piping to vessels are either threaded, flanged, or welded, although flanged joints are probably the most common for larger process vessels with connections over two inches. Faulty installation or repair may be the most common causes of failure at a flanged attachment. Incorrect gaskets may be used. Flanges may be bent by excessive bolt tightening, which results in insufficient pressure on the gasket to assure a complete seal. Gasket faces may be scratched or dented. A gasketed joint may be bolted with the gasket half in the grove and half out. Mechanical stress due to inadequately supported loads or vibrations may overstress bolts and result in a leak. The bolts around the perimeter of a gasketed joint often require a specific tightening sequence at a specific tightening torque. A failure to follow these specifications can result in a poor seal. Flanges or bolts of incorrect specifications may be substituted during a repair, but substitutions of this kind could result in a premature failure of the flanged joint.

Flanged or threaded joints are sometimes used to join equipment and piping of different materials of construction. A joint of this type is potentially subject to galvanic corrosion. The threads on a threaded attachment can be destroyed by galvanic corrosion, resulting in a complete failure of the joint.

Vessel Seam--

Generally, two types of seams are used on vessels: welded and riveted. Riveted vessels are not recommended for use with toxic chemicals because the overlap at a riveted seam is subject to both internal and external corrosion and each rivet is the potential source of a leak due to corrosion. Stresses can be concentrated around each rivet and stress or stress corrosion cracks can form around rivets. Vessels for toxic chemicals should have welded seams, but even these may fail under certain circumstances.

With vessel seams, the primary preventive measure is proper initial welding. Sloppy welding may result in a porous or incomplete weld that does not provide the necessary structural strength. Also, a welded seam may leak because of corrosion. The metal surrounding a weld is heated during the welding process, and this heating can result in unrelieved stress in the metal which may contribute to the formation of stress corrosion cracking. Heating can also create the precipitation of complex carbides, which can result in intergranular corrosion.

Sometimes welding is used to join two different metals. This type of weld is subject to thermal fatigue when it is subjected to repeated heating and cooling cycles caused by differential expansion and contraction of the different metals. Cracks from thermal fatigue will initially form slowly but will progress more rapidly with time.

Some welds are actually designed to fail in the event of an overpressure. An atmospheric storage vessel will often be designed to fail at the shell-to-head joint to prevent the entire contents of the tank from emptying.

Vessel Shell Failure--

The nozzles attached to a vessel shell are sites for potential vessel failure. Inadequately supported loads on fittings connected to a nozzle may result in failure of the nozzle. Thermal expansion and contraction of piping may create excessive stress on a nozzle. A tank settling under subsidence of the ground beneath it can cause connection and nozzle failure. Vibration of support piping or equipment may also cause a nozzle failure. Unrelieved stress formed during fabrication may contribute to stress corrosion cracking. Galvanic corrosion is a potential wherever a fitting made of a different metal is attached to a nozzle.

External forces may cause the main vessel shell to be susceptible to failure. Unsupported loads such as support piping, platforms or walkways can cause the collapse of a vessel shell. Loads created by foundation settling can also cause a vessel shell to collapse.

Thermal shock caused by a rapid temperature swing will result in unequal expansion or contraction of different parts of a vessel, which may result in damage. Thick walled vessels are more susceptible to cracking due to thermal shock than are thin walled vessels. Some vessel coatings may pull loose and fittings may pull away because of thermal shock. A vessel shell (particularly around a nozzle) may crack because of repeated heating and cooling cycles. The effect is similar to thermal shock except that the fatigue of the metals and of the joints between different metals develops more slowly.

A vessel shell is subject to several types of corrosion. Corrosion in the plate body of the shell, as opposed to seam welds, can result in thinning of the wall and cracks which lead to shell plate failure. A bottom plate in contact with a pad is particularly susceptible to external corrosion. An insulated vessel may be subject to external corrosion in situations where moisture has soaked into insulation and is trapped next to the vessel. A vessel head is susceptible to corrosion that accompanies evaporation and condensation of trace quantities of water. Numerous additional varieties of corrosion specific to the chemicals and materials of construction are possible.

Vessel Failure Prevention Controls--

The primary prevention control for vessel failure related to the equipment itself is proper design and construction and the avoidance of the kinds of conditions, discussed above, that can lead to failure. The ASME Boiler and Pressure Vessel Code, which specifies design standards for boilers, and pressure vessels, was referred to in Subsection 3.2 of this manual. Most states have incorporated all or part of the Code into their legal requirements. Where toxic chemicals are involved, it may be appropriate to use the ASME Code as a minimum standard for the design of vessels. This may not be practical for large storage vessels, however, where at a minimum, design might be according to API (19,20,21). For all vessels, more stringent design may be appropriate than these minimum standards where toxic chemicals are involved.

Most processing vessels are designed in accordance with ASME Code Section VIII, Division 1. Subsection B of Section VIII provides some specific instructions for the fabrication of vessels used in "lethal service," which is defined as a vessel used to handle poisonous gases or liquids of such a nature that a very small amount of the gas or of the vapor of the liquid mixed or unmixed with air is dangerous to life when inhaled. The Code specifies that all vessels in lethal service shall have all butt-welded joints. These joints must be fully radiographed to test their integrity before use. All vessels made of carbon or low-alloy steel shall be postweld-heat-treated. The ASME Code stipulates that brazed vessels cannot be used in lethal service. In general, butt-welded or flanged joints should be used throughout a process handling hazardous materials. Threaded or riveted joints should be avoided.

A storage vessel usually has less strength than a process vessel (such as a reactor). For example large atmospheric storage vessels are often designed to withstand only 8 inches of water and will burst at about three times this pressure. These same vessels can only tolerate about 2.5 inches of vacuum. Because they are easy to rupture or collapse, this type of tank may not be suitable for storing large quantities of toxic chemicals. Care must be taken when sizing vents and inlet and outlet lines on these vessels. The vent should be sized to accommodate the maximum possible flowrates into and out of the vessel. To prevent an overflow from overpressuring the tank, the vent line and overflow lines must not rise more than eight inches above the side of the vessel. Rapid heating or cooling of these vessels can cause a rupture or collapse due to expansion or contraction in the vapor space.

Other storage vessel designs commonly used in the chemical industry may not be appropriate for storing toxic chemicals. For example, because of vapor loss, external floating roof tanks are probably not appropriate.

The American Petroleum Institute has stated that corrosion is the prime cause of deterioration in vessels (14). Because this deterioration is often slow, it may go undetected. The primary defense against corrosion is the selection of the correct materials of construction. Very rarely will a material be completely resistant to corrosion where it is being used. In most cases, an allowance for corrosion is provided by increasing the thickness of the material.

Protective coatings are often used to slow the rate of corrosion. These coatings should be used with care when hazardous materials are involved. Very thin coatings (less than 0.75 mm thick) should not be relied on for corrosion protection because coatings of this thickness may have very small flaws that will allow material to contact the surface below the coating (3). Most coatings are permeable to some chemicals. When a coated material is used, it should be resistant to all components to which it will be exposed. If a coating is permeable to some species in a mixture, then this species will penetrate the coating and contact the surface below. Even if the permeable species is not corrosive to this surface it may damage the bond between the coating and the support surface and ultimately result in a failure of the coating.

Vessels should be constructed to minimize crevices and allow for complete drainage, cleaning, and easy inspection. The internals of a vessel should be so arranged that all exposed surfaces can be cleaned and so that liquid is not held up when the vessel is emptied. External corrosion can be reduced by protecting all outer surfaces with a coating. If insulation is used, care must be taken to seal the insulation so that moisture does not get in and become trapped against the vessel wall.

Care must be taken when dissimilar metals must be contacted with each other as galvanic corrosion can result. The degree to which galvanic corrosion will occur will depend on the metals involved, on the area of contact, and on the electrolytic character of the liquid contacting the metals. Table

3-4 shows the Galvanic Series of Metals and Alloys. The farther apart two metals appear on this table, the more susceptible to galvanic corrosion they will be when in contact with each other. The American Petroleum Institute has compiled a list of design recommendations to combat galvanic corrosion (21):

- Select combinations of metals as close together as possible in the galvanic series.

- Avoid making combinations where the area of the less noble material is relatively small. It is good practice to use the more noble metals for fastenings or for other small parts in equipment built largely of less resistant material.

- Insulate dissimilar metals wherever practical. If complete insulation cannot be achieved, materials such as paint or plastic coatings at joints will help to increase the resistances of the circuit.

- Apply coatings with caution. For example, do not paint the less noble material without also coating the more noble, otherwise greatly accelerated attack may be concentrated at imperfections in coatings on the less noble metal. Keep such coatings in good repair.

- In cases where the metals cannot be painted and are connected by a conductor external to the liquid, increase the electrical resistance of the liquid path by designing the equipment to keep the metals as far apart as possible.

- If practical, add suitable chemical inhibitors to the corrosive solution.

TABLE 3-4.   GALVANIC SERIES OF METALS AND ALLOYS

Corroded end (anodic, or least noble)

Magnesium
Magnesium alloys
Zinc
Aluminum alloys
Aluminum
Alclad
Cadmium
Mild steel
Cast iron
Ni-Resist
13% chromium stainless (active)
50-50 lead-tin solder
18-8 stainless type 304 (active)
18-8-3 stainless type 316 (active)
Lead
Tin
Muntz metal
Naval brass
Nickel (active)
Inconel 600 (active)
Yellow brass
Admiralty brass
Aluminum bronze
Red brass
Copper
Silicon bronze
70-30 cupronickel
Nickel (passive)
Inconel 600 (passive)
Monel 400
18-8 stainless type 304 (passive)
18-8-3 stainless type 316 (passive)
Silver
Graphite
Gold
Platinum

Protected end (cathodic, or most noble)

Source: Reference 3.

- If dissimilar materials well apart in the series must be used, avoid joining them by threaded connections, since the threads will probably deteriorate excessively. Welded joints are preferred, and the use of filler material more noble than at least one of the metals to be joined.

- If possible, install relatively small replaceable sections of the less noble material at joints, and increase its thickness in such regions. For example, extra-heavy wall nipples can often be used in piping, or replaceable pieces of the less noble material can be attached in the vicinity of the galvanic contact.

- Install pieces (sacrificial anodes) of bare zinc, magnesium, or steel to provide a counteracting effect that will suppress galvanic corrosion.

Control Effectiveness--

The control effectiveness of various vessel design factors is measured in terms of the reliability of vessel components. Some reliability data related to vessels are summarized in Table 3-5. The overall reliability of multi-component control measures depends on the reliability of the individual components and on the specific design or procedure.

Costs--

Costs for each vessel control measure are summarized in Table 3-6, which lists vessel hazards, corresponding control technologies, and both typical capital and total annual costs associated with each control measure. Costs are in 1986 dollars. Table 3-7 presents design bases for the costs.

Case Examples (1)--

A cylindrical LNG storage tank ruptured and discharged its entire contents over the plant and nearby areas. The LNG vapor ignited and an intense

TABLE 3-5.   TYPICAL FAILURE RATES FOR VESSEL COMPONENTS

Component	Failure Rate (Failures/Year)
Vessel shell	
Complete failure	3×10^{-6}
Rupture equivalent to 6 inch opening	7×10^{-6}
Flanged head joint	N/A
Flanged nozzle piping connections, flange leak or rupture	0.0026
Seam welds	N/A

Source: Adapted from References 1, 23 and 24.

TABLE 3-6. VESSEL HAZARDS AND CONTROL TECHNOLOGIES

Basis: Costs represent incremental costs over the baseline system presented in Table 3-7.

Hazard	Control Technology	Capital Cost ($)	Total Annual Cost ($/Yr) (1986 Dollars)
General Failure	Corrosion resistant materials.	30,000	5,200
	Higher pressure rating.	16,000	2,800
	Greater corrosion allowance.	6,000	1,100
	Vibration control in attached piping systems, and on agitators.	630	55
	Extra heavy foundations and extra support structures.	1,500	280
	Fire protection.	$6.60/ft^2$	$0.60/ft^2$
	Increased inspection frequency.	—	800
	Pressure/temperature cycling control.	9,500	1,400
	Adequate pressure relief.	600	50
	Protection from external physical damage (curbing or barrier).	6.00/ft	0.50/ft
Flanged Joint Failure	Correct flange/gasket combinations for intended service.	--	320
	Full diameter, correctly torqued, and full circle bolting.	--	320
	Leak monitoring.	--	1,000
Welded Joint or Seam Failure	Correct weld type and materials.	--	320
	Correct weld procedure.	--	320
Vessel Wall Rupture	Double-walled vessel.	17,000	3,000
	Wall thickness testing.	--	1,200

Source: Adapted from references 25, 26 and 27.

TABLE 3-7.   DESIGN BASES FOR VESSEL CONTROLS

Vessel baseline - 10,000 gallon carbon steel pressure, storage vessel, 50 psig rating, ellipitical heads.

Increased corrosion resistance - same as vessel baseline except 316 stainless steel.

Higher pressure rating - same as vessel baseline except 100 psig rating.

Greater corrosion allowance - same as vessel baseline except 1/8" thicker wall and heads.

Vibration control for attached piping - assume increase in piping supports from one per 14 ft. to one every 7 ft. for 100 ft. of pipeline, 4 inch pipe.

Extra heavy foundations and extra support structures - Assume 50% increase in cost over that required for baseline vessel, where cost for the vessel is assumed to be 10% of the baseline vessel allowed, without any peripheral equipment.

Fire protection - Assume the addition of a water deluge system over the baseline vessels.

Increased inspection frequency - Assume an increase of 40 hours per year at an hourly total labor cost of $20/hour.

Pressure/temperature cycling control - Assume the installation of a better control loop somewhere in the process unit of which the vessel is a part. Assume a conventional control loop.

Adequate pressure relief - Assume the addition of a 4-inch rupture disk to the vessel. The baseline vessel is assumed to already have a safety relief valve.

Protection from external physical damage - Assume the addition of vehicle curbing around the vessel process area.

Correct flange/gasket combination - Assume increased supervisory labor at 8 hours per year and $40 per hour.

Full diameter, correctly torqued bolts, and full circle bolting - Assume increased supervisory labor at 8 hours per year and $40 per hour.

Leak monitoring - Assume monitoring costs as primarily labor at an extra 40 hours per year at $25 per hour.

Correct weld type, materials, and procedure - Assume increased supervisory labor at 8 hours per year and $40 per hour.

(Continued)

TABLE 3-7 (Continued)

Double-walled vessel - Same basic specification as baseline vessel except with double wall allowing for space between inner and outer wall.

Wall thickness testing - Assume inspection labor requirement at 40 hours per year and $30 per hour.

fire burned at the plant, causing great loss of life and extensive damage. More LNG flowed from the plant as liquid down the storm sewers, where it mixed with air and exploded. Though the cause of the ruptures is uncertain, an investigation concluded that the low carbon steel used in the vessel construction may have been unsuitable and that failure may have been caused by vibration or seismic shock. The final death toll was 128 and the number of injured was estimated to be between 200 and 400.

In another incident, an estimated 30 tons of ammonia escaped from a ruptured anhydrous ammonia storage tank. A gas cloud 492 feet in diameter and 66 feet deep formed, killing 18 people both inside and outside the plant. The cause of the failure was brittle fracture of the dished end of the tank. Evidence suggested there had been no overpressure or overtemperature of the tank contents, and no other triggering event was determined. Ultrasonic investigation of the dished end of the tank revealed numerous subsurface fissures, perhaps caused by progressive cold-forming of the dished end.

3.3.4 Piping

The term "piping" is used here to describe three closely associated categories of equipment: piping, fittings and valves. Piping can be drawn, rolled and welded, or cast. Fittings are used to connect various pipe sections or connect pipe to another piece of equipment, to change the direction of flow, or to provide a branch. Valves are used to regulate the flowrate or direction of flow through lines. Valves use a number of different mechanism to regulate flow. Some common varieties include, ball valves, gate valves, globe valves, plug valves, diaphragm valves, butterfly valves, and check valves. Check valves as a control technology for preventing backflow have been discussed in Section 2 of this manual. Materials of construction for all of these devices commonly include both metals and plastics.

Piping failures may occur at the following locations:

- Piping wall (including fittings);

- Joints;

- Valve stems;

- Valve bodies;

- Internal valve mechanisms; and

- A hose or hose connections.

Pipe Wall Failure--

Piping can be drawn, rolled and welded, or cast. Welded piping is fabricated with a welded seam running the length of the pipe. Welded seams are subject to stress corrosion cracking, intergranular corrosion and defective welds, which can cause a pipe wall failure. For toxic chemicals, seamless rather than welded piping is preferable.

A pipe wall is subject to the same failure modes as a vessel shell, including excessive stress. Piping can be subjected to excessive stress by:

- Vibration;

- Misalignment; or

- Thermal expansion and contraction.

Vibration can be a very destructive force to process piping. For example, 18-8 stainless piping subjected to a stress of 30,000 psi at a peak deflection failed after approximately one million cycles. This same pipe subjected to a vibration frequency of 60 cycles/sec. failed from fatigue in five hours (26). Causes of vibration include periodic mechanical motion, pressure pulsation associated with reciprocating machinery, high-pressure drop across a control valve or unstable two-phase flow conditions.

Misalignment of pipe may result in failure by fatigue or in the formation of stress corrosion cracking. Misalignment may be caused by an error in installation of by shifting equipment. Sometimes during construction, two piping ends come together and are not completely in line. These out of line ends are usually forced into position. Forcing the pipes into position will subject them to a stress probably unanticipated in the original design specifications. This unanticipated stress can result in premature pipe failure. Even if the initial installation was done correctly, equipment can settle or poorly placed foundations and stress a pipe run in the same way as would a misaligned pipe. Broken or defective anchors and hangers can create stresses on a pipe run that can result in premature failure of the pipe.

Thermal expansion and contraction that is not properly accounted for in the initial design can cause internal stress. Even if the original design provided for thermal expansion, broken or defective sliding saddles or rollers could cause excess stress during expansion and contraction. In addition to stress caused by thermal expansion, repeated thermal temperature cycles can result in failure caused by thermal fatigue.

Piping is often insulated. Moisture can get trapped beneath the insulation and cause external corrosion. Piping will be particularly susceptible to this near joints and fittings where it is difficult to ensure a good seal for the insulation. The potential problem of atmospheric corrosion under insulation applies for all insulated equipment. It is discussed here under process piping because piping systems are often complex and it is difficult to monitor all portions of the piping system for corrosion problems.

Pipe Joint Failure--
Some of the potential failure modes for pipe joints were discussed in Subsection 3.3.4 on vessels.

Most flanges are attached to the end of a pipe in the form of a flanged-end fitting (a few varieties of pipe may be purchased as flanged-end pipe). Some types of flanged-end fittings have a weaker tolerance for cyclic stress loads than the piping to which they are joined. These varieties include slip-on, socket-welded, and lap-joint flanges. Subjecting these flanges to cyclic stresses may cause an unanticipated failure. Flanges may fail as a result of thermal shock. Flanges must be made of a heavier gage of metal than the surrounding pipe because they may be more susceptible to thermal shock cracking than the surrounding pipe. Selecting flange bolts with inappropriate strength and corrosion resistance can lead to a flanged joint failure.

Pipes can be joined by welding. One variety of welded joint is the socket-weld joint, which cannot resist bending stress as well as the pipe which it joins. This joint may fail prematurely under bending stresses. Additionally, a crevice between the two pipe ends is formed in a socket-weld joint. Liquid can enter this crevice and result in corrosion. A second variety of welded joint is the butt-weld joint. For most types of piping, this joint is as strong as the pipe to which it is joined and may, therefore, be preferred for toxic chemicals. The exception is where a butt-weld joint is used to connect work-hardened pipes annealed by the welding (3). All varieties of welded pipe can fail because of sloppy welding.

A threaded joint can be used to connect pipe sections. A threaded joint is not as strong and resistant to fatigue as the pipe it is joining. Threaded joints have other disadvantages. Turbulence will form at the contraction caused by a threaded joint. This turbulence can contribute to corrosion at a point where the pipe is already thinned by threading. A threaded joint may keep its seal even after most of the threads or pipe wall are destroyed. Such a joint may fail completely before any leakage occurs and before any external decay is detected. Threaded joints can be crushed by the force of the pipe wrench used to tighten the joint, which will weaken the connection and possibly result in later failure. A threaded joint can loosen when subjected to torque. Such torque could be caused by thermal expansion or contraction or by shifting equipment. A loosened joint may not seal completely and may not have as much resistance to stress as a tight joint.

Valve Stem Failure--

A common source of leaks in a piping system is valve stems. Most valves are constructed with some type of packing material around the valve stem that allows the valve stem to rotate while preventing the process fluid from escaping around the stem. Exceptions are diaphragm valves where the process liquid is separated from the stem by the diaphragm and ball valves where face seals are used.. Valve packings are made of a number of materials including plastics, metals, and metal composites. Because the valve packing experiences wear every time the valve is opened or closed, it is usually the weakest point of the piping system and will be the first point to fail in the event of an overpressure.

Leaks around a valve stem are frequent but are rarely serious unless the leak results in a fire or explosion. It is possible to tighten the packing to stop a leak when the valve is in operation. However, the life of valve packing and the number of times that the packing can be compressed is limited, requiring periodic replacement. A valve packing failure on a very large valve might be sufficient to discharge a significant amount of toxic chemical, but in most cases a release from a valve steam blowout would not be catastrophic. Leakage around valve stems could indirectly lead to major equipment or process failures, however. For example, small amounts of leakage over a long period of time could result in corrosion that makes the valve inoperable. Such a failure at a critical time in a critical line could lead to a major equipment or process failure.

Valve Body Failure--

The failure of a valve body would be comparable to a pipe failure, and could occur for the same reasons; excessive stress, corrosion or erosion.

The effect of excessive stress is of special concern with valves. Valve bodies are constructed in irregular shapes that may not withstand bending or twisting stresses. Where valves consist of two or more pieces forming a body, the seal between the parts is a weak point, especially if the assembly bolts are improperly tightened. Overstressing could also occur in the body parts themselves if not joined properly to the piping. A valve that is rapidly closed will experience a tremendous surge in pressure. This surge is known as a "water-hammer" or "hammer blow." This surge in pressure can result in failure because of overpressure of the valve body, of the pipe-to-valve fittings, or of the pipe near the valve.

Since valves disrupt the fluid flow, considerable turbulence occurs, which can contribute to erosion.

Internal Valve Mechanism Failure--
The potential cause of failure of an internal valve mechanism depends on the type of valve. Basic types include:

- Gate valves;

- Globe valves;

- Plug valves or cocks;

- Ball valves;

- Diaphragm valves; and

- Check valves.

A failure of the internal valve mechanism can result in low flow or excess flow with all their attendant hazards (see Section 2 of this manual).

A gate valve is composed of a body containing a gate that interrupts the flow. The valve is normally used in the fully open or fully closed positions. Gate valves must usually be constructed of metals, which could limit their usefulness in some corrosive environments.

When handling high-velocity flow of dense fluids, the gate assemblies can shake violently. This vibration can result in failure of the gate assembly, the valve body or the attached piping and equipment. Gates have been known to fall off gate valve stems (27).

Some portion of the stem on a gate valve must be exposed to the process fluid. For some varieties the threaded portion of the stem is exposed. This could result in corrosion or erosion of the stem. Liquid entering the bonnet of this valve when it is opened can be trapped in the bonnet when the valve is closed. Thermal expansion or contraction could rupture the bonnet.

Globe valves are commonly used to regulate fluid flow. The valve is composed of a disk (or plug) that moves axially to seat in the valve body. These valves have a large pressure drop and can be the source of vibrations. In most designs the disk is free to rotate on the stems, which prevents galling between the disk and the seat. However, the ability to rotate can result in a tilted and misaligned disk on closure and the disk on a globe valve can detach from the stem. Process fluid solids, scaling, or corrosion of the seat of disk can result in an inadequate seal.

A plug valve is composed of a tapered or cylindrical plug fitted snugly into a seat in the valve body. When in the open position, the plug has an opening in line with the flow openings in the valve body. A plug valve can be constructed from metals, coated metals, or plastics.

The surface area between the plug and the valve body is large, which can make the plug difficult to rotate. Some types of plug valves use a lubricant to act as a seal and to allow the plug to be more easily rotated. A common way these valves can fail is by seizure of the plug. Temperatures above 500° can cause differential expansion between the plug and the body and result in seizure. A loss of lubrication between the plug and body can result in seizure.   Seizure of the valve could contribute to an accidental release if the inability to open or close the valve results in a loss of control of the process.   An opposite problem is that a small plug valve could vibrate open or shut if the fit between the plug and valve body became too loose (through wear, for example).

Plug valves can trap a small quantity of the process fluid in the plug cavity whenever the valve is closed.   If the process fluid has a high coefficient of thermal expansion, this trapped fluid could result in a ruptured valve if it were heated.   An example of such a fluid is liquid chlorine. Since water expands as it freezes, this could also occur with water that is frozen when trapped in the valve.   Some valves have a bypass built into the plug, which relieves the plug cavity to the high pressure side of the valve when it is closed.   As long as the relief bypass does not become plugged, these valves solve the fluid expansion problem.   These valves are unidirectional and are likely to leak if installed backwards.

A ball valve is similar to a plug valve except the plug is in the shape of a ball.   Plastic seats are used to seal the ball.   Though not as prone to seizure as plug valves, ball valves can seize as a result of thermal expansion, corrosion or fouling.   The ability of a ball valve to seal depends on the performance of the plastic seals.   Thus, once these seals are damaged or worn, the valve could fail to function properly.   Ball valves are usually constructed of metals, which could limit their usefulness in some corrosive environments.   A ball valve also has the same disadvantage as a plug valve in trapping process material in the ball when the valve is closed.   Again, unidirectional ball valves are available with a vent from the ball cavity to the upstream side of the valve to prevent liquid entrapment.

A diaphragm valve is usually a packless valve composed of a diaphragm made of a flexible material which functions as both a closure and a seal.   A compressor mechanism closes the valve by forcing the diaphragm against a seat. Failure of the diaphragm can result in the leakage of the process fluid around the stem, since some varieties of diaphragm valves have no packing around the stem. A diaphragm rupture in a packless valve could result in a significant release of a toxic chemical.

A butterfly valve is composed of a disc mounted on a stem in the flow path. A 90 degree turn of the stem changes the valve from closed to completely open. The butterfly valve forms a complete seal around the perimeter of the pipe. However, minor corrosion or erosion of the disk and seal may cause them to not seal properly. The presence of solid deposits around the sealing surface also may prevent a tight seal upon closure.

A number of varieties of check valves are available. Common varieties include swing, lift (piston check or ball check), tilting-disk or spring-loaded wafer-type check valves. Check valves are also discussed under backflow prevention in Section 2 of this manual.

The common varieties of check valves vary slightly in their mode of action; however, they all incorporate certain common features necessary for their performance.

Successful function of all check valves depends on a reliable seal. The two surfaces that form the seal must remain clean and smooth. The sealing surfaces could be marred by corrosion, erosion, and solids in the fluid stream that could lodge between the sealing surfaces and prevent a complete seal. The portions of a check valve that form the seal have a coefficient of thermal expansion typically 24 to 45 % greater then that of cast or forged carbon steel (4). This could cause unequal thermal expansion of portions of the valve and prevent a tight seal.

All common varieties of check valves have moving parts that must remain flexible and function under the influence of slight pressure changes. Corrosion, solids in the fluid stream, or formation of scale or polymer buildup could prevent these moving parts from operating correctly. Corrosion could also weaken springs and prevent valve closure.

Hoses and Hose Fittings--

Hoses are a frequent source of leaks and may result in large releases.
Two common causes of hose failure are physical wear and poor connections.
Because hoses are usually applied for intermittent use or quick connection, a
release can result from an incorrectly attached hose. Threaded hoses could be
incorrectly secured by only a few threads, by different thread types incorrectly combined, or by the attachment made with a gasket missing.

A release could result from the use of a hose-to-end connection that is
not strong enough. An example would be the use of a clamp or metal compression band connection of the type used to attach automobile hoses. These are
not suitable for use with toxic materials.

An operating error could result in an accidental release from a hose that
is disconnected before the line is properly isolated from the process.

Release Prevention Controls for Piping--

The American National Standards Institute has published standards for
chemical plant piping. ANSI standard B31.3 applies to most of the piping that
would be present at a facility handling hazardous chemicals. The standards
prescribe minimum requirements for the materials of construction, design,
fabrication, assembly, support, erection, examination, inspection, and testing
of piping systems subject to pressure or vacuum.

Where possible, seamless piping should be used for hazardous chemicals.
All joints should be either welded or flanged and threaded joints and fittings
avoided. As far as possible, all joints should be butt-welded. Properly
done, this provides joint strength equivalent to the original pipe. Flanges
should be used only where a removable joint is necessary. If a threaded joint
or fitting must be used, then it should be seal-welded. Because of their
superior strength, welding-neck flanges may be preferred over other kinds of
flanges. An exception must be made for some kinds of coated or specialty
piping that can only use lap-joint flanges. The potential for cyclic stresses
should be considered when an alternate kind of flange (other than

welding-neck) is used. Flange bolts must be consistent in strength and
material of construction with the design requirements of the system.

Metal-polyfluorinated ethylenes and metal-graphite spiral-wound gaskets
should be used instead of metal-asbestos gaskets where hazardous chemicals are
involved. These gaskets will often provide a better seal than the
metal-asbestos -type. A metal-ring-joint facing may sometimes be preferred
over a conventional gasket. In the event of a fire, this type of seal is less
likely to leak. The sealing surface is less likely to be damaged in handling
than a conventional gasketed seal. The metal-ring-joint facing can be more
resistant to the fluid being handled than a conventional gasket. This type of
seal is more difficult to disassemble because the flanges can be separated
only in the axial direction.

As shown in the previous sections, stress is a common failure mode for
all portions of a piping system. The potential for stress-related failures
can be reduced by proper design. Some considerations are:

- Shorten connections to reduce bending moment and raise the
  fundamental frequency of vibration (destructive frequencies tend to be less than 60 cycles/sec).

- Use pulsation dampeners where applicable.

- Provide engineered anchors and braces.

- Account for the potential for two-phase flow conditions by
  providing extra bracing.

- Use pipe loops to control movement and vibration at
  equipment connections. Avoid the use of bellows where
  hazardous materials are involved.

- Allow for thermal expansion by using sliding supports.

- Do not support a valve body and use it to support the attached piping but support the attached piping separately.

- Do not allow a vessel or pump fitting to support attached piping but support the attached piping separately.

- Consider the destructive force of "water-hammer" or "hammer blow" when installing automatic valves. Valves should not be allowed to close more rapidly than the system can tolerate.

- During installation or repair, do not force piping ends into alignment. This creates extra stress not planned for in the original piping design.

- Use a heavier grade of flange than required by typical practice.

A discussion of the applicability and design of each valve type for use where hazardous materials are involved is presented below.

For reasons discussed previously, it may be preferable to avoid gate valves when using hazardous materials.

Although globe valves have several drawbacks, they are usually the only alternative when automatic flow control is required. When globe valves are used, they should be designed with built-in precautions to prevent misalignment of the disk with the seat. Such precautions could include guides above and below the disk or using a spherical seat. Large globe valves must be installed with the stems vertical.

Valves should be installed with the high-pressure side connected to the top of the disk. This would result in the valve closing if the disk were to separate from the stem.

Plug valves and ball valves probably provide the most reliable form of off/on flow control. However, care must be taken to select a valve with the appropriate overpressure protection where materials with high coefficients of expansion could be trapped and heated or frozen in the valve when it is closed. To ensure adequate sealing, ball valves should be of the fixed ball variety with spring-loaded seats.

Diaphram valves are well-suited for applications where a high level of solids are present in the process fluid. If solids are not a concern, it may be advisable to avoid using diaphragm valves. If the diaphragm fails, then the valve will leak with little or no constriction. These valves are generally limited to low pressure applications and probably should not be used anywhere near their rated maximum pressure. If they are used they should be equipped with packing around the stem.

If there is any potential for corrosion, erosion or fouling, and if the system operates at anything other than moderate pressure, then butterfly valves should be avoided.

Whenever possible, hoses should not be used where hazardous materials are involved, although in some operations it may be impossible to eliminate them. Each hose should be designed to withstand the conditions to which it will be subjected. They should be joined to end connections that can be securely fastened to form joints with other piping. Where more than one type of hose is in use, each variety of hose should use a different type of end connector. The corresponding connectors should only be installed at the process equipment locations where each hose may be used. This will help prevent the use of hoses in applications for which they were not intended.

The above discussion has focused on the individual elements in a piping system. The following discussion lists some design considerations for overall piping systems handling hazardous materials.

- Avoid dead ends or unnecessary and rarely used piping branches. Minimizing piping will reduce the potential for operating errors. Rarely used piping tends to be neglected. Process materials can become trapped in dead ends and result in corrosion, plugging, or unwanted mixing with future incompatible process material.

- Take precautions against liquid holdup. All pipe runs should be pitched to allow liquid to drain. Drains should be at the lowest point in the system. Often this is not the case when piping is installed or when future changes are made. Provisions should be made to prevent a liquid-full condition in a blocked section of line. A liquid-full condition can result in a pipe rupture if the liquid expands as it warms or cools to ambient temperatures.

- Minimize locations in main arteries outside high-integrity block valves. This decreases the likelihood of a main artery leak where rapid isolation could be more difficult.

- Bottom drain lines on vessels handling large volumes of hazardous materials should be limited in diameter. This will limit the rate of material release in the event of a failure in drain line valves. A flow limiter valve which automatically closes if flowrate exceeds a prescribed rate can also be used. Drain lines should always be equipped

with two valves in series. A remotely operated emergency
isolation valve may be appropriate.

● The line and valve sizes into and out of a vessel should
be of the same capacity. If the inlet line has a higher
capacity than the outlet, then an overflow is inevitable
once control is lost.

● Piping systems should never be designed so that the
failure of a single valve will result in a significant
accidental release.

Control Effectiveness--

The control effectiveness of various piping design considerations is
measured in terms of the reliability of piping system components. Some reliability data are summarized in Table 3-8. The overall reliability of multicomponent control measures depends on the reliability of the individual
components and on the specific system design or procedure.

Costs--

The costs for each control measure applicable to piping are summarized in
Table 3-9, which lists piping hazards, corresponding control measures, and
both typical capital and total annual costs associated with each control
measure. Table 3-10 presents the design bases for these costs.

Case Examples (1)--

Dead-ends in pipes can be the source of pipe failure when water collects
in the dead-end and freezes, breaking the pipe, or when corrosive matter
dissolves in the water and corrodes the line. In one case, water and impurities collected in and corroded a dead-end branch of a 12-inch diameter
natural gas pipeline operating at a gauge pressure of 550 psi. When the
dead-end failed, the escaping gas ignited immediately, killing three men who
were looking for a leak.

TABLE 3-8. TYPICAL FAILURE RATES FOR PIPING COMPONENTS

Component	Failure Rate (Failures/Year)
Pipe wall, under 3 inches diameter rupture	8.8×10^{-6}
over 3 inch diameter, rupture	8.8×10^{-7}
Flanged joint, leak or rupture	0.0026
Gasket leak	0.026
Welded joint, leak	2.63×10^{-5}
Valve casing	N/A
Valve stem seal	N/A
Manual valve	
- failure to operate	0.365
- failure to remain open (plug)	0.0365
- leak or rupture	8.8×10^{-5}
Solenoid valve - failure to operate	0.365
Automated valves	
- failure to operate	0.1095
- failure to remain open (plug)	0.0365
- leak or rupture	3.65×10^{-6}
Check valves	
- failure to open	0.0026
- reverse flow leak	1.10×10^{-4}
- rupture	8.8×10^{-5}

Source: References 23, 24 and 28.

TABLE 3-9. EXAMPLES OF PIPING HAZARDS, CONTROL TECHNOLOGIES, AND COSTS

Basis: Costs represent incremental costs over the baseline system presented in Table 3-10.

Hazard	Control Technology	Capital Cost ($)	Total Annual Cost ($/Yr) (1986 Dollars)
General Failure	Corrosion resistant materials.	16,000	2,800
	Higher pressure rating.	2,500	430
	Greater corrosion/erosion allowance.	23,000	4,000
	Monitoring/extra supports.	630	1,000
	Vibration control.	630	55
	Thermal expansion allowance.	750	65
	Heat tracing per 100 ft. of line for freeze protection.	1,000	200
	Fire protection.	200	30
	Increased inspection frequency.	--	800
	Additional piping supports.	630	15
	Avoidance of long unsupported piping runs.	--	--
	Pressure/temperature cycling control.	9,500	1,400
	Valve selection.	--	--
	Control of "hammer blow." (Valve closure rate control)	500	90
	Protection from external physical damage.	60	5
Flanged Joint Failure	Correct flange/gasket combination for intended service.	--	320
	Full diameter, correctly torqued, and full circle bolting.	--	320
	Installation/maintenance supervision.	--	320
	Leak monitoring.	--	1,000
Welded Joint Failure	Correct weld type and materials.	--	320
	Correct weld procedure.	--	320
Pipe Wall Rupture	Double-walled pipe.	7,500	1,300
	Wall thickness testing.	--	1,200
Valve Failure	Installation/maintenance supervision.	--	320
	Correctly sized and torqued body bolding.	--	320

Source: Adapted from References 25, 26 and 27

TABLE 3-10.   DESIGN BASES FOR EXAMPLE PIPING CONTROLS

Piping baseline - 100 ft. of 4 inch, Schedule 40, carbon steel piping with average number of fittings.

Increased corrosion resistance - Same as piping baseline except 316 stainless steel.

Higher pressure rating - Substitute Schedule 80 for Schedule 40.

Greater corrosion/erosion allowance - Substitute Schedule 80, 316 stainless steel for Schedule 40, carbon steel.

Vibration control - Assume an increase in piping supports from one per 14 ft. to one every 7 ft. for 100 ft. of pipeline.

Thermal expansion allowance - Assume addition of an extra loop of 10 ft. of piping to the baseline.

Heat tracing - Assume electrical wrap heat tracing added per 100 ft. of baseline piping.

Fire protection - Assume fireproofing added to baseline piping.

Increased inspection frequency - Assume an increase of 40 hours per year at an hourly total labor cost of $20/hour.

Pressure/temperature cycling control - Assume the installation of a better control loop somewhere in the process unit of which the piping is a part.

Valve selection and control of "hammer blow" - Assume substitution of different valve type.

Protection from external physical damage - Assume 10 ft. of curbing placed around pipe bridge supports.

Correct flange/gasket combination - Assume increased supervisory labor at 8 hours per year and $40 per hour.

Full diameter, correctly torqued bolts, and full circle bolting. Assume increased supervisory labor at 8 hours per year and $40 per hour.

Leak monitoring - Assume monitoring costs as primarily labor at an extra 40 hours per year at $25 per hour.

Correct weld type, material, and procedure - Assume increased supervisory labor at 8 hours per year and $40 per hour.

(Continued)

TABLE 3-10 (Continued)

Double walled pipe - Same basic specification as baseline piping except with double wall allowing for space between inner and outer wall.

Wall thickness testing - Assume inspection labor requirement at 40 hours per year and $30 per hour.

Installation/maintenance supervision - Assume increased supervisory labor at 8 hours per year and $40 per hour.

Correctly sized and torqued body bolting - Assume increased supervisory labor at 8 hours per year and $40 per hour.

In another incident, at a pumping station on a liquid propane pipeline a sudden increase in throughput indicated a major line break. The pipeline failure occurred on high ground and the gas flowed down a sparsely inhabited valley, which was evacuated. The cloud ignited in a sudden flash, creating an overpressure and a fire storm which rolled up the hillside to a highway. The pipeline pressure was believed to have been about 942 psig, and the amount of liquid propane estimated to have escaped in the first 24 minutes was 750 barrels.

3.3.5 Process Machinery

Process machinery is process equipment that contains moving parts and is used to physically treat, handle, or modify process streams. Examples include pumps, compressors, agitators and refrigeration units. The discussion in this section focuses on features common to most types of process machinery and highlights some design considerations for pumps and compressors. Process machinery failures may contribute to accidental releases by directly releasing a toxic chemical contained therein, or by causing a process upset that in-directly leads to a release elsewhere in the process system.

Direct releases from process machinery may occur at the following loca-tions:

- Rotating shaft seals;

- Machinery casing; and

- Machinery attachments.

Causes of releases at these locations are discussed below. Erosion and corrosion are not discussed in any detail since these conditions have been discussed in the previous sections of this manual.

Rotating Shaft Seals--

Process machinery comprises both rotating and reciprocating equipment. A seal is required wherever a rotating or reciprocating device contacts the process fluid. Seals include ring seals on pistons in reciprocating compressors and rotating shaft seals. Rotating shaft seals are far more common in the process plant, and only these are discussed further. Two methods used to make this type of seal are a "packing" or "stuffing" box, or a mechanical seal.

The seal is made in a stuffing box by wrapping the shaft in a woven material and enclosing the wrapped section in a box to hold the packing in place. The advantage of a stuffing box is that the packing is inexpensive and may be easily replaced. The disadvantage is that a packing box does not form a complete seal. At high speeds a packing box must leak to allow for lubrication between the shaft and the packing. These devices are not usually appropriate for use with toxic materials.

A mechanical seal is a prefabricated assembly that forms a running seal between flat, precision-finished surfaces. A mechanical seal can leak because of wear, incorrect installation, erosion, or system overpressure. A mechanical seal depends on a precise contact of the two sealing surfaces. A scored, corroded or in any way deformed shaft can result in a seal failure. The presence of solids in the pumping fluid can destroy the integrity of a seal, which is more of a problem when pumping gases with solids than liquids with solids.

A seal that is not properly installed can fail; all components of the seal must be precisely aligned. Most older pumps were designed for stuffing and the stuffing box may not be big enough to accommodate the seal. A seal installed with too small a clearance could fail because of overheating (29). Even a properly installed seal will fail eventually as a result of wear between the sealing surfaces.

Machinery Casings--

A release from a machinery casing may be caused by stress, corrosion, or erosion.

There are both internal and external sources of stress on a machinery casing. An internal source of stress would be overpressure of the process fluid. External sources of stress include vibration, unsupported weight of attached piping, or improperly torqued, especially overtorqued, bolting that holds casing parts together. These stresses can cause deformation or cracking of the casing.

Corrosion and erosion may be caused by many of the factors listed in previous sections. A pump or compressor casing is particularly susceptible to corrosion and erosion because of the high velocity and turbulence of the fluid. The presence of solids can result in significant wear on the casing and the internals of the pump or compressor. The combined action of corrosion and erosion can significantly reduce the life of a casing.

The casing on positive displacement pumps and compressors can fail because of flow blockage at the discharge (deadheading) or flow blockage at the inlet (resulting in cavitation). Deadheading will result in an overpressure, and if not relieved, can cause a ruptured casing.

The casing on centrifugal pumps and compressors can also fail because of flow blockage at the discharge. The liquid or gas trapped in the pump or compressor will be heated if the device is left running after the discharge has been blocked. This heating can cause thermal decomposition of the process fluid or thermal expansion if both the inlet and discharge are blocked. Both may result in an overpressure and failure of the casing.

Cavitation can result in vibration, erosion, and overheating resulting from the formation and collapse of vapor cavities in the liquid at the metal-liquid interface. Impellers are most frequently affected by cavitation;

however, damage to an impeller can cause excess vibration, which can result in a failure of the casing.

The forces that can cause the failure of a machine casing can cause the failure of a machinery attachment. Process machinery is often the source of most of the vibrations in a process. This vibration is often amplified as it is transmitted to piping and supports and can be very destructive.

Some machinery attachments may fail because of neglect, abuse or improper design. A pump case drain line will usually require a long nipple before mounting the drain valve. An excessive force (operator standing on the drain) may break the line. Sometimes the machine attachments provided by the manufacturer will be improperly designed for the specific application for which the machine is being purchased. A compressor, pressure relief device that comes attached to a compressor should be carefully evaluated. Unless specifically communicated, the assumptions under which the manufacturer sized the pressure relief may not be compatible with the requirements of the actual process.

Process Machinery Failure Prevention Controls--

Stuffing or packing boxes may not be appropriate for toxic chemicals because these devices do not form a tight seal and small amounts of liquid, vapors, or gases are released during normal operation. Enclosing the pump in a ventilated enclosure connected to a control device (such as a scrubber) may be possible but may not be an acceptable alternative in many situations.

A properly installed single mechanical seal will typically operate for about a year without any detectable release of material (29). A seal can fail prematurely for any number of reasons, and for toxic chemicals, any failure may be considered unacceptable. A number of levels of control can be achieved by altering the seal design. Seal performance might be improved by using a larger seal, which will result in a longer wear life. A cartridge seal comes preassembled and may be used to avoid some of the potential installation errors of standard seals. A bellows seal is a mechanical seal designed in a way that reduces the potential for seal plugging over conventional spring designs.

The most effective method for controlling accidental releases from mechanical seals is to use two seals. Further protection may be added by using a sealing fluid between the two seals. Where hazardous materials are involved, the sealing fluid must be compatible with the process fluid and it may be necessary to monitor the sealing fluid for the presence of the process fluid.

In addition to seals, there are many other specific design considerations for pumps and compressors which are discussed in the remainder of this subsection.

Some kinds of pumps do not require mechanical seals and are less susceptible to failures that might directly result in an accidental release. A canned-motor pump is designed so that the motor rotor and pump casings are interconnected; the motor bearings run in the process liquid and all seals are eliminated. Eliminating the seal reduces the potential for an accidental release from these pumps. Because the motor is exposed to the process fluid, their use is restricted to nonabrasive and relatively noncorrosive fluids. Repairing these pumps may require employee exposure to the pumping fluid and thus they may not be suitable where toxic materials are involved.

A diaphragm pump is another type that does not employ a mechanical seal. This pump uses the reciprocating action of a flexible diaphragm to move the process fluid. A diaphragm pump can be mechanically or pneumatically driven. The advantage of this type of pump is that only the diaphragm contacts the liquid. The disadvantage is that eventually the diaphragm will break as a result of wear or overpressure. If these pumps are used for toxic materials, precautions must be taken to avoid an accidental release when a diaphragm fails, which can be accomplished by totally enclosing the pump in a ventilated enclosure; the vent gas going to a control device. An alternative is to connect just the nonprocess side of the diaphragm to a controlled vent. The pumping system design must account for the potential of explosive mixture formation when a diaphragm ruptures on a pneumatically actuated pump.

A final variety of sealless pump is one where the impeller and motor are linked magnetically. The impeller is totally enclosed and leakage is possible only by failure of the impeller casing. These pumps are generally used only for very small, low pressure pumping operations.

Overpressure protection is important for pumps and compressors. The potential for deadheading a positive displacement pump can be reduced by using a pump return loop which connects the discharge directly to the intake and allows liquid to continue to circulate when flow is stopped downstream of the pump loop. The loop should be sized to accommodate the minimum flowrate required by the pump. The danger of using a pump return loop is that the loop creates a route for potential backflow when the system is shut-off. A system's design must provide sufficient backflow protection in such a situation.

Where hazardous materials are involved, pressure relief from pumps and compressors must be piped to a total containment system. The backpressure created by the containment system must be accounted for in sizing the pressure relief device.

It may often be appropriate to install remotely controlled emergency isolation valves at the intake and discharge of a pump or compressor that is handling a hazardous material. This combined with a remote pump shutoff switch allows flow to be stopped in the event of an accidental release from the pump or compressor. Where two pumps are parallel, an alternative or addition to the above recommendations would be to add the ability to remotely switch from one pump to the other.

Care must be taken whenever encasing a pump or compressor in a vented enclosure. While providing some accidental release protection, such

enclosures can be a trap for flammable mixtures. To avoid this, vented
enclosures should be purged with an inert gas, or with sufficient quantities
of air to keep the level of flammable vapors below the flammable limit. In
either case, the contents should be monitored for the presence of flammable
mixtures.

A number of pump parameters can be continuously monitored to help detect
potential failures before they occur. The following pump or compressor parameters can be monitored: overspeed, low oil pressure, high jacket-water
temperature, high gas discharge, high oil temperature, high level on knockout
drums, overload relays, vibration monitors, bearing load cells, and temperature indicators imbedded in bearings. Incipient failure detection is discussed in more detail in Section 4 under maintenance and modification practices.

Control Effectiveness--

The control effectiveness of various process machinery design considerations depends on the reliability of system components. Some reliability data
are summarized in Table 3-11. The overall reliability of multicomponent
control measures involving machinery depends on the reliability of the individual components and on the specific system design or procedure.

Costs--

Table 3-12 summarizes costs for some example control measures applicable
to process machinery. This table lists process machinery hazards, corresponding control measures, and both typical capital and total annual costs
associated with each control measure. Design bases for the costs are presented in Table 3-13.

Case Examples (31)--

In a polyethylene plant, fatigue failure of a vent connection on a compressor suction line allowed ethylene to escape. A gas cloud formed and later
ignited. Highly directional pressure waves caused widespread damage. Six
people were killed and thirteen injured.

TABLE 3-11.   TYPICAL FAILURE RATES FOR PROCESS MACHINERY COMPONENTS

Component	Failure Rate (Failures/Year)
Pumps	
— Failure to start	0.365
— Failure to run	0.011
— Shaft seal, major leak	5×10^{-3}
Compressors	N/A

Source: Adapted from References 23 and 24.

TABLE 3-12. EXAMPLE PROCESS MACHINERY HAZARDS, CONTROL TECHNOLOGIES, AND COSTS

Basis: Costs represent incremental costs over the baseline system presented in Table 3-13.

Hazard	Control Technology	Capital Cost ($)	(1986 Dollars) Total Annual Cost ($/Yr)
General Failure	Corrosion resistant materials.	2,500	430
	Higher pressure rating.	3,000	520
	Greater corrosion/erosion allowance.	3,000	520
	Vibration monitoring and control.	--	800
	Freeze protection.	250	50
	Increased inspection frequency.	--	800
	Installation/maintenance supervision.	--	320
	Full bolting with correctly torqued bolts.	--	320
	Pressure/temperature cycling control.	9,500	1,400
	Control of hammer blow.	500[a]	90
	Adequate overpressure protection.	--[a]	--
	Protection from external physical damage.	120	10
Shaft Seal Failure	Proper seal type for service.	1,000[b]	90
	Proper seal materials and construction.	--[b]	320
	Adequate seal flushing, lubrication, and cooling.	--	320
	Adequate overpressure protection.	--	--
	Vibration monitoring and control.	--	800
	Increased inspection frequency.	--	800
Bearing Failure	Specialized acoustic monitoring.	--	800
	Preventive maintenance.	--	160

Source: Adapted from references 25, 26 and 27.

[a] Costs depend on specific systems used which are not directly related to the process machinery itself.
[b] Proper seal material depends on specific service.

TABLE 3-13.   DESIGN BASES FOR PROCESS MACHINERY CONTROLS

Baseline equipment – Centrifugal pump, carbon steel, 250 gpm, 150 psig rating, stuffing box seal.

Corrosion resistant materials – Same as equipment baseline except 316 stainless steel.

Higher pressure rating, greater corrosion/erosion allowance – Same as equipment baseline except heavier casing.

Vibration monitoring – Assume equivalent cost of 40 hours additional operating labor per year at $20 per hour.

Vibration control – Assume heavier foundations and supports.

Freeze protection – Assume 25 ft. of heat tracing.

Increased inspection frequency – Assume an increase of 40 hours per year at an hourly total labor cost of $20/hour.

Full bolting with correctly torqued bolts – Assume increased supervisory labor at 8 hours per year and $40 per hour.

Pressure/temperature cycling control – Assume the installation of a better control loop somewhere in the process unit of which the pump is a part.

Control of "hammer blow" – Assume substitution of different type of valve or valve closure mechanism in liner connected to pump for a 4-inch pipeline.

Adequate overpressure protection – No general estimate possible. Costs depend on specific systems used which are not directly related to the process machinery itself.

Protection from external physical damage – Assume curbing around pump area for perimeter of 20 ft.

Proper seal for type of service – Double mechanical seal substituted for stuffing box seal on baseline centrifugal pump.

Proper seal materials of construction – Assume increased supervisory labor at 8 hours per year and $40 per hour. Actual seal materials depend on specific service.

Adequate seal flushing, lubrication, and cooling – Assume increased supervisory labor at 8 hours per year and $40 per hour. Actual seal flush system depend on specific service.

Specialized acoustic monitoring for bearings – Assume equivalent cost of 40 hours additional operating labor per year at $20 per hour.

Preventive maintenance – Assume bearing replacement once a year.

Failure of a pump in a butadiene processing unit spilled an estimated 27,000 lb of butadiene that resulted in a vapor cloud release which drifted 600 feet before igniting 10 minutes later. One person was killed, six were injured, and the plant suffered major damage.

3.3.6 Instrumentation and Control System Hardware

Mechanical failure of the components of a facility's instrumentation and control system can cause a loss of process control which could result in an accidental release. The relationship between process control and the potential for an accidental release was discussed in Section 2 of this report. This section focuses on potential mechanical failures for the various components of a control system. It does not address the problems that may result from poor control system logic, which was also addressed in Section 2.

Instrumentation equipment may fail as a result of the following conditions:

- Corrosion,

- Erosion,

- Mechanical deterioration, and

- Fouling.

The relative importance of these potential causes of instrumentation failure depend on the kind of instrumentation involved. For example, corrosion may be a significant consideration for temperature measurement devices, while fouling is a well-known problem with pH control systems. The various listed conditions are discussed below.

Corrosion--

Both internal and external corrosion can damage instrumentation and
control system components. Any components that contact a process stream can
be subject to internal corrosion. The nature and concentration of the corrosive agents in the stream and the materials of construction of the component
determine the potential for internal corrosion. Corrosion damaging to system
operation is most often encountered in such control components as
control-valves, orifice plates, analyzers, chambers of certain level instruments, gage glasses, probes, or any other device in direct contact with a
corrosive process stream. Internal corrosion can cause moving parts to seize,
and membranes or sensing elements can be destroyed. Such failures can make
sensing and other functions impossible and cause a loss of control. Corrosion
can also cause leaks through instrument connections and housings.

External corrosion can be caused by moisture, salt air, fungi or corrosive vapors in the atmosphere. The American Petroleum Institute states that
"in hot, humid climates, these factors are the major causes of instrument
component failure" (29). Electrical components are susceptible to failure by
external corrosion. The formation of oxides, and the growth of fungus are
examples of external corrosion that can alter electrical properties and
prevent some portion of a system's electronics from functioning properly. In
more severe cases, external corrosion can destroy the physical integrity of
casings, conduit, or other structural portions of a control system. Such
deterioration could result in a failure of the control system by exposing the
internal portions of the system to the same corrosion that destroyed the
external portions.

Erosion--

The portions of a control system exposed to flowing process streams are
susceptible to erosion, especially streams containing significant amounts of
suspended solids. The high pressure drop and rapid fluid motion through an
orifice plate make it susceptible to erosion. Erosion of the orifice would
result in an altered pressure-to-velocity relationship and hence in failure of
the device. Any sensor the protrudes into a flowing stream, such as a

thermowell, can be subject to failure by erosion. Failure occurs when erosion has physically destroyed the sensing element. The operation of a control valve and its ability to accurately regulate flow can be impaired by erosion.

Mechanical Deterioration—

Several kinds of mechanical deterioration can affect a control system. Examples of cases of mechanical deterioration include: fatigue failure and wear, mechanical abuse, exceeding instrument limits, and overheating or freezing.

Any instrument with moving parts is subject to fatigue failure and wear. Instruments that oscillate over a narrow portion of the operating range are most susceptible to wear. This type of wear can result in cracked, bent, or broken moving parts. Seals and bearings will eventually wear out. Fatigue failure can also be caused by external vibration or by the weight of unsupported attached equipment. Such forces can result in cracked, bent, or broken casings or internals.

Mechanical abuse is damage that results from careless maintenance or operation. Examples of mechanical abuse damage are broken glass, damaged housings, and bent valve stems. Damage can occur when personnel use instruments attached to equipment as steps to climb on equipment. Control-valve stems can be distorted from excessive torque if valve wrenches on the handwheel are used to free binding valves. Equipment may be damaged by collision with moving maintenance equipment.

Each instrument has a specific acceptable operating range and exceeding this range may result in permanent damage to the instrument. Acceptable operating limits are usually placed on the variable being monitored by the instrument, on some of the chemical and physical properties of the process stream and on the electrical power supply. For example, a stainless steel bourdon tube pressure gauge may be equipped to monitor pressures from 0 to 500 psi. Such a gauge may fail if subjected to greater than 750 psi pressures, to temperatures over 500°F, or to acidic process streams, depending on the materials of construction.

Exposure of instrumentation to external sources of heat or to freezing temperatures can cause instrument malfunctions. External sources of heat could include fires, adjacent hot equipment or even direct sunlight in some cases. Heat can alter or impair the performance of or damage the electronic components of the system, such as heat-sensitive sensing devices. Extreme heat (such as fire) can cause total destruction of the instrument. Low ambient temperatures can also have adverse effects. For example, freezing may impair free movement of parts.

Fouling--

Portions of a control system exposed to process streams are subject to fouling from direct deposition of contaminants or by or reaction by products in the streams. Fouling may block a sensing element from exposure to the stream it is intended to monitor, thus preventing the system from controlling the condition of the stream.

Design Considerations--

In designing control system hardware, proper initial instrument selection may be the most effective method for reducing the potential for subsequent failures. For almost every process variable at least two kinds of sensing devices are available, each kind functioning on a different principle. Different ranges are usually available for each device. For each range several different materials of construction are likely to be available, which means that for every specific application there are opportunities to select combinations of specifications suitable for any particular service.

Selecting an appropriate sensor may not be a simple matter. The more information available about the operating conditions, the easier the selection will be. Since vendors will be quick to point out the advantages of their device, it will often be up to the designer to explore all of the alternatives available and use engineering judgement to make a selection. Guidance is available from the technical literature that describe how various sensing devices work and that aids in selection (29). From a practical perspective, it is not always possible to pick a sensing device clearly superior to the alternatives. For critical applications it may be necessary to install redundant sensors that operate on different principles. Such redundancy should be considered wherever the failure of a sensor would significantly contribute to the potential for an accidental release.

Devices that protect a sensing element from severe process conditions are also available. Such special protection should be considered whenever the process involves a hazardous material. These devices are designed to lengthen the life and therefore improve the reliability of the instrumentation. Many of these protective devices can reduce the accuracy or increase the response time of the sensor they are protecting, which must be taken into consideration when incorporating such a device into a control system. Examples of some protective devices are discussed below.

Pressure sensors are almost always protected from direct contact with the process stream by various types of diaphragms. They may also be protected from damage by sudden shock pressure or rapidly fluctuating pressure by pulsation dampeners. Temperature sensors are often protected from the process stream by a thermowell. Other probes can be protected from physical damage by perforated casings. Where solids are present, a sensor may be installed in a filtered slip stream. Where fouling is likely to occur, automatic or manual cleaning devices are available to periodically remove solid buildup on a sensing surface.

Extra support, shielding and vibration suppression equipment may be in order for various components in a control system. Where possible, sensitive electronic portions of the system should be protected. Often, the sensor, transducers and control valves are the only parts of the control system that need to be out in the process itself. The rest of the system can often be placed in the protected environment of the control room.

Operator training will help prevent failures caused by abuse. Regular inspection will help reduce the potential for an accidental release by correcting instrument problems before they occur. The operation of most control equipment can be tested on line without shutting down the process. Control systems that oscillate around the set point should be returned before such oscillation results in a failure. Control systems should be protected from damage during shutdown or repair. Sensors may need to be isolated before pressure or leak testing is performed, or before equipment is cleaned. Control systems may also need protection during sandblasting or painting.

3.4 REFERENCES

1. Lees, F.P. Loss Prevention in the Chemical Process Industries. Butterworth's, London, 1980.

2. Hazard and Survey of the Chemical and Allied Industries. American Insurance Association, New York, NY, 1979.

3. Green, D.W. (ed.). Perry's Chemical Engineers Handbook, 6th Edition. McGraw-Hill Book Company, New York, NY, 1984.

4. Piping Component Standards for Refinery Service, Std. 593-609. American Petroleum Institute, Washington, D.C.

5. NFPA 30, Flammable and Combustible Liquids Code, 1984. National Fire Protection Association, Quincy, MA, 1984.

6. Battelle Columbus Division. Guidelines for Hazard Evaluation Procedures. American Institute of Chemical Engineers. The Center for Chemical Plant Safety, New York, NY, 1985.

7. Kletz, T.A. Plant Layout and Location: Methods for Taking Hazardous Occurrences into Account. Loss Prevention. American Institute of Chemical Engineers. Volume 13, 1980.

8. Lewis, D.J. The Mond Fire, Explosion and Toxicity Index Applied to Plant Layout and Spacing. Loss Prevention. American Institute of Chemical Engineers. Volume 13, 1980.

9. Munson, R.E. Safety Considerations for Layout and Design of Processes Housed Indoors. Loss Prevention. American Institute of Chemical Engineers. Volume 13, 1980.

10. National Electric Code. National Fire Protection, 1984.

11. McNaughton, K.J. Selecting Materials for Process Equipment. Materials Engineering I. McGraw-Hill Publications Company, NY, 1980.

12. Guidelines to the Selection and Use of Corrosion Resistant Metals and Alloys. CS79-10. Pfaudler Company, Rochester, NY, 1979.

13. Failure Analysis and Prevention. Metals Handbook, 8th Edition. Volume 10. American Society for Metals, 1975.

14. Kletz, T.A. What Went Wrong. Gulf Publishing Company, Houston, TX, 1985.

15. Guide for Inspection of Refinery Equipment. Chapter VI, American Petroleum Institute. Washington, D.C. December 1982.

16. Fontana, M.G., and N.D. Greene. Corrosion Engineering. McGraw-Hill, Incorporated, 1978.

17. Vhlig, I.H. Corrosion Handbook. John Wiley and Sons, Incorporated, New York, New York, 1948.

18. Proceedings of the International Corrosion Forum. 4 volumes. National Association of Engineers, Boston, MA, March 1985.

19. Spec 12D, Specification for Field Welded Tanks for Storage of Production Liquids, 9th Edition. American Petroleum Institute, January 1982.

20. Spec 12f, Specification for Shop Welded Tanks for Storage of Production Liquids, 8th Edition. American Petroleum Institute, January 1982.

21. Std 620, Recommended Rules for Design and Construction of Large, Welded, Low Pressure Storage Tanks, 7th Edition. American Petroleum Institute, 1982.

22. Guide for Inspection of Refinery Equipment. American Petroleum Institute, 1973. Chapter II.

23. Reactor Safety Study. U.S. Nuclear Regulatory Commission, WASH-1400 (NUREG-75/014), October 1985. Appendices III and IV.

24. Anyakora, S.N., G.F.M. Engel, and F.P. Lees. Some Data on the Reliability of Instruments in the Chemical Plant Environment. The Chemical Engineer. Number 255, 1971.

25. Peters, Max, and Klaus Timmerhaus. Plant Design and Economics for Chemical Engineers. McGraw-Hill Book Company, New York, NY, 1980.

26. The Richardson Rapid Construction Cost Estimating System, Richardson Engineering Services, Inc., San Marcos, California,

27. Radian Corporation, internal cost files, 1986.

28. Tomfohode, J.H. Design for Process Safety. Hydrocarbon Processing. December 1985.

29. Guide for Inspection of Refinery Equipment. American Petroleum Institute, 1981. Chapter 15.

30. Liptak, B.G. Instrument Engineers' Handbook. Chilton Book co. Radnor PA, 1982.

31. Davenport, J.A. A Survey of Vapor Cloud Incidents. Loss Prevention. American Institute of Chemical Engineers. Volume 11. September 1977.

4. Procedures and Practices

A well-designed and operated process, quality hardware and mechanical equipment, and protective devices all increase plant safety; however, they must be supported by the safety policies of management and by clear specifications on their operation and maintenance. This section describes how management policy and training, operation, and maintenance procedures relate to the prevention of accidental chemical releases. Within the chemical industry, these procedures and practices vary widely because of differences in the size and nature of the processes and because any determination of their adequacy is inherently subjective. For this reason, the following subsections focus primarily on fundamental principles and do not attempt to define specific policies and procedures.

4.1 MANAGEMENT POLICY

Management is a key factor in controlling industrial hazards and preventing accidental releases. Management establishes the broad policies and procedures that influence the implementation and execution of specific hazard control measures. It is important that these management policies and procedures be designed to match the level of risk in the facilities where they will be used. Most organizations have a formal safety policy. Many make policy statements to the effect that safety must rank equally with other company functions, such as production and sales. The effectiveness of any safety program, however, is determined by a company's commitment to it, as demonstrated throughout the management structure. Specific goals must be derived from the safety policy and supported by all levels of management. Safety and loss prevention should be an explicit management objective. Ideally, management should establish the specific safety performance measures, provide incentives for attaining safety goals, and commit company resources to safety and hazard control.   The advantages of an explicit policy are that it sets the standard by which existing programs can be judged, and it gives evidence that safety is viewed as a significant factor in company operations.

In the context of accident prevention, management is responsible for (1, 2, 3):

- Ensuring worker competency;

- Developing and enforcing standard operating procedures;

- Adequate documentation of policy and procedures;

- Communicating and promoting feedback regarding safety issues;

- Identification, assessment, and control of hazards; and

- Regular plant audits and provisions for independent checks.

Because human error is a common cause of accidental chemical releases, personnel selection and maintenance of a qualified, experienced workforce is a significant part of release prevention.   Managers are responsible for ensuring that personnel selection procedures adequately assess worker competency and match worker skills to job responsibilities.   One author identified such abilities as signal detection, signal filtering, probability estimation, manual control, and fault diagnosis as important process operator skills (1). In addition, however, personal characteristics, practical training and recent relevant experience should be considered in the selection process.   The ideal process operator should be responsible, conscientious, reliable, and

trustworthy (1). Temperament is important since it affects a previous response to monotony and stress. Motivation and communication skills are other
personality traits that could be assessed during personnel selection. Particular care should be taken when selecting operators for facilities handling
hazardous materials. The qualifications and capabilities of personnel in high
hazard facilities should be higher than for those in other process facilities.

Many accidental release incidents and/or major industrial disasters have
been caused in part by a failure to properly assign responsibilities (1, 2).
It is just as important to avoid a situation where positions critical to
safety maintenance are left unoccupied; accidents have been attributed to
situations in which key safety responsibilities were vacated and left unas-
signed. An effective management policy sees that the responsibilities of such
positions are carried out even in the event of resignations and absences.

The responsibility for safety issues such as release prevention should be
shared by all management and workforce staff. The staff must be supported by
procedures for staff training in operating and maintenance practices, handling
upset and emergency conditions, using safety equipment, and performing facility audits. Establishing standard practices for these activities is necessary
to ensure uniformity, avoid confusion, and make enforcement possible. The
enforcement of standard procedures is one of management's most fundamental
responsibilities in the area of facility safety and accident prevention.

Management is also responsible for ensuring that an adequate description
of safety policy and standard procedures is available. In the chemical industry, a great deal of documentation is produced covering areas from design and
layout, operation, maintenance, and inspection, hazard identification and
assessment, and emergency planning. So that procedure and policy statements
will not be overly cumbersome and monotonous, these documents should:

●   Be easy to use;

- Be complete;

- Establish the importance and benefits of the procedures;

- Be clear and concise.

The document should be written to a specific audience. In other words, the users of the policy and procedures document must feel it addresses them. These documents should explain not only how release prevention activities are conducted, but also why it is important that they be conducted in the speci- fied manner, so that the benefits of proper safety practices will be clear. Finally, attention to readability and style should be an important priority in document preparation, since if personnel find the documents cumbersome and wordy, they won't read them and they will not be easy to use as references.

Publicizing safety material is not enough, since personnel often think these bulletins are dull (1, 2, 3). Encouraging communication, promoting feedback, and emphasizing facility safety as a matter of professionalism are examples of ways of countering "safety boredom." Active communication stres- sing the priority of accident prevention should come from all levels of management, so that personnel will feel that discussion of the effectiveness of current safety standards and ideas for improvement is an important part of their job. When management demonstrates a willingness to respond to initia- tives from below and participates directly with workers in improving safety, worker morale increases, increasing the degree to which standard procedures are followed. Assigning safety and loss prevention responsibility to compe- tent people and giving them specific, meaningful objectives and adequate resources also encourages a professional attitude about the importance of safety.

Hazard identification, assessment, and control is another area that should be addressed by management to minimize the potential for accidental chemical release incidents. In the past, these activities have been handled relatively informally; however, many companies now use a more systematic approach (3). In most instances, safety hazards are identified by the experienced staff. Safety review committees that evaluate processes by a "what if" approach are an example of using staff experience to identify hazards. The objective of hazard assessment techniques is to gather information on how frequently accidents could occur, how severe the consequences could be, and what changes could be made in processes, equipment and procedures to reduce the accidents and their consequences. Management must have a mechanism for obtaining this type of information so that they can rank potential problems and decide how to allocate hazard control resources. Various analytical techniques, both qualitative and quantitative, are available for performing hazard assessments (1, 2, 3). Managers are responsible for hazard control decisions during the design, construction, startup, operation, maintenance, and shutdown of chemical process facilities. Hazard control has typically been aimed at minimizing the potential for incidents such as fires and explosions; however, prevention of accidental chemical releases should also be a fundamental consideration.

A facility safety audit is a frequently used activity associated with hazard identification, assessment, and control. A total facility safety audit involves a thorough evaluation of a facility's design, layout, equipment and procedures, and is aimed specifically at identifying and correcting potentially unsafe conditions. Audits are often conducted during facility startup and repeated a year after startup. Thereafter, the frequency of safety audits varies, typically ranging from every two to five years. Specific objectives of these audits include alerting operating personnel to process hazards, determining whether or not safety procedures need to be changed, screening for equipment or process changes that may have introduced new hazards, assessing the feasibility of applying new hazard control measures, identifying additional hazards, and reviewing inspection and maintenance programs (4).

In-house safety audits are generally the responsibility of senior management and are conducted by teams selected by these managers. One review of
audit procedures recommended that audit committees include the plant manager,
a safety department representative, a project or design engineer, a maintenance engineer, a process engineer, and an instrumentation/control engineer
(5). Another system is the independent safety audit, which is conducted by
qualified consultants or insurers. One of the primary advantages of hiring
independent entities to perform safety audits is that in addition to
evaluating facility safety, they check management organization, policies, and
attitudes.

4.2 OPERATOR TRAINING

The performance of operating personnel is a significant factor in preventing accidental chemical releases. Many case studies documenting industrial incidents note the contribution of human error to accidental releases
(1,2). Causes of release incidents include improper routine operating procedures, insufficient knowledge of process variables and equipment, insufficient
knowledge of emergency or upset procedures, failure to recognize critical
situations, and, in some cases, a direct physical mistake (e.g., turning the
wrong valve). A comprehensive operator training program can decrease the
potential for accidents resulting from such causes.

Operator training can include a wide range of activities and a broad
spectrum of information. Training, however, is distinguished from education
in that it is specific to particular tasks. While general education is
important and beneficial, it is not a substitute for specific training. The
content of a specific training program depends on the type of industry, the
nature of the processes used, the operational skills required, the characteristics of the facility management system, and tradition.

Some general characteristics of quality industrial training programs
include:

- Establishment of good working relations between management and personnel;

- Definition of trainer responsibilities and training program goals;

- Use of documentation, classroom instruction, and field training (in some cases supplemented with simulator training);

- Inclusion of procedures for normal startup and shutdown, routine operations, and upsets, emergencies, and accidental releases; and

- Frequent supplemental training and the use of up-to-date training materials.

In many instances, training is carried out jointly by facility managers and by a training staff selected by management. In others, management is solely responsible for maintaining training programs. In either case, responsibilities should be explicitly designated to ensure that the quality and quantity of training provided is adequate and that there is accountability for the training. Training requirements and practices can be expected to differ between small and large companies, partly because of resource needs and availability, and partly because of differences in employee turnover.

The job of the trainer is essentially to ensure that the right type and amount of information is supplied at the right time. To do this the trainer must not only understand the technical content of a job, but also those aspects of the job where operators may have difficulty. It is therefore advantageous for trainers to spend time observing and analyzing the tasks and skills they will be teaching. This is referred to by Lees as prior-task analysis, and is important for finding out which factors affect the learning of skills and how training programs can be organized to best address difficult areas (1).

Factors such as pace, motivation, and feedback are standard considerations in educational and training applications (1). Training programs should be conducted at a pace that allows operators to become thoroughly familiar with the material. Maintaining motivation is essential; interest in the material can be fostered by focusing not only on how tasks are conducted but on why they are performed the way they are (e.g., pointing out the hazards associated with improper procedures can motivate workers to do the job properly and safely). Finally, performance feedback and reinforcement during training is needed to correct poor technique, resolve misunderstandings, reward good technique, and encourage further skill development.

Training programs may use a combination of classroom training, where lectures can and should be supplemented with written material (e.g., specific training documents as well as copies of standard operating procedures), simulator training, and training within the facility itself. In-facility training is particularly effective because it allows the operator to become familiar with the actual process layout and to learn directly how to operate equipment and instrumentation. During facility startup, valuable in-facility training exercises can be conducted in hazardous areas using safe fluids instead of actual process chemicals (2). Digital simulators are extremely useful training devices, particularly for training in high-hazard areas. Using these devices to closely simulate process operation allows training to be conducted without endangering either a process or an inexperienced operator. Although simulators are useful in routine operator training, they are perhaps even more important for upset and emergency training. Simulated emergencies are an effective way to allow operators to witness a variety of situations and make judgements in a non-stressful environment (6).

The objectives of normal operator training are to familiarize operators with the nature of processes, equipment, chemicals, and standard operating

procedures. A list of the aspects typically involved in the training of
process operators is presented in Table 4-1.

Upset and emergency training typically augments the information presented
during routine training. Ideally, emergency training should be integrated
with routine operation training (2). This kind of presentation allows operators to become familiar with processes as a whole and illustrates how both
process characteristics and operator responses evolve from normal operation to
upset and on to emergency conditions. Emergency training includes topics such
as:

- Recognition of alarm signals;

- Performance of specific functions (e.g., shutdown
  switches);

- Use of specific equipment;

- Actions to be taken on instruction to evacuate;

- Fire fighting; and

- Rehearsal of emergency situations.

Safety training includes responses to emergency situations, but is also
concerned with preventive measures. Safety courses are typically required of
new operators at chemical process facilities, and frequently these courses are
later supplemented by more thorough safety training. Aspects specifically
addressed in safety training include (1, 3):

- Hazard recognition and communication;

- Actions to be taken in particular situations;

TABLE 4-1.   ASPECTS OF TRAINING PROGRAMS FOR ROUTINE PROCESS OPERATIONS

Process goals, economics, constraints, and priorities
Process flow diagrams
Unit operations
Process reactions, thermal effects
Control systems
Process materials quality, yields
Process effluents and wastes
Plant equipment and instrumentation
Equipment identification
Equipment manipulation
Operating procedures
Equipment maintenance and cleaning
Use of tools
Permit systems
Equipment failure, services failure
Fault administration
    Alarm monitoring
    Fault diagnosis
    Malfunction detection
Communications, recordkeeping, reporting

Source: Reference 1.

- Available safety equipment and locations;

- When and how to use safety equipment;

- Use and familiarity with documentation such as:
  - plant design and operating manuals,
  - company safety rules and procedures,
  - procedures relevant to fire, explosion, accident, and
    health hazards,
  - chemical property and handling information; and

- First aid and CPR.

Although emergency and safety programs traditionally focus on incidents such as fires, explosions, and personnel safety, the prevention of accidental chemical releases and release responses should also be addressed. Release incidents are often overlooked because, except for infrequent catastrophes, their impact can be less dramatic than that of fires or explosions. Combined training with responding agencies is also a good practice because it allows personnel from both sides to learn how to adapt their procedures to increase the benefits of a joint emergency response effort.

The frequency of training and the frequency with which training materials are updated (1-5) are also important in maintaining strong training programs. One of the main problems associated with industrial training programs is that the level of training provided during facility startup (or new employee orientation) is often not followed up during the life of the facility (or in the case of the employee, during his career) (2). Additional training programs, such as refresher courses, offer many advantages, particularly when they cover procedures that may not be used very often (e.g., startup, shutdown, upset, emergency). In some instances, experienced operators may become careless or complacent, and additional training reminds them of the importance of following proper procedures. Chemical processes may be modified to the extent that equipment changes require operational changes. Supplemental training ensures that operators are aware of the changes and of the safety considerations that accompany them. Another advantage of training experienced operators is that these workers can often point out the weaknesses and in-adequacies in documented procedures and suggest alternate techniques. Differences in turnover in different companies also affect the experience profile of the operating staff and influence the frequency of training.

Maintaining updated training materials and updated process design and operating manuals is the responsibility of managers and training personnel. These materials should be reviewed regularly, especially when plant modifications are made, to ensure that workers have access to up-to-date descriptions and procedures.

Much of the type of training discussed above is also important for management personnel. Safety training gives management the perspective necessary to formulate good policies and procedures, and to make changes that will improve the quality of facility safety programs. Lees suggests that training programs applied to managers include or define (2):

- Overview of technical aspects of safety and loss prevention approach;

- Company systems and procedures;

- Division of labor between safety personnel and managers in with respect to training; and

- Familiarity with documented materials used by workers.

The training programs in effect in industries using, manufacturing, and storing toxic chemicals represent an effort to minimize the potential for accidental release and, therefore, should be considered as important as physical containment. The training fundamentals presented here are only an introduction to some features of the types and content of specific training that can be provided, but they highlight the importance of a good training program in fostering a staff genuinely concerned with safety and aware of how their actions can affect safe operation of the facility.

4.3 MAINTENANCE AND MODIFICATION PRACTICES

Maintenance practices include both general practices that involve overall maintenance policy and procedures, and specific practices for specific maintenance objectives.

4.3.1 General Practices

Proper facility maintenance ensures the functional and structural integrity of chemical processing equipment, ancillary equipment, and services. Modifications are often necessary to allow more effective production, reduce costs, or enhance safety. However, since these activities can be a primary source of accidental release incidents, proper maintenance and modification practices are an important part of accidental release prevention. Use of a formal system of controls is perhaps the most effective way of ensuring that maintenance and modification are conducted safely. In many cases, formal control systems have had a marked effect on the level of failures experienced (2).

Maintenance refers to a wide range of activities, including preventive maintenance, production assistance (e.g., adjustment of settings), servicing (e.g., lubrication and replacement of consumables), running maintenance, scheduled repairs during shutdown, and breakdown maintenance. These activities in turn require specific operations such as emptying, purging, and cleaning vessels, breaking pipelines, tank repair or demolition, welding, hot tapping (attaching a branch to an in-service line), and equipment removal (2).

The potential for accidents during these types of procedures is quite high. Two of the more common maintenance problems have been identified as equipment identification and equipment isolation (2). Accidents frequently result from incorrect identification of the equipment on which work is to be done. This can have disastrous effects, especially if connections associated with high temperature, high pressure, or hazardous material operations are involved. Failure to adequately isolate equipment is another major source of maintenance accidents. To protect process and maintenance personnel, it is essential to have positive isolation of both process materials and moving parts during maintenance activities. Other potential sources of maintenance accidents are improper venting to relieve pressure, insufficient draining, and not cleaning or purging systems before maintenance activities begin.

Permit systems and up-to-date maintenance procedures minimize the potential for accidents during maintenance operations. Permit-to-work systems control maintenance activities by specifying the work to be done, defining individual responsibilities, eliminating or protecting against hazards, and ensuring that appropriate inspection and testing procedures are followed. Such permits generally include specific information such as (2):

- The type of maintenance operations to be conducted;

- Descriptions and identifying codes of the equipment to be worked on;

- Classification of the area in which work will be conducted;

- Documentation of special hazards and control measures;

- Listing of the maintenance equipment to be used; and

- The date and time when maintenance work will be performed.

Maintenance permits originate with the operating staff. In this way, operators and operations supervisors are aware of impending maintenance activities and are instructed to make any required operating changes and properly isolate the equipment to be serviced. Permits may be issued in one or two stages. In one-stage systems, the operations supervisor issues permits to the maintenance supervisor, who is then responsible for his staff. Two-stage systems involves a second permit issued by the maintenance supervisor to his workforce (2).

Permit-to-work systems offer many advantages. They explain the work to be done to both operating and maintenance workers. In terms of equipment identification and hazard identification, they provide a level of detail that significantly reduces the potential for errors that could lead to accidents or releases. They also serve as historical records of maintenance activities. To make sure that these systems give the desired protection, various authors recommend that they be reviewed as part of facility audit procedures (2, 5).

Another form of maintenance control is the maintenance information system. Such a system can generate information on facility incidents, failures, and repairs (2). Ideally, these systems should log the entire maintenance history of equipment, including preventive maintenance, inspection and testing, routine servicing, and breakdown or conditional maintenance. This type of system is also used to track incidents caused by factors such as human error, leaks, and fires, including identification and quantification of failures responsible for hazardous conditions, failures responsible for downtime, and failures responsible for direct repair costs. This information is used to assess current maintenance practices and to develop maintenance procedures and schedules that increase facility safety and reduce operating costs.

Maintenance of the integrity and safe operation of process equipment depends on proper facility modification practices. To avoid confusion with maintenance activities, a modification is defined as an intentional change in

process materials, equipment, operating procedures, or operating conditions
(2). Industrial facilities frequently undergo modification; these activities
sometimes start during the design phase and nearly always take place during
commissioning and operation. Modifications may be temporary or may involve
permanent changes to equipment or operation.

Accidental releases frequently result from some aspect of facility
modification. Accidents happen when equipment integrity and operation are not
properly assessed following modification, or when modifications are made
without updating corresponding operation and maintenance instructions. In
many instances, equipment is changed under the stress of attempting to get a
facility operating as soon as possible. In these situations, it is important
that careful assessment of the modification results has a priority equal to
that of getting the facility on line.

Frequently, hazards created by modifications do not appear in the exact
location of the change. Equipment modifications can invalidate the arrangements for system pressure relief and blowdown and can invalidate the function
of instrumentation systems. Even relatively minor modification can introduce
hazards if proper precautions are not taken (e.g., improperly supported bypass
lines) (2).

For effective modification control, there must be established procedures
for authorization, work activities, inspection, and assessment, complete
documentation of changes, including the updating of manuals, and additional
training to familiarize operators with new equipment and procedures (2, 3).
Several factors should be considered in reviewing modification plans before
authorizing work. According to Lees, these include (2):

•   Sufficient number and size of relief valves;

•   Appropriate electrical area classification;

- Elimination of effects which could reduce safety standards;

- Use of appropriate engineering standards;

- Proper materials of construction and fabrication standards;

- Existing equipment not stressed beyond design limits;

- Necessary changes in operating conditions; and

- Adequate instruction and training of operation and maintenance teams.

Following authorization, established procedures should be used to ensure that work is conducted according to appropriate codes and standards and that systems are fully inspected before commissioning. In many cases, it is advantageous that the people involved with authorizing modifications be involved with the pre-startup inspection.

Formal procedures and checks on maintenance and modification practices must be established to ensure that such practices enhance rather than adversely affect plant safety. As with other facility practices, procedure development and complete documentation are necessary. However, training, attitude, and the degree to which the procedures are followed also significantly influence facility safety and release prevention.

4.3.2 Equipment Monitoring and Testing

The potential for an accidental release may be reduced by repairing or replacing equipment that appears to be headed for failure. A number of

testing methods are available for examining the condition of equipment. Some
of the most common types of tests are listed below:

- Metal thickness and integrity testing;

- Vibration testing and monitoring; and

- Relief valve testing.

All of the above testing procedures are nondestructive; they do not
damage the material or equipment that they test. Only a few of the most common
test methods are discussed here. Additional methods and further detail on the
methods mentioned here may be found in references on testing procedures
(1,2,3,4).

4.3.3 Metal Thickness and Integrity Testing

Metal thickness and integrity tests are used to determine the thickness
of metal in vessels and piping and the presence of general corrosion, cracks,
pitting, or other defects. These methods are also used to inspect the integrity of welds. The two most common test methods are radiographic testing and
ultrasonic testing.

Radiographic testing uses an X-ray type of photograph of the equipment of
interest. A radiation sensitive film is attached to one side of a metal
surface. A radiation source (usually gamma rays) is opened up on the other
side of the metal surface. The radiation penetrates the metal and exposes the
film. The end result is an image of the surface and interior of the metal.
If properly taken and interpreted, the presence of defects in the metal as
well as the thickness of the metal can be determined. This method is often
applied to welds, and small diameter pipes, valves and fittings may be
radiographed without opening the line or taking it out of service. For larger

diameter equipment, the radiation source must be placed inside the equipment, requiring that it be taken out of service. Proper handling of the radiation source and proper interpretation of the radiographic picture require special training. Most companies hire outside specialists to perform their radiography, although some companies have trained in-house personnel.

The advantages of radiography are that it provides a reliable and detailed picture of the condition of the item of interest, and for small equipment that does not have to be taken off line it is a fairly quick testing method. There are several disadvantages of radiography. Many pieces of equipment must be taken out of service and opened up to allow access for the radiation source. In these situations, more time will probably be required to prepare the equipment for testing than will the actual test. Additionally, operating personnel must be cleared from an area of the process when the radiation source is in use. This could require additional process downtime. The radiation from the testing may also interfere with certain types of instrumentation.

In ultrasonic testing, a probe that generates an ultrasonic pulse is held against the bare metal. The thickness of the metal is determined by the instrument measuring the time required for the ultrasonic signal to travel from the probe to the opposite side of the metal surface and back to the probe. Small pieces of metal of known thickness must be available as a reference for each material to be tested. The pulse and echo from the probe are displayed on a cathode ray screen. Special training is required to be able to properly interpret the output from this test. A signal sent perpendicular to the metal surface is used to measure the wall thickness at that spot. By placing the signal probe at various angles, and moving the probe to various locations on the surface, a trained tester can determine the presence of cracks, pits or other irregularities in metal equipment and welds.

The advantages of ultrasonic testing are that it often requires less preparation than radiography. No radiation is involved so testing areas do not have to be cleared of personnel. Equipment does not have to be opened for an ultrasonic test to be performed. But there are several disadvantages to ultrasonic testing. Test results may not be easy to interpret, and it is difficult to generate a permanent copy of the results for future examination. Most ultrasonic equipment is not explosion proof and flammable materials must be eliminated from the testing area. Since the ultrasonic probe cannot be held directly against high temperature surfaces, an insulated material must be used. The insulation, however, reduces the sensitivity of the probe.

Other test methods are available but less commonly used. These include; magnetic-particle examination, liquid-penetrant examination, eddy-current test method, thermography and electrical-resistance test method. The magnetic-particle and liquid-penetrant examination methods are simple proced- ures used for detecting the presence of surface cracks. The eddy-current test method is often used to test the integrity of tubing. The electrical-resis- tance test method, which is used to locate flaws in metal structures, has been used for years by railroads to locate transverse cracks in rails. Additional details of these and other testing methods may be found in the technical literature (1,2).

4.3.4 Direct Corrosion Monitoring

Direct corrosion monitoring is accomplished by visual inspection, using corrosion coupons, or various instrumental methods that measure corrosion products in process streams. The shortcomings of these methods, especially the coupon method, is that special circumstances, such as corrosion in a specific area of equipment caused by localized stress conditions, for example, might not be detected.

4.3.5 Vibration Testing and Monitoring

Vibration testing and monitoring is used to detect incipient equipment failure and to diagnose what part of the equipment is failing. The basic premise of vibration testing and monitoring is that defects in piping, structural supports or machinery are characterized by corresponding abnormal vibrations.

Vibrations are measured using a variety of methods. A common vibration measuring probe is the accelerometer, which measures the motion of the probe relative to a motionless reference mass. A proximity probe is used to measure the motion of one point relative to another.

The science of interpreting vibration information is called vibration signature analysis. The method used in conventional vibration signature analysis is as follows (5):

"Instantaneous mechanical motion (vibration) is converted into an electronic signal that is then analyzed, filtered, or otherwise manipulated. In mechanical vibration theory, this complex signal is the result of summing many discrete sources, each vibrating at a single frequency. Through a process called spectrum analysis, this summing effect is mathematically reversed so that the energy content at individual frequencies can be determined. Therefore, spectrum analysis helps one to identify the specific components in a machine that are vibrating."

Lower frequency testing is good for obtaining a detailed diagnostic picture of different components in a given piece of machinery, but it is not usually a good method for detecting incipient failures. Low frequency tests are often run once a problem has been detected. Higher frequency monitoring is good for detecting incipient equipment failure but is not well-suited for diagnosing what the specific problem might be. A permanent high frequency

probe is often attached to a piece of machinery with an alarm to warn when
abnormal vibrations occur.

4.3.6 Relief Valve Testing

Two types of tests are carried out on relief valves. In one test, the
valve open and reseat pressures are measured. In another test, the capacity
of the relief valve is measured. The test for determining opening and
reseating pressures involves a very simple bench top apparatus that uses
hydraulic pressure. The valve is pressurized with water until it opens, at
which point a small amount of water is ejected and the valve closes. The
opening and closing pressures are recorded. Most plant maintenance shops have
this type of capability. Measuring valve capacity is a much more complicated
procedure and almost always must be done at one of a few labs that is equipped
for this type of test.

The opening and reseating pressures of a relief valve should be tested
periodically as a part of routine maintenance. Testing a valve's capacity
should be done whenever any corrosion, fouling or scaling has occurred.

4.4 OPERATING AND MAINTENANCE MANUALS

As discussed in Sections 2 and 3 of this manual, accidental chemical
releases are often caused by either improper equipment or equipment failure or
process operation. Good operating and maintenance practices are the key to
preventing such releases. In both cases, clearly and correctly defined
procedures in maintenance and operating manuals contributes to good practices.

Well-written instructions give enough information about a process that
the worker with hands-on responsibility for operating or maintaining the
process can do so safely, effectively, and economically. These instructions
not only document day-to-day procedures, but also are the basis for most

industrial training programs (7, 8). In the chemical industry, operating and maintenance manuals vary in content and detail. To some extent, this variation is a function of process type and complexity; however, in many cases it is a function of management policy. Because of their importance to the safe operation of a chemical process, these manuals must be as clear, straightforward, and complete as possible. In addition, standard procedures should be developed and documented before plant startup, and appropriate revisions should be made throughout plant operations.

Documentation of operation and maintenance procedures may be combined or documented separately. Procedures should include startup, shutdown, hazard identification, upset conditions, emergency situations, inspection and testing, and modifications (1, 2). Several authors think operating manuals should include (1, 3, 7, 8):

- Process descriptions;

- A comprehensive safety and occupational health section;

- Information regarding environmental controls;

- Detailed operating instructions;

- Sampling instructions;

- Operating documents (e.g., logs, standard calculations);

- Procedures related to hazard identification;

- Information regarding safety equipment;

- Descriptions of job responsibilities; and

●   Reference materials.

Equipment sketches and process flow diagrams are a useful feature of the
process descriptions included in operating manuals. Since safety and occupational health information is an integral component of operating manuals, this
section should include information on the properties and hazards of materials,
special precautions to prevent exposure, and spill and fume control measures
(8).

Operating instructions must include those for normal operations, normal
startup and shutdown (including variations based on length of shutdown), and
abnormal or emergency shutdown (1). Diagnostic guides and a listing of normal
(safe) operating limits specific to each type of operation should also be
provided (3). Sampling instructions can include the location and identification of sampling points, sampling frequency, sampling methods, and safety
precautions. Operating documents include operator's logs, manager's logs, and
standard calculations. Safety equipment descriptions generally include a
guide to locations, the equipment inspection schedule, and manuals for their
use and maintenance. Descriptions of job responsibilities should explicitly
define the operator actions expected in both routine and upset situations, and
should explain who has authority over specific functions (3, 7).

Serious industrial accidents have occurred during startup and shutdown
periods (2). Formal startup and shutdown procedures should be described in
the operating manual or in separate but readily available documentation.
Startup manuals, or manual sections, may include descriptions of the overall
startup strategy, pre-startup conditions, required utilities, and material,
process design, and equipment performance characteristics. Frequently this
type of documentation will include checklists summarizing the sequence of
steps that must be taken during startup (7,9). Hazard identification, accident prevention, and emergency actions specifically related to startup activities should also be addressed. Shutdown procedures should define practices to
be used for both normal and emergency shutdown. In many instances, process

facilities cannot be shut down rapidly, and tasks must be conducted in an orderly progression to avoid creating hazardous conditions. For this reason, sequential checklists summarizing activities are often included in shutdown procedures. Some of the areas typically addressed by shutdown procedures are (9):

- Cooling and depressurizing equipment;

- Pumping out vessels (particularly if vessel contain flammable, corrosive, or toxic materials);

- Purging flammable vapors; and

- Inspections and testing before entering equipment.

The potential for chemical releases during startup and shutdown activities is significant. Fittings and equipment placed on line (i.e., placed in contact with process fluids) during startup are potential leakage sources. During shutdown, operations such as purging, depressurizing, and equipment disassembly are potential chemical release sources. The potential for, prevention of, and proper response to accidental chemical releases should be specifically addressed in startup and shutdown procedures.

Upset and emergency procedures instruct workers on how to handle non-routine operations. Situations may range from operation outside the bounds defined as normal to serious situations involving fire, explosion, health hazards, or pollution episodes. Other examples of hazards for which pre-scribed response actions should be available include:

- Leaks of materials from equipment;

- Defective or damaged equipment;

- Detection of unusual odors or sounds;

- Abnormal conditions such as high or low temperature or
  pressure,

- Infractions of operating procedures or safety regulations,

- Unauthorized hazardous work, and

- Unauthorized personnel or vehicles in hazardous areas.

Maintenance manuals typically contain procedures not only for routine
maintenance, but also for inspection and testing, preventive maintenance, and
facility or process modifications. These procedures include specific items
such as codes and supporting documentation for maintenance and modifications
(e.g., permits-to-work, clearance certificates), equipment identification and
location guides, inspection and lubrication schedules, information on lubri-
cants, gaskets, valve packings and seals, maintenance stock requirements,
standard repair times, equipment turnaround schedules, and specific inspection
codes (e.g., for vessels and pressure systems) (2). Full documentation of the
maintenance required for protective devices is a particularly important aspect
of formal maintenance systems. These devices include:

- Pressure relief valves;

- Rupture discs;

- Tank vents and filters;

- Other pressure relief devices;

- Non-return valves (e.g., check valves);

- Mechanical trips and governors;

- Instrument trips; and

- Alarm, sprinkler, and fire water systems.

Maintenance manuals should also describe the way to keep equipment records. Such records are usually maintained for individual equipment items, and include identification, location, engineering descriptions, operating conditions, inspection intervals and maintenance history (2). In some instances, equipment records are incorporated as part of the maintenance manual; in others these types of records are maintained separately. These records, and the frequency with which they are reviewed, are important to release prevention because they have the data needed to evaluate equipment integrity.

The preparation of operating and maintenance manuals, their availability, and the familiarity of workers with their contents are all important to safe operations. The objective, however, is to maintain this safe practice throughout the life of the facility. Therefore, as processes and conditions are modified, documented procedures must also be modified. The written documentation used to control operating and maintenance activities on a day-to-day basis must be up-to-date and include the latest process changes. One way of ensuring that documentation is properly maintained is to institute a regular procedure review process (1, 5). This type of review is often conducted as part of facility safety audits, but can also be performed independently of total system checks. Examples of the questions often asked during procedure reviews are presented in Table 4-2.

Documented operating and maintenance procedures form the backbone of effective safety and loss prevention policy. Adherence to these procedures is the primary standard against which management judges the worker's performance in safe operations, and effort put into the preparation, maintenance, and enforcement of these procedures is a measure of management's commitment to

TABLE 4-2.   EXAMPLES OF QUESTIONS ASKED DURING PROCEDURE REVIEWS

Are manuals written in a clear and concise manner?

Are they available to all operators and maintenance personnel?

Are they kept up to date and reviewed at regular intervals?

Are the documented procedures obeyed?

Are the procedures reviewed regularly with each employee?

Are emergency plans included?

Are hazards and preventative measures spelled out?

Source: Reference 1.

accident prevention. In addition, these procedures are the building blocks of
training programs, of communications between operating staff and supervisors,
and finally, of improved procedures.

4.5 REFERENCES

1. Lees, F.P. Loss Prevention in the Chemical Industries — Hazard, Identi-
    fication, and Control. Volume 1. Butterworth and Company, 1983.

2. Lees, F.P. Loss Prevention in the Chemical Industries — Hazard, Identi-
    fication, and Control. Volume 2. Butterworth and Company, 1983.

3. Process Safety Management (Control of Acute Hazards). Chemical
    Manufacturers' Association. Washington D.C., May, 1985.

4. Kubias, F.O. Technical Safety Audit. Paper presented at the Chemical
    Manufacturers' Association Process Safety Management Workshop.
    Arlington, VA. May 7-8, 1985.

5. Conrad, J. Total Plant Safety Audit. Chemical Engineering. May 14,
    1984.

6. Training Plant Operators — New Digital Simulators Make It Easier and
    Cheaper. Chemical Week. September 21, 1983.

7. Stus, T.F. On Writing Operating Instructions. Chemical Engineering.
    November 26, 1984.

8. Burk, A.F. Operating Procedures and Review. Paper presented at the
    Chemical Manufacturers' Association Process Safety Management Workshop.
    Arlington, VA. May 7-8, 1985.

9. Hazard Survey of the Chemical and Allied Industries. American Insurance
    Association, New York, NY, 1979.

5. Protection Technologies

This section discusses the various types of technologies that may protect against the incipient accidental air release of a toxic chemical. Protection technologies, as defined in this manual, apply when the chemical is still contained; it has escaped from its primary containment but has not yet escaped to the atmosphere. An example of an incipient accidental release is a relief valve discharge still contained within its manifold piping. The technology of protection involves equipment and systems that capture or destroy a toxic chemical before it is released to the environment. Protection technologies include the following:

- Flares,

- Scrubbers, and

- Enclosures.

Each of these technologies represents an add-on to the basic process system it protects. Flares and scrubbers are technologies in common use for ordinary pollution control with many well-developed applications. In the context of accidental toxic chemical releases, however, there are special design problems and considerations that may differ from the applications of these technologies to ordinary process or vent streams. An enclosure provides temporary containment of a released chemical until it can be released to the atmosphere at a controlled, non-threatening rate or treated by one of the other technologies discussed above. While the purpose of the first two protection technologies is to reduce the quantity of chemical ultimately released, enclosures control the rate of release to the atmosphere or to treatment technology.

The appropriate application and proper design of these systems for toxic chemicals must be evaluated on a case-by-case basis. At present, a major limitation of both applicability and design are that fundamental design and performance data under the severe and unstable operating conditions en-countered in an emergency are not well developed. The basic characteristics of flares and scrubbers, including a brief process description and a discus-sion of the applicability of the technique, performance, and typical costs, are discussed below.

5.1   FLARES

Flares are devices routinely used in the chemical process industries to burn intermittent or emergency emissions of flammable waste gases. Sources of such emissions are generally process vessels. Flares are distinguished from other process combustion devices such as incinerators by their design to handle extreme flow rate variations and their unenclosed combustion zone. A flare is basically a stack which burns a flammable gas at the discharge.

5.1.1   Process Description

A total flare system consists of collection piping, a seal pot, a liquid knock-out vessel, and the flare itself. Figure 5-1 is a conceptual illustra-tion of a typical flare system. The flare is a section of vertical piping with a specially designed combustion tip. The tip consists of a pilot light to ignite flammable gases flowing out the end of the flare pipe. Tip designs vary according to the specific application and are an important aspect of flare design. The system shown in the figure is for an elevated flare.

Two basic types of flares are elevated flares and ground flares. Figure 5-2 shows a typical ground flare. Elevated flares are long vertical pipes which may range to heights as great as 600 feet, while ground flares usually do not exceed 150 feet in height (1). Ground flares are generally surrounded by a refractory enclosure. Elevated flares are usually designed to handle larger flows than ground flares, and is more likely to be used for high volume upset or emergency flaring situations. The ground flare is more likely to be

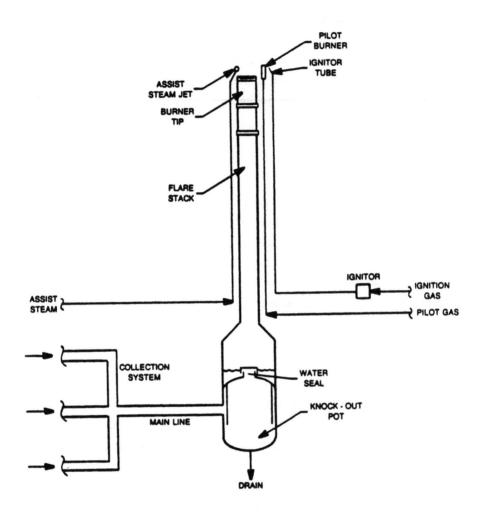

Figure 5-1. Conceptual diagram of elevated flare system (steam assisted).

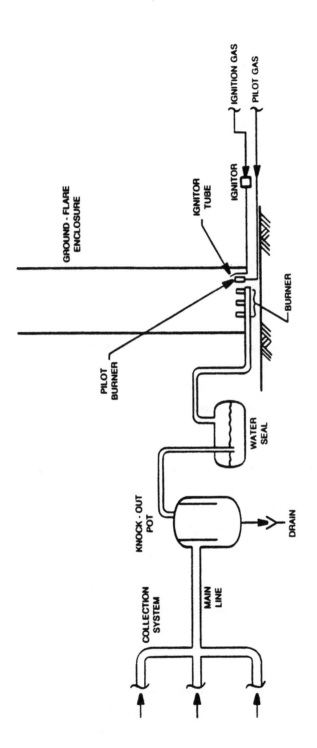

Figure 5-2.   Conceptual diagram of ground flare system.

used for smaller volume, routine process venting. Elevated flares may be more common, however. A survey of 21 petroleum refineries in California indicated that elevated flares made up 87 per cent of the total (1).

Gases are collected through the collection piping and routed to the flare. The collection piping consists of two basic parts: the main flare line and lateral lines to individual process units or vessels. Proper operation of a flare system depends on sizing the collection system to accommodate possible simultaneous discharges of different processes at different pressures into the flare system under widely varying flow rates. Pressure drop considerations in different parts of the collection system are, therefore, quite important. The maximum allowable line pressure is limited by the lowest pressure relief valve discharging into the collection system.

A seal pot is often installed between the collection piping and the flare. This water seal maintains some positive pressure on the flare piping collection system when there is no discharge in order to prevent backflow from the flare into the collection system. Its primary purpose is to prevent air from getting into the collection piping. Some systems use mechanical air restriction seals.

A knockout drum is used between the collection piping and the flare stack to prevent liquid hydrocarbons entrained with the gas from entering the flare. Entrained liquid or solid particles may not totally burn which would degrade flare performance or even cause a "raining" fire.

Other features of flares include air or steam injection at the flare type to ensure smokeless operation, and the use of a purge gas flow at low flow rates to maintain gas velocity and ensure flame stability.

5.1.2 Applicability

Flares may be applicable as protection against accidental releases of toxic chemicals when the release is initially contained within piping or ducting or can be routed to the same, for example, from a release within a

building, and when the toxic chemical is also combustible. The heating value
of the chemical must be sufficient to ensure its combustion at flare temperature conditions.

Basic design requirements for flares burning toxic chemicals are:

● The ability to operate safely over a wide range of flow rates
  and varying compositions,

● Have acceptable emissions of radiant heat and noise, and

● Achieve acceptable destruction efficiency for the toxic
  chemical.

Flow rates for flares are highly variable because flare systems are
designed for intermittent routine process or upset and emergency venting.
Compositions vary because gases are usually controlled from multiple sources
within a process facility. Elevated flares, which are used for higher capacities than ground flares, are designed with capacities as high as 2 million
pounds per hour with turn-down ratios of as much as a thousand. There are
estimates that over 95 per cent of the time, these flares operate at less than
five per cent capacity (2). The ability of flares to accommodate large variations in flow rate is an important consideration in using them for protection
against accidental chemical releases. Depending on the size of a flare relative to a potential release, the accidental release flow rate may constitute a
large or small fraction of the total flare flow rate and, correspondingly,
could have a significant or relatively minor effect on the instantaneous total
flow rate and flare performance. For example, in a dedicated flare system,
where the accidental release would constitute the entire flow, significant
variability and performance fluctuations could occur. In a large shared flare
system, on the other hand, the accidental release flow might not have a large
effect on variability and performance if it is a small fraction of the total
flow.

Whether a dedicated flare or shared flare is used depends on site specific considerations. A shared flare may be advantageous for several reasons. A dedicated flare system for infrequent emergency use may be difficult to safely maintain in full working condition. A shared flare may allow connection to an existing system. With a shared flare, however, design for emergency conditions must ensure that a large release of toxic gas does not overwhelm the flare and lead to flame blowout, or that the pressure from the toxic emergency release does not cause a backflow into other process units tied into the flare collection piping.

The primary compositional requirement of flares is that the vented gases are easily ignitable and have adequate heating value to minimize supplementary fuel requirements.

A fundamental flare design variable is exit velocity. Typical flare exit velocities range from 0.2 to 400 ft/sec. Excessive exit velocities cause flame detachment from the burner tip or flare quenching (flame-out). A typical limiting exit velocity is 400 ft/sec or less, depending on the heat content of the flare gas. This criterion has been recommended by the EPA to ensure 98 percent destruction efficiency of flared chemicals using a steam-assisted flare (3). The addition of an accidental release discharge to an existing flare must not cause this maximum flow to be exceeded. The normal maximum gas flow is determined by an inventory of all contributors to the gas collection system and an analysis of the probability of various streams venting simultaneously. From this total flow and from the maximum velocity limitation the flare diameter can be calculated. The diameter of commercial flares varies from several inches to about 3 feet.

Empirical correlations have been developed for sizing the remaining flare components. Flame length can be estimated from the gas molecular weight (which is related to heat content), temperature, and flow rate. The flame length, combined with limitations on allowable heat radiation to personnel and equipment, is used to determine the height of the flare and the required clear area around the base. Numerous references present these calculations (4,5,6).

Two other important design considerations for flares include air infiltration and flashback. Water seals or mechanical seals are two means used to
prevent these conditions. At low process flows a purge gas may be used to
maintain adequate flow for proper flare operation. In a dedicated flare it
might even be necessary to maintain a slight positive nitrogen pressure on the
system to preclude air infiltration.

An existing flare system collects vent gases from process units as
needed, as well as emissions from pressure relief valves and other emergency-
generated sources. Some units may not be tied into the flare system for any
of several reasons. If a process unit does not require periodic normal
venting or in an old facility relief valves may vent directly to the atmosphere. Certain precautions should be observed when connecting multiple
discharges to a flare system. The precautions are avoidance of incompatible
chemicals, corrosive materials, and chemicals prone to fouling. Concern for
these chemical characteristics is significantly greater in a chemical plant
than in a petroleum refinery where much traditional experience on flaring is
based.

For example, some processes in chemical plants contain oxygen in the vent
stream. Releasing hot gases containing oxygen into a fuel-rich mixture could
create a flashback or explosion. Similar results could occur with other
mutually reactive materials.

Besides these incompatibilities between gases, fouling caused by venting
a polymerization reaction or other viscous substance into the collection
network could occur. Plugging could eventually cause a high back pressure in
the collection system, preventing relief valves from functioning properly.
Organic liquids entering the flare could overload the knockout drums and
create rain fires. Low vapor pressure liquids could flash, condensing and
freezing water vapor in the line, which would cause high back pressure.

Another concern is corrosion of the pipe network, which is normally made
of mild steel. Chlorides and acids are detrimental, although this is not a
critical concern in an emergency situation.

accidental release could damage the pipes or potentially affect the venting of other process units. Table 5-1 summarizes the factors that need to be considered to prevent accidental chemical releases when using a flare system.

5.1.3 Control Effectiveness

Flaring protects against accidental toxic chemical releases by reducing the quantity of toxic chemical released. This reduces the overall consequences of the release. Ideally a flare would destroy all of the toxic chemical so none would be released. It is difficult to estimate the destruction and removal efficiency (DRE) of flares because of the many variables associated with their operation. Numerous studies have been conducted to determine the operational performance of flares. EPA has published a set of flare requirements that are meant to ensure 98 percent or greater destruction of the gases (3). These requirements include a gas heating value of at least 900 Btu/scf and a maximum gas velocity related to the heating value. In an emergency condition, however, these conditions might not be met. For example, a large release of carbon tetrachloride could significantly reduce the lower heating value of the flare gas since carbon tetrachloride is nonflammable. A screening study on some 25-30 compounds performed on 1/16 and 1/8 inch diameter test flares generally obtained over 99 percent destruction efficiency except for high concentrations of carbon monoxide and ammonia (7). A 100 ppm hydrogen cyanide stream was only 85 percent controlled.

It is difficult to measure the efficiency of an operating flare because of the intense heat and the instability of the flame. Even pilot scale facilities encounter these difficulties. A properly run elevated flare uses steam or forced air to increase turbulence and allow complete combustion. The gas flow can vary instantaneously, and since the flame life is 2-3 seconds, steady state conditions are reached very quickly. As long as the accidental released toxic is combustible and the volumetric rate does not exceed the capacity of the flare nozzle, one would expect a high DRE.

In the case of noncombustible vapors, the use of a flare would dramatically increase the dispersion. The gas velocity in the flare of up to 400

TABLE 5-1.   IMPORTANT CONSIDERATIONS FOR USING FLARES TO PREVENT
ACCIDENTAL CHEMICAL RELEASES

- Maximum flow rate - will it cause a flame blowout?

- Possibility of air, oxygen, or other oxidant entering system?

- Is gas combustible - will it smother the flare?

- Will any reactions occur in collection system?

- Can liquids enter the collection system?

- Will liquids flash and freeze, overload knockout drum or cause rain fire?

- Is the back pressure of the collection system dangerous to the releasing vessel?

- Is releasing vessel gas pressure or temperature dangerous to collection system?

- Will acids or salts enter the collection system?

- Will release go to an enclosed ground or to an elevated flare?

- If toxic is not destroyed, what are the effects on surrounding community?

steam or forced air to increase turbulence and allow complete combustion. The gas flow can vary instantaneously, and since the flame life is 2-3 seconds, steady state conditions are reached very quickly. As long as the accidental released toxic is combustible and the volumetric rate does not exceed the capacity of the flare nozzle, one would expect a high DRE.

In the case of noncombustible vapors, the use of a flare would dramatically increase the dispersion. The gas velocity in the flare of up to 400 feet per second associated with the buoyancy of the thermal plume could easily dilute ambient concentrations by several orders of magnitude. Some noncombustible compounds could possibly undergo a pyrolysis reaction in the oxygen-lean portion of the flame, especially if hydrogen is present.

Because of the variable flow capacity, high temperature, high gas velocity, and usual remote locations, using a flare to prevent accidental chemical releases can be a highly effective technique. Small or isolated vessels that contain a process that does not require normal venting of a flammable gas, and that employ relief valves and rupture disks which may have never been used, can be connected to a flare system to prevent accidental releases.

The number of potential hazards associated with flaring include:

• Explosions in the system;

• Obstruction in the system;

• Low temperature embrittlement of the pipework;

• Heat radiation from the flame;

• Liquid carry over from the flare; and

• Emission of toxic materials from the flare.

In designing and operating a flare system, precautions must be taken to prevent these hazards from occurring. A rigorous examination of all probable accident conditions should be performed with respect to the hazards listed above to ensure that using a flare system to prevent accidental chemical releases does not create a greater environmental hazard.

5.1.4 Costs of Flare Systems

As protection systems, flares have relatively low capital costs. Elevated flares, constructed mainly of pipe, are very inexpensive. The flare tip has a special design for efficient steam, air and fuel mixing. Enclosed ground flares are an order of magnitude more expensive because of the numerous burner nozzles, refractory material, acoustical insulation, etc.

The operating costs of elevated flares, however, are high because of the need for a purge gas and steam injection. Enclosed flares have much smaller pilot and purge gas difference becomes a significant factor, which explains the use of ground flares for normal process flaring in conjunction with an elevated flare for emergencies.

Table 5-2 compares the costs (1986 dollars) of elevated and ground level flares (8). The costs of two applications are presented, ethylene waste gas, and a low-Btu waste gas. As can be seen, initial costs for the elevated flare are significantly lower than for the the ground flare. Costs for connecting a new source to an existing flare system would depend primarily on the piping and engineering costs for the modification. The capital cost of the elevated system is only moderately sensitive to flowrate. The operating costs, however, for the elevated flare are proportional to the flowrate. The operating costs of the ground fare depend little on size. The costs presented in the table do not include costs for the collection system, which could easily exceed the flare cost. Also, the design basis for the systems in this table may not meet recently promulgated EPA requirements for a fuel gas heat content minimum of 300 Btu/scf. This requirement could raise the operating costs presented for the low-Btu waste gas example.

TABLE 5-2.   COST COMPARISON OF ELEVATED AND ENCLOSED GROUND FLARING SYSTEMS (May 1986 Dollars)

Waste Gas Flow Rate, lb/hr	Flare-Tip Diameter, in.	Capital Cost* ($)	Total Annual Cost** ($/yr)	Capital Cost ($)	Total Annual Cost ($/yr)
I. Systems designed for smokeless flaring of high-Btu (ethylene) waste gases**					
	3	8,000	16,000	30,000	980
25,000	8	20,000	150,000	100,000	1,500
250,000	24	50,000	1,450,000	450,000	3,200
II. Systems designed for flaring of low-Btu waste gases***			pilots, purge and assist gas		pilots and purge (no assist gas)
2,500	3	8,000	6,500	8,000	980
25,000	8	14,000	52,000	26,000	1,500
250,000	24	26,000	480,000	60,000	3,200

* Complete elevated flaring system with stack of sufficient height to ensure maximum grade-level radiation of 1,500 Btu/hr/ft^2, ladders and platforms are all painted and ready for erection.

** Operating costs are based on 10 percent running time for flaring ethylene waste gases with continuous natural gas pilots and continuous natural gas purge at $2.00/1000 ft^3, steam at $0.02/lb.

*** Operating costs are based on 10 percent running time for flaring low-Btu/ft^3 and molecular weight of 24-with natural gas pilots, continuous purge and assist gas at $2.00/1000 ft^3.

Source: Adapted from Reference 8.

5.2   SCRUBBERS

Absorption is the transfer of a soluble (or vapor) into a relatively nonvolatile liquid. Absorption of a gas or vapor by a liquid can be physical only, or can result in a chemical reaction. Regardless of the absorption mechanism, intimate mixing of the gas and liquid is needed to achieve a high absorption efficiency.

Scrubbers or gas absorbers that remove both organic and inorganic compounds from gas streams are routinely used in many process facilities for raw material and/or product recovery and as air pollution control devices (9). There are several standard types as well as special designs for specific applications. Scrubbers can be used in some situations to control emergency releases of toxic chemicals if they are properly designed and operated.

5.2.1   Process Description

Mass transfer equations that describe the absorption process have been presented in numerous other works and are not shown here (10,11,12). The removal efficiency of a gas is a complex function of equilibrium-related factors (temperature, pressure, gas phase concentration, liquid phase composition) and kinetic factors (gas-liquid interfacial area, contact time, liquid and gas rates, etc.). In general, the removal effectiveness of a gas by absorption can be estimated from the vapor pressure of the gas at equilibrium with the gas in the liquid. If the gas equilibrium vapor pressure of the gas is low, the gas can be readily absorbed in a properly designed system.

As protection devices for accidental chemical releases, scrubbers can be used to control toxic gas releases from emergency vents and pressure relief discharges from process equipment or vents from secondary containment enclosures. All scrubbers require a liquid feed system and some type of contacting mechanism to provide high surface area contact between the gas and liquid. A schematic of common absorber types is shown in Figure 5-3. The internals of column or tower absorbers vary with with application. Formed packings of

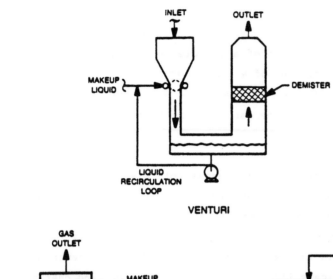

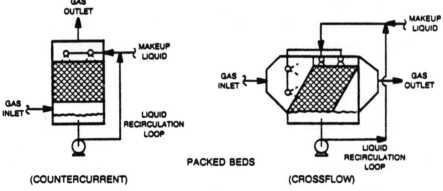

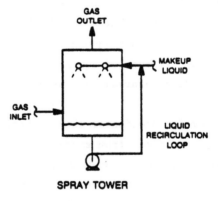

Figure 5-3.   Three common types of scrubbers.

plastic and other materials, perforated trays, valve trays, and open columns with spray nozzles have all been used. The types of absorbers most applicable to accidental chemical releases are probably spray towers, packed towers, and venturis. Spray towers have the advantage of low pressure drop and high liquid-to-gas (L/G) ratios, but the disadvantage of low efficiencies. The open spray tower is used to reduce plugging problems for liquids that may contain solids. The other tower internals are used with clean liquids. Venturis have the advantage of simplicity, but the disadvantage of high back-pressure and limited removal efficiencies. Packed towers have higher effi-ciencies than spray towers but limited L/G ratios and a moderate backpressure. Selection of specific tower internals is based on trade-offs between the surface area for mass transfer and gas side pressure drop. Vendors are continually striving to develop high area packings with low pressure drop characteristics.

The size of an absorber is determined by flow rate, the gas-liquid system involved, and the removal efficiency required. Flow rate determines diameter, (or cross-sectional area) and the specific system and removal efficiency determine the height (or time in the contact zone) of an absorber. In general for packed beds or spray towers, the gas velocity is limited to a maximum of about 10 feet per second, which establishes the diameter for a given flow rate. The liquid rate is set based on the L/G ratio required to achieve a specific removal rate for the particular gas and solvent combination.

5.2.2 Applicability

In general, absorbers are appropriate for protecting against accidental releases of toxic chemicals when the substance released is readily soluble in some non-flammable absorbing medium, and the size of the release allows an economically reasonable scrubber size.

As discussed earlier, an absorber is sized to handle certain flow rates. Significant variations in the flow can cause severe operating problems. High gas rates cause flooding, weeping, and pressure surges in the column. The flow characteristics of the accidental release may be steady or it may be a sudden surge load on the absorber. The absorber must be designed to accommodate these operational demands. Absorbers for emergency relief applications typically operate under a slight positive pressure. A high pressure gas surge beyond the absorber vessel design pressure could damage the vessel. The absorber must also be designed so that the pressure drop does not create a backpressure that would cause a destructive overpressure at the source of release. Such an occurrence would cause a greater release than the one the scrubber was trying to prevent. Additionally, scrubbers must be designed to accommodate hot gases, or those that generate heat when absorbed, to prevent thermal damage to the absorber, vessel liner, or other equipment.

In the event of an accidental release, the concentration of the gas being absorbed could be much higher than typically encountered in most industrial applications. Scrubbers designed specifically for emergency discharges will have taken this into account. Ordinary vent scrubbers may not be able to handle the high inlet gas concentrations. For some gases, absorption is an exothermic reaction. At high concentrations and flow rates, the heat generated may be sufficient to boil all of the scrubbing liquid, which could damage the internals (if plastic), or perhaps cause a fire.

Another consideration is the selection of scrubbing solution. Both aqueous and organic solutions can be used as absorbents. The absorbent selected depends on the absorption characteristics of the toxic material being absorbed. For example, while water may be satisfactory for hydrogen chloride, an alkaline solution is required for substantial removal of chlorine form a gas stream. It might be possible to use organic liquids in some situations as long as they are not flammable or toxic.

Compared with a flare system, which may handle waste gases from multiple sources, absorbers are often dedicated to specific units. Total plant capacity absorbers are not common because of the diverse nature of gas contaminants and the poor turndown ratio of absorbers. A typical turndown ratio is about 10:1. Also, absorbers are not readily adapted to handling flammable gases.

Absorbers generate waste material in the form of blowdown liquids and solids. Even the regenerable systems require periodic replacement of the scrubbing solution as it degrades and loses activity. The wastes may be hazardous, depending on the contamination present in these streams, which is another disadvantage when absorber systems are compared to flares. However, for infrequent emergency use these disadvantages may not be critical.

5.2.3   Control Effectiveness

The ability of an absorber to remove contaminants from gas streams is a function of the parameters discussed previously. Absorbers in process applications or pollution control with steady flow rates and constant compositions are capable of very efficient control. The effectiveness of a scrubber for protecting against accidental releases depends on whether it is dedicated to the emergency system or is used continuously with an emergency tie-in. In general, unless the occurrence of an accidental release significantly changes the composition and flow rate of a gas normally being scrubber, the performance of the non-dedicated absorber during a release incident would be expected to diminish only slightly. As long as the liquid phase chemistry is not disturbed, absorption should reduce the effects of an accidental release. For critical applications, dedicated scrubbers are preferable to common scrubber units.

As an indicator of the performance capabilities of scrubbers in typical industrial applications, Table 5-3 shows the removal efficiencies for various organic compounds and solutions. As can be seen, most of the compounds are effectively controlled. These types of efficiencies are for commercial scrubbers where economics may be limiting the efficiency feasible for long term operation. For emergency scrubbers, very high efficiencies may be achievable that would not be economical for sustained, long-term operation. Many other compounds, including H_2S, HCl, SO_3, Cl_2, SO_2, NH_3, and other inorganics, can also be removed effectively by absorption. Properly designed absorbers are capable of economically removing over 90% of these compounds. Operating costs usually limit the efficiency that is economically feasible for long-term operation. For an emergency scrubber, where intermittent use makes

TABLE 5-3.   SUMMARY OF SELECTED TYPICAL COMMERCIAL ABSORPTION
EFFICIENCIES FOR VARIOUS INDUSTRIAL CHEMICALS

Compound	Scrubbing Liquid	% Control Efficiency
Acetone/phenol	water	97
Acrylonitrile	water	99
Aniline	dilute sulfuric acid	99.4
Chloroprene	oil	100
Chloroprene/neoprene	oil	97
Cyclohexanol/cyclohexanone	oil	99
Formaldehyde	water	74
Methyl chloroform	water	90
Nitrobenzene	water	99+
Per/tri-chloroethylene	water	90
Terephthalic acid	water	95.6
Dimethyl terephthalate	xylene	97
Toluene diisocyanate	water	60
Toluene diisocyanate	caustic solution	98

Source: Reference 13

operating cost considerations less important a very high efficiency may be feasible.

Table 5-4 provides an example of the physical size and control capabilities of a scrubber for a hypothetical 50% ammonia stream. For a highly soluble gas like ammonia, relatively high efficiencies appear to be achievable in a "reasonably" sized system.

5.2.4 Costs

For a given application, the cost of a scrubber system depends primarily on the size, as determined by flow rate and removal efficiency required. Materials of construction based on corrosivity and pressure conditions also determine costs. Specific items included in the system battery limits can vary between applications. These include ductwork, fans or compressors, scrubber internals, reagent feed systems, instrumentation requirements, and degree of redundancy. Without considering a specific application, it is impossible to present a system cost. Total capital costs (including installation, indirect costs, and other fees) for absorber vessels have been presented in the literature in terms of column weight (14). The weight is related to the length and diameter of the absorber. The developed correlation was presented on a log-log graph. The data convert to the following equation:

$$\log (\text{June 1981 \$}) = 0.89 * \log (\text{column weight, lbs}) + 0.62$$

This corresponds to about $64,000 for a 5,000 pound column and does not include packing, platforms, ladders, ducting, fans, or other equipment. It does include face piping, some instrumentation, painting and insulation.

Operating costs of absorbers are highly variable. They are affected by the type of system and the reagents consumed. Tail end absorbers often produce throwaway materials which cause high reagent costs. The regenerable systems, although they recycle chemicals, incur costs for the regeneration step in the form of steam, reducing gases, or other commodities. Some costs are also associated with pumping the liquid through the column and fans or

TABLE 5-4.   EXAMPLE OF PERFORMANCE CHARACTERISTICS FOR A
PACKED BED SCRUBBER

Basis: Inlet stream of 50% NH_3 in 50% air. Constant gas flow per unit
cross-sectional area of 455 $scfm/ft^2$.

Packing: 2 inch plastic Intalox® saddles.

Pressure Drop: 0.5 inch water column

Removal Efficiency, %	50	90
Liquid to Gas Ratio		
(gal/thousand scf)		
— at flooding	160	160
— operating	80	80
Packed Height, ft.	3.1	11.4

Column Diameter and Corresponding Gas Flow Rates for Both Removal Efficiencies

Column Diameter (ft)	Flow Rate (scfm)
0.5	90
1.0	360
2.0	1,400
6.3	14,000

compressors to move the gas. Absorbers typically generate a gas side pressure drop on the order of 0.1 to 0.3 psi. Venturi absorbers can have pressure drops as high as 3 psi, depending on Venturi throat size and L/G ratios.

For emergency scrubbers, which are used rarely except for periodic testing, operating costs other than maintenance are not particularly important. Selection of a specific system will be determined primarily by the performance and capital cost.

5.3 ENCLOSURES

Enclosing equipment and vessels that store or process toxic chemical are another way of preventing or decreasing the effects of an accidental chemical release. Enclosures are containment structures that capture toxic chemicals spilled or vented from storage or process equipment, thereby preventing immediate discharge to the environment. The enclosures contain the spilled liquid or gas until it can be transferred to other containment, discharged at a controlled rate that would not be injurious to people or to the environment, or transferred at a controlled rate to a flare or scrubber. While enclosures may offer some protection, there may also be disadvantages or secondary hazards, which are discussed below.

5.3.1 Process Description

Enclosures can be constructed around individual equipment items, process units, and entire plants, depending on the nature of the hazard being controlled. For example, many facilities using solvents might vent all of the air in the manufacturing building through an activated carbon system. In laboratories and other small-scale facilities, small absorbers or filters may be used on the exhaust gases. These type of systems are effective in controlling smaller emissions of toxic compounds in a normal atmosphere.

Reaction vessels may also be enclosed in buildings. However, several factors should be considered in the design of such enclosures, including potential flammability and explosive hazards. Pressures over 5 psi are capable of destroying conventional buildings.

Another enclosure scheme is to use underground storage tanks. Although it is not common practice to bury pressure vessels, many tanks containing volatile liquids are underground. The overburden acts as an effective damper on emissions from such tanks. The use of underground storage tanks for preventing a sudden release of toxic chemicals is an attractive concept. Regulations concerning the design and operation of underground tanks for hazardous wastes are designed to prevent any release to the environment (15). Although regulations have not been promulgated for hazardous products, the same concern would be expected. These requirements are discussed further below.

5.3.2 Applicability

Enclosures are useful for preventing the release of small amounts of a toxic chemical because an enclosure can temporarily contain the chemical while it is vented at a controlled rate to a control device (see protection technologies) suitable for removing the chemical from the vented stream. A secondary hazard is explosion. Enclosures must be used with caution, however. For example, a small leak of a flammable vapor in an enclosure could be concentrated to the LEL, causing an explosion if a ventilation system is not properly designed. Also, the rupture of a pressure vessel within an enclosure could overpressurize the enclosure, resulting in it destruction. When enclosures are considered for reducing the potential of an accidental release, safety aspects must be carefully reviewed to prevent greater hazards. Enclosures are more appropriate for non-flammable toxic chemicals than for toxic ones.

A modified form of enclosure is the underground storage tank. Hazardous product and reactants stored in underground tanks will have to conform to regulations for these tanks. Such tanks require approved secondary containment which include external liners, vaults, and double wall tanks (15). For tanks of hazardous waste substances currently regulated, monitors capable of detecting a leak within 24 hours must be used, and all underground piping must

be double walled. Additional requirements are specified for ignitable,
reactive, and incompatible wastes in Part 264.198 and 199. Regulations
governing release detection, prevention, and correction are expected in the
near future. Congress has specified the following dates by which EPA regulations must be promulgated for underground tanks (16):

- February 9, 1987 - regulations for existing and new
  petroleum tanks

- August 9, 1987 - regulations for new hazardous chemical
  product tanks

- August 9, 1988 - regulations for existing hazardous
  chemical product tanks

5.3.3 Control Effectiveness

Enclosures may be a very effective means of controlling accidental
releases. Depending on the type and design of the enclosure, emissions could
be almost totally controlled since they could be confined until vented to a
destruction or capture device such as a flare or scrubber. These types of
systems are used to control exothermic reactions that use pyrophoric reactants
(17).

The effectiveness of enclosures may be compromised by two things:
flammable materials and pressurized vessels. With flammable materials,
enclosures can create a secondary hazard both without and with an accidental
release. During routine operations, small leaks of a flammable substance in
an enclosed area can lead to the accumulation of a flammable vapor-air mixture
which, if ignited, could damage the enclosure and process equipment, thus
causing the catastrophe it was intended to prevent. Proper ventilation could
reduce this hazard. During a release, a properly designed enclosure should be
effective if the release occurs at a low to moderate rate. A sudden release
from a pressurized vessel, however, could overpressure the enclosure itself,
causing damage and releasing the chemical. The enclosure would have served no

causing damage and releasing the chemical. The enclosure would have served no purpose. Because the sudden catastrophic failure is less likely than a large uncontrolled leak, however, an enclosure still might be appropriate if designed with appropriate explosion relief venting.

5.3.4 Costs

The variety of enclosures that can be used makes it difficult to estimate their costs. Total containment systems can exceed the cost of the controlled vessel, since they must be larger than the vessel and still be designed to meet the same criteria. Costs for buildings with directed ventilation systems can be estimated easily; however, a total cost can not be assigned to the control system. Small enclosures around process vessels or equipment can also vary in cost because of safety considerations (explosion risk, temperature limitations, atmosphere requirements, etc.). Additional costs are associated with the cleanup equipment needed when the enclosure becomes contaminated.

5.4 REFERENCES

1.   Product Line Reference Catalog. National Air Oil Burner Company, Inc., Philadelphia, Pennsylvania, 1984.

2.   Joseph, D., et al. Evaluation of the Efficiency of Industrial Flares: Background--Experimental Design--Facility. EPA-600/2-83-070 (NTIS PB83-263723) August 1983.

3.   Federal Register. Volume 50. April 16, 1985. pp. 14,941 - 14,945.

4.   Evans, F. L., Jr. Equipment Design Handbook for Refineries and Chemical Plants. Volume 2. Gulf Publishing, Houston, TX, 1974.

5.   Tan, S. H. Simplified Flare System Sizing. Hydrocarbon Processing. October 1967. p. 149.

6.   Kent, G. R. Practical Design of Flare Stacks. Hydrocarbon Processing and Petroleum Refiner. August 1964, p. 121.

7.   Pohl, J. H. and N. R. Soelberg. Evaluation of the Efficiency of Industrial Flares: Flare Head Design and Gas Composition. EPA-600/2-85-106 (NTIS PB86-100559) September 1985.

8.   Straitz, J. F. III. Flaring with Maximum Energy Conservation. Pollution Engineering. Volume 12, February 1980. p. 47.

9. Kohl, A.L. and F.C. Riesenfield. Gas Purification, 3rd Edition. Gulf Publishing Corporation, September 1979.

10. Perry, R.H. Chemical Engineers' Handbook, 5th Edition. McGraw Hill, New York, NY, 1973.

11. Sherwood, T.K. and R.L. Pigford, Absorption and Extraction, 2nd Edition. McGraw Hill, New York, NY, 1952.

12. Treyball, R.E. Mass-Transfer Operations, 2nd Edition. McGraw Hill, New York, NY, 1968.

13. Shareef, G.S. et al. Hazardous/Toxic Air Pollutant Control Technology: A Literature Review. EPA-600/2-84-194 (NTIS PB 85-137107) December 1984.

14. Vatavuk, W.M. and R.B. Neveril. Costs of Gas Absorbers. Part XIII. Chemical Engineering. October 4, 1982, pp. 135-136.

15. Lees, F.P. Loss Prevention in the Process Industries. Butterworth and Company, Boston, MA, 1980.

16. Federal Register. July 14, 1986. 40 CFR Parts 260, 261, 262, 264, 265, 270, and 271.

17. Welding, T.V. Operational Experience with Total Containment Systems in Protection of Exothermic Reactors and Pressurized Storage Vessels. (EFCE Event 292). Chester England, April 1984.

Part III

Mitigation Measures

The information in Part III is from *Prevention Reference Manual: Control Technologies—Volume 2. Post-Release Mitigation Measures for Controlling Accidental Releases of Air Toxics,* prepared by D.S. Davis, G.B. DeWolf, K.A. Ferland, D.L. Harper, R.C. Keeney, and J.D. Quass of Radian Corporation for the U.S. Environmental Protection Agency, January 1989.

Acknowledgments

This manual was prepared under the overall guidance and direction of T. Kelly Janes, Project Officer, with the active participation of Robert P. Hangebrauck, William J. Rhodes, and Jane M. Bare, all of U.S. EPA. In addition, other EPA personnel served as reviewers. Radian Corporation principal contributors were Graham E. Harris (Program Manager), Glenn B. DeWolf (Project Director), Daniel S. Davis, D.L. Harper, R.C. Keeney, Jeffrey D. Quass, and Sharon L. Wevill. Contributions were also made by other staff members. Secretarial support was provided by Roberta J. Brouwer and Sonita Cross. Special thanks are due many other people, both in government and industry, who served in the Technical Advisory Group and as peer reviewers.

1. Introduction

Increasing concern about the potentially disastrous consequences of major accidental releases of toxic chemicals resulted from the Bhopal, India, methyl isocyanate release on December 3, 1984, which killed approximately 2,000 people and injured thousands more. Concern about the safety of process facilities that handle hazardous materials increased further after the accident at the Chernobyl nuclear power plant in the Soviet Union in April of 1986.

While headlines of these incidents have created the current awareness of toxic release problems, other, perhaps less dramatic, incidents have occurred in the past. These accidents contributed to the development of the field of loss prevention as a recognized specialty area within the general realm of engineering science. Interest in reducing the probability and consequences of accidental toxic chemical releases that might harm workers within a process facility and people in the surrounding community prompted preparation of this manual and a series of companion manuals.

This manual, which addresses the post-release mitigation measures for toxic chemical releases, is designed to give the reader useful information about the state of the art in this area. Post-release mitigation measures are defined as any measures that may reduce the consequences of an accidental toxic chemical release after it has occurred. Mitigation measures decrease the quantity of a released chemical that can reach receptors, decrease the area exposed to the chemical, and/or reduce the duration of exposure. The emphasis is on post-release mitigation measures for facilities that handle toxic chemicals so that the potential risk to surrounding communities can be minimized.

Companion volumes to this manual include:

Prevention Reference Manual: User's Guide Overview for Controlling
                              Accidental Releases of Air Toxics (1)

Prevention Reference Manual: Control Technologies, Volume 1:
                              Prevention and Protection Measures for
                              Controlling Accidental Releases of Air
                              Toxics (2)

The series also contains several chemical-specific manuals on controlling accidental releases of specific chemicals. An example is the Prevention Reference Manual: Chemical Specific, Volume 8: Control of Accidental Releases of Hydrogen Fluoride (3).

1.1 FUNDAMENTAL CONCEPTS

The physical release of a chemical is the final event in a sequence of events leading to the release. Such a sequence of events begins with the initiating, or primary, event and propagates through enabling events. The final event can be prevented by preventing the initiating event or the enabling events. Success in preventing a release depends on correctly identifying primary and enabling events and event chains, on knowing the relative probability of the events, and on the skill and knowledge of the individuals charged with conducting the analysis. If prevention and protection meausures fail and a release occurs, mitigation measures must be invoked.

Figure 1-1 illustrates the place of mitigation, among other measures, in reducing the ultimate consequences of a potential release. As seen in the figure, mitigation can be viewed as one of several barriers between a release hazard and realization of its consequences. Before appropriate mitigation

Figure 1-1.   The role of various accidental release control measures
in reducing the consequences of an accidental release.

measures can be determined, the potential consequences of a release must be determined. The mitigation measures addressed in this manual include: 1) emergency planning; 2) siting and layout; 3) dispersion modeling; 4) detection and warning systems; 5) meteorological instrumentation; and 6) technical mitigation measures that can effectively control a release.

Emergency planning and training ensure the rapid and correct response of the people charged with applying other mitigation measures and define what mitigation measures should be employed.

Plant siting and layout concerns the placement of hazardous facilities relative to sensitive receptors in the surrounding community and within plant boundaries. Important considerations in this area, besides the obvious distance factor, include taking advantage of terrain features such as hills that might act as natural barriers, and avoiding the funneling effects of valleys.

Detection and warning systems give advance notice that a release is incipient or has occurred and define the magnitude and location of the release so that other mitigation measures can be taken. Meteorological instrumentation serves a similar purpose by providing data needed to monitor the physical location and movement of a release and the ambient characteristics that affect the movement and dispersion of a vapor or gas plume or cloud.

Vapor dispersion modeling is used to predict the extent, duration, and concentration of the plume or cloud of released toxic vapor or gas. Numerous dispersion models of varying levels of sophistication, accuracy, and verification by actual field data are available. The results predicted by these models depend on a source term which describes the characteristics of the initial release, and a dispersion term that describes the characteristics of the resultant cloud or plume. Characteristics of the release are primarily process and process equipment related, while dispersion characteristics are related to the properties of the chemical vapor or gas and to meteorological conditions. Such models can inform decisions about plant siting and layout, the placement of detection and warning systems and meteorological instrumentation, and the selection of technical mitigation measures.

Containment systems reduce the area that could be exposed to vapors from a release and confine the liquid portion of a release until measures can be taken to recontain or destroy the released chemical. While some containment may be successful with gases, containment is probably more applicable to spilled volatile liquids.

Spray systems disperse, dilute, and absorb a released airborne chemical. Spray systems rely on fixed or mobile equipment that applies a spray of water, other materials, or a condensing cloud of steam directly to the cloud or plume of noxious chemical. Some spray systems for toxic gas releases are similar to fire fighting systems.

Foam systems are primarily used to contain volatile liquid evaporation from pools of spilled liquid. These systems are based on specialty chemical materials that generate foams whose specific characteristics are tailored to the chemical properties of the material to which they are applied. Foams act as a physical barrier to prevent or decrease evaporation from liquid surfaces.

1.2   ORGANIZATION OF THE MANUAL

The remainder of this manual is divided into 8 sections, beginning with Section 2. Sections 2 through 6 deal with mitigation measures that enable a plant to prepare for an accidental release and to respond appropriately if one occurs. Sections 7 through 9 discuss the technologies used to control the liquid and vapor potions of an accidental release.

Section 2 addresses emergency planning and training, which are essential components of the mitigation effort. In Section 3, facility siting and layout as related to mitigation are discussed. Detection and warning systems within the facility and outside the facility boundary are discussed in Section 4.

Section 5 covers the basic types of vapor dispersion models and discusses the applicability of these models to the mitigation problem. Meteorological instrumentation, as described in Section 6, allows real-time estimates of where, when, and to what extent a release will affect the community.

Section 7 introduces the topic of secondary containment measures, or those measures that can control the spread of a toxic release once it occurs. The last three sections of the manual discuss three technologies that may be used to reduce hazardous vapor concentrations, thereby minimizing their effect on surrounding communities. These technologies are spray and steam systems, covered in Section 8, and foam systems, covered in Section 9.

1.3 REFERENCES

1.   Davis, D.S., G.B. DeWolf, and J.D. Quass. Prevention Reference Manual: User's Guide Overview for Controlling Accidental Releases of Air Toxics. EPA-600/8-87-028 (NTIS PB87-232112), U.S. Environmental Protection Agency, AEERL, Research Triangle Park, North Carolina, 1987 (Part I, this book.)

2.   Davis, D.S., G.B. DeWolf, and J.D. Quass. Prevention Reference Manual: Control Technologies, Volume 1. Prevention and Protection Technologies for Controlling Accidental Releases of Air Toxics. EPA-600/8-87-039a (NTIS PB87-229656), U.S. Environmental Protection Agency, AEERL, Research Triangle Park, North Carolina, 1987 (Part II of this book.)

3.   Davis, D.S., G.B. DeWolf, and J.D. Quass. Prevention Reference Manual: Chemical Specific, Volume 8. Control of Accidental Releases of Hydrogen Fluoride. EPA-600/8-87-034h (NTIS PB87-234530), U.S. Environmental Protection Agency, AEERL, Research Triangle Park, North Carolina, 1987.

2. Emergency Planning and Training

Emergency planning and training are an integral part of mitigation. Reducing the consequences of an accidental release depends on a timely, effective response by people at the scene who know the what, when, where, and how of various countermeasures. This includes the application of various technical mitigation measures discussed elsewhere in this manual, as well as calling for assistance, notifying authorities, and, if necessary, evacuating people from the affected areas.

2.1 BACKGROUND

Numerous recent publications on emergency preparedness include sections on emergency planning and training. The U.S. EPA has published a guidance manual on the Chemical Emergency Preparedness Program (1). The Chemical Manufacturer's Association has sponsored the Community Awareness and Emergency Response (CAER) program (2,3). Other publications also deal with the issue (e.g., Reference 4).

2.2 EMERGENCY PLANNING

2.2.1 Description

The details of emergency planning and training will vary from facility to facility and community to community. In general, however, any program should address certain fundamental elements:

- Program initiation;

- Hazard evaluation;

- The identification, evaluation, selection, and implementation of countermeasures;

- Resource requirements and availability;

- Organization; and

- Mobilization and demobilization.

Emergency programs may be initiated by facility or corporate personnel or by an outside authority. Once the program is started, an overall plan can be developed to address each of the other fundamental elements listed above. The plan should address community involvement as well as on-site emergency planning and response activities. This is discussed further in Subsection 2.2.2.

Evaluating the potential hazards of a particular facility, which involves both hazard identification and analysis, is an essential starting point in actual plan development. Hazard identification begins with a list of designated chemicals, a determination of which ones are present at the facility, the quantities, and their physical states. The identification of potential release hazards is based on physical and toxicological data and on how the chemicals are used at the facility. The potential hazards are then analyzed to determine how release incidents might occur and what the consequences might be. These topics are examined more fully in other manuals in the Prevention Reference Manual series, as well as in the technical literature. Various formal procedures can be followed to more effectively determine the relative probability and consequences of various release scenarios.

Once the potential accidental release scenarios are known, countermeasures can be proposed for incorporation into the emergency plan. Since there may be more than one acceptable response to a given emergency situation, the various response measures must be evaluated to select those that appear most effective for the specific scenarios under consideration. The final step is to implement the measure(s) when an emergency occurs.

Two other important subjects to consider when designing an emergency plan are resource requirements and availability, and emergency organization. Resource requirements include personnel, equipment, and money to fund both the plan and its implementation. Personnel requirements and equipment needs should be explicitly noted, while funds should be allocated in management budgets but not explicitly addressed in the plan. An on-going financial commitment will be required for periodic reviews and updates of the plan, as well as for training programs. Availability addresses what resource additions may be required that are not already present in a facility, company, or community. The emergency organization must be clearly defined. The function of the emergency organization is to establish clear lines of communication and define responsibilities and authority so that confusion, delays, and inappropriate actions can be avoided during an emergency.

Finally, a complete plan must address mobilization--how the plan will be activated in an emergency--and demobilization--how the plan will be concluded after an emergency.

The plan should be documented; the emergency response plan (ERP) documents the planning activities and provides a ready reference for use during emergencies. The ERP must be a "living document," that is, frequently updated to reflect changes in equipment, staff, and the facility. The ERP should include the following major topics:

- General facility description, including plot plans and area maps;

- Description of hazards associated with facility operations (detailed assessments should be included or attached);

- Description of the organization of personnel involved in emergency response;

- Functions, responsibility, and authority of key personnel;

- Location and equipment of each emergency control center;

- Alarm systems and notification;

- Description of all communication systems and emergency backups;

- Ranked list of notifications that must be made for each type of accident;

- Evacuation and/or sheltering plans for non-essential personnel;

- Assembly points and procedures for accounting for all employees, visitors, contractors, and others;

- Locations of and equipment stored at each on-site first aid center;

- Transportation facilities and likely locations of vehicles; including specialty vehicles like trackmobiles, cranes, front-end loaders, and others;

- Security precautions;

- Continuation of utilities and services;

- Local emergency response agencies;

- Mutual aid organizations;

- Possibility of and procedures for community evacuation;

- Community shelter-in-place potential;

- Community emergency education;

- Public relations in an emergency;

- Procedures and equipment for responding to
  -- fires and/or explosions,
  -- toxic materials releases,
  -- floods,
  -- hurricanes (if applicable to location),
  -- ice storms,
  -- other natural phenomena (earthquakes, tornados, etc.),
  -- civil disturbances;

- Restoration of normal operations;

- Training in emergency response procedures;

- Drilling in emergency response procedures;

- Emergency systems training and maintenance; and

- Emergency response plan updates.

The appropriate depth of planning for each of these topics depends on the nature/severity of the potential hazard, the size and complexity of the plan, and the use of the facility relative to the use of outside agencies for fire fighting, first aid, etc.

One sample outline of a community emergency response plan published by the U.S. EPA is shown in Table 2-1 (1). The details of this approach are discussed in the reference. Comparison of this approach with the previous discussion illustrates that, while certain topics are common to both industrial and community plans, individual details may be addressed in different ways.

2.2.2 Implementation

Implementation of an emergency response plan, including coordination between the facility and the community, has been addressed by both the U.S. EPA and the chemical industry (1,2,3).

One formal, collective chemical industry initiative is the Community Awareness and Emergency Response program developed under the auspices of the Chemical Manufacturers Association (3). Its purpose is to prepare communities to respond effectively to man-made or natural disasters. The goals of the program are to improve community awareness and to integrate industrial emergency response plans with those of the community. Some objectives of CAER are to:

- Inform the public about hazardous chemicals;

- Review, renew, or establish emergency response plans;

- Integrate chemical facility emergency response plans with those of the community to form an overall plan for handling all emergencies; and

- Involve members of the local community in developing and implementing overall emergency response planning.

TABLE 2-1.   EXAMPLE OUTLINE OF A COMMUNITY EMERGENCY RESPONSE PLAN

---

   i.   Emergency Response Notification Summary
  ii.   Record of Amendments
 iii.   Letter of Promulgation
  iv.   Acknowledgment
   v.   Table of Contents

   I.   Introduction

      A.   Abbreviations and Definitions
      B.   Purpose
      C.   Relationship to Other Plans
      D.   Assumptions/Planning Factors
      E.   Concept of Operations
          1.   Governing Principles
          2.   Organizational Roles

  II.   Emergency Response Operations

      A.   Notification of Release
      B.   Initiation of Action
      C.   Coordination of Decision-Making
      D.   Public Information/Community Relations
      E.   Personal Protection/Evacuation
      F.   Resource Management
      G.   Personnel Safety
      H.   Acutely Toxic Chemicals
      I.   Countermeasures
      J.   Response Action Checklist
      K.   Attachments
          1.   Emergency Assistance Telephone Roster
          2.   Siren Coverage
          3.   Emergency Broadcasting System Messages
          4.   Evacuation Routes
          5.   Traffic Control Points
          6.   Access Control Points
          7.   Evacuation Routes for Special Populations

 III.   Appendices

      A.   Basic Support Documents
          1.   Legal Authority and Responsibility for Responding
          2.   Acutely Toxic Chemicals Information
          3.   Hazards Identification and Analysis
          4.   Response Organization Structure/Coordination
          5.   Laboratory, Consultant, and Other Technical Support
             Resources
          6.   Computer Utilization
      B.   Post-Emergency Operations
          1.   Documentation of Accidental Releases
          2.   Investigative Follow-Up

A CAER Handbook has been prepared that gives detailed information on organizing and implementing a CAER Program (3). Major steps in a CAER program are:

- Community status review and program organization;

- Facility status review;

- Implementation;

- Community involvement; and

- Emergency exercises.

Community Status Review--

This review involves a survey of the status of local emergency planning in communities within a few miles around a facility. It seeks to determine whether integrated community emergency response planning is a complex activity (multifacility/hazard) or a local chemical facility function. Factors that should be considered are:

- Are there significant natural disaster risks in the area and do emergency plans exist for these risks?

- Are there other hazardous industrial activities, such as nuclear power generation, hazardous waste disposal, or significant transportation activities, and do emergency plans exist for these hazards?

- Are there other chemical facilities in the area?

- Does the facility's own emergency response plan clearly define the roles and responsibilities of community officials and off-site responders?

Based on this information, a status report can be prepared that describes the facility, the key elements of the facility emergency response capability, the objectives of CAER, and other pertinent issues. It should stress the need for off-site emergency response planning to be integrated with the existing facility emergency response capability, and that an integrated facility-community plan can be used to improve emergency response to all hazards.

- As a facility initiative in getting the CAER program underway, this document can be distributed and discussed with selected community contacts. The CAER program is explained to other people who should be involved in community emergency response planning as they are identified. These people will probably include first responders such as the fire chief, police chief, ambulance service workers, mutual aid program participants, local officials, business leaders, and facility employees, who may also be elected officials. Local disaster preparedness officials may be helpful as initial contacts.

- These initial contacts will identify the appointed and elected officials whose cooperation will be necessary to complete the integrated community emergency planning process. Meetings with key officials should be used to enlist their support in establishing a coordinating group. In many locations, the equivalent of this group or of a well-defined emergency response planning agency may already be in place.

A more formal preliminary organizational meeting for the coordinating group is a next step. Participants should include members of the groups listed above. Other participants could be public health officials, hospital administrators, Red Cross personnel, or local civic leaders. In most areas, chief elected officials have a general charter to protect public safety. Gaining official recognition from the local community is an important prerequisite for the coordinating group. This gives the group the necessary authority to begin the integrated planning process described later in this discussion.

Facility Status Review--

A number of activities are required. A first step is to review the facility emergency response plan to ensure that interactions with off-site agencies and responders are clearly defined. Responsibilities and notification procedures should be clearly defined.

- Employee awareness of emergency response planning must be provided for by ensuring that basic emergency concepts are included in initial training and periodic safety training. The facility news-letter or bulletin boards can announce that the facility has completed a new or revised company plan and has started an inte-grated emergency response planning process.

- A written community relations plan should be prepared. This can begin with a list of local organizations, government agencies, community groups (business, civic, public interest) and local media who should be briefed on the CAER program. Past local media cover-age of the facility and national industry issues should be reviewed. Past and ongoing facility communications efforts should also be reviewed and it should be determined whether the community and its leaders are likely to be knowledgeable about the facility. Using this information, assign priority to community awareness efforts. Pay special attention to relations with local officials, off-site responder groups, neighborhood associations and labor groups. It may be appropriate to obtain expert assistance in community rela-tions if this is a new activity for the facility.

Assured that the internal emergency planning activities are in order and that the appropriate coordinating group has been established, the facility's first contribution should be a presentation of a ten-step implementation process to the coordinating group.

Implementation Steps--
    Ten implementation steps are defined under the CAER program:

- Identify the emergency response participants and establish their roles, resources and concerns;

- Evaluate the risks and hazards that may lead to emergency situations in the community;

- Have participants review their own emergency plan for adequacy relative to a coordinated response;

- Identify the required response tasks not covered by existing plans;

- Match these tasks to the resources available from the identified participants;

- Make the changes necessary to improve existing plans, integrate them into an overall community plan and achieve agreement;

- Commit the integrated community plan to writing and obtain approvals from local governments;

- Educate participating groups about the integrated plan and ensure that all emergency responders are trained;

- Establish procedures for periodic testing, review, and updating of the plan; and

- Educate the general community about the integrated plan.

2.3 TRAINING

The emergency response plan, which reflects the character of a specific facility, determines the basic training needs, which will vary with different groups within the facility according to their designated roles in an emergency. The general objectives of training are to increase the awareness, knowledge, and skills of management, and of operating, maintenance, and specialized emergency response personnel. Emergency exercise activities, suggested by CAER, involve integrated training with community emergency personnel.

2.3.1 Training Overview

As defined by one author, training must address employee roles in three stages of an emergency (4):

- Raising the alarm;

- Declaring that an emergency exists; and

- Implementing the emergency plan.

The training must ensure that personnel understand their roles in the general areas of:

- Communications and control procedures;

- Individual responsibilities;

- Specific emergency operating and countermeasure procedures;

- Coordination with outside services; and

- Public relations.

There are two distinct functional aspects of training: procedures and equipment. Procedures include the safe shutdown of a facility, communications, first aid and medical response, evacuation, traffic control, rescue, fire fighting, and the implementation of other counter measures.

Equipment-oriented training deals with both protective and remedial equipment, including: protective clothing and breathing apparatus, facility shutdown and control, fire fighting, leak control, spill control, meteorological monitoring, movement of equipment, emergency construction, and actuation of hardware safety systems.

To be effective, a training program must be ongoing, periodically reviewed, updated and revised, and have full management support to ensure that the necessary training resources are available. Management provides the environment for adequate training development and presentation through guidelines and policies and also the enforcement and incentives to make the program work. The initial training of personnel should give a thorough grounding in fundamentals and be followed by subsequent enhanced training in critical emergency areas. The development of an effective training program depends on a thorough task analysis of the various personnel roles defined in the emergency response plan.

The conduct of training should include on-the-job training, seminar/discussion sessions, formal classroom training, and field exercises. The support that vendors of equipment and technical services can provide should not e overlooked. Such outside influences can inspire new interest in otherwise routine in-house programs. Full advantage should also be taken of newer training tools such as video tapes and computer simulation.

In-depth training should be given to employees soon after they are hired and should include a review of facility hazards, general emergency response activities, the emergency response plan, and the specific duties of the new employee. This level of training should be repeated each time an employee moves to a new position in the facility with different hazards and different emergency response duties.

Refresher training should be given regularly. The frequency of refresher training should be based on the nature of the hazards and on the type of emergency response duties assigned to the employee.

Drilling is an important part of training. While the objective of training is to teach employees the appropriate response, the objective of drilling is to improve response time. Drills should be held for each type of hazard present at the facility. A combination of announced and surprise drills should be used. Outside agencies should be included in some drills. The employees should be given feedback on drill performance and supplemental training when performance is seriously deficient.

Decisions about the frequency of drills should be based on the nature/severity of the hazards, the size/complexity of the facility, and the amount of reliance on outside emergency aid.

Drills are discussed further as part of emergency exercise programs in Subsection 2.3.3.

2.3.2 Training Coordination with Outside Agencies

Fire departments, police, emergency medical services, mutual aid groups, and commercial emergency response contractors can help substantially during a facility emergency. The effectiveness of these outside agencies depends, however, on their knowledge of the facility, the processes, and the chemicals

involved. Representatives of these agencies should regularly tour the facility and be trained in its hazard-related features. A spirit of open
communications and cooperation should be fostered.

2.3.3 Emergency Exercise Program

An emergency exercise program is an essential part of CAER. It completes
the cycle begun when the facility, along with community and industry representatives, formed a community coordinating group to manage emergencies. The
coordinating group integrated all emergency plans into one community emergency
response plan. Emergency responders have their tasks, and resources are
committed. An emergency response system is in place, but will it work?

Besides a real emergency, only the simulation of an emergency can answer
that question. Simulation and evaluation exercises, by testing all or part of
a system against the plan, complete the cycle the facility and the community
started. There are several types of exercises and each has its use, but no
single exercise can adequately test all of the elements of an emergency
management system. Properly implemented, an ongoing exercise program will:

- Help evaluate emergency plans and response capability;

- Provide the information necessary to improve plans and procedures;

- Train the participants;

- Improve coordination between on-site and off-site personnel;

- Ensure the continued involvement of key community organizations;

- Provide a way to inform and involve the public and the media; and

- Serve as a visible demonstration of industry and community commitment to protect the public.

Facility Management Role--

Facility management has a pivotal role. Exercises are complex events that require the cooperation of many people and organizations. Without the active support and encouragement of the chief executives of these organizations, there is little chance for success. The most important parts of the task are committing the necessary time and resources of the organization and gaining the cooperation of the chief officials in the community.

Components of the Program--

Four types of exercises should be the major components of the program:

- Tabletop;

- Emergency operations simulation;

- Drill; and

- Field exercise.

They range from the simple to the complex and each has a definite function.

Most industrial managers are familiar with the concept of a drill or field exercise. Such exercises typically involve actual response by personnel and the use of protective equipment, emergency apparatus, and field communications. A drill focuses on a single aspect of an emergency response system, while a field exercise tests all or most of the response system.

The tabletop and emergency operation simulation exercises are less commonly used and are characterized by the simulation of all or most of the actions to be performed in emergency response. They provide substantial benefits within an overall exercise program because they are usually less complex and require less time and effort to plan and conduct. They are also more flexible as learning experiences and can build confidence and capability for more involved exercises.

- A tabletop is primarily a learning exercise that takes place in a classroom or meeting room setting. Situations and problems generate discussion of plans, procedures, policies, and resources. Tabletop exercises are an excellent way to familiarize a group of organizations and agencies with their assigned roles. This type of exercise is also a good method for testing the logic and content of the plan. It is sometimes referred to as a "What if" exercise.

- An emergency operation simulation (EOS) is a full-scale emergency simulation that uses various forms of message traffic (telephones, radios, message forms, etc.) to create a realistic, high-pressure, emergency environment. EOS exercises stimulate decision-making and interaction by emergency managers in response to simulated emergency conditions. An EOS is conducted in an Emergency Operating Center or in another suitable facility and involves command, control, and decision making functions. All field response activities are simulated, with activity limited to the controlled environment of the facility used. They are very useful for testing direction and control functions and for evaluating how well the total response system is coordinated.

- The drill is a supervised activity that tests, develops or maintains skills in a single emergency response function (i.e., communications drills, fire drills, command post drills, medical emergency drills, etc.). These frequently involve actual field response, activation of emergency communications networks, and equipment and apparatus that would be used in a real emergency. The effectiveness of a drill is its focus on a single, or relatively limited, portion of the overall response system in order to evaluate and improve it.

- A field exercise practices all or most of the basic functions of the response system simultaneously and the ability of the different organizations to work together. A full field exercise involves at least a partial mobilization of each organization's resources and a demonstration of their response actions. It also includes activation of all emergency facilities, including emergency operations centers, communications centers, command posts, hospitals, media centers, first response units, etc. A field exercise may exclude some specific functions previously tested, but it should include all response functions that require significant coordination with any others being tested.

Table 2-2 summarizes some of the advantages and disadvantages of each type of exercise.

TABLE 2-2.   EXERCISE TYPES--ADVANTAGES AND DISADVANTAGES

Advantages	Disadvantages

1. Tabletop Exercise

Modest time, cost, and resource commitment. Effective method of reviewing plans, procedures, and policies. Good training for key personnel in responsibilities and procedures. helps build coordination and consensus.	Lacks realism. Does not provide a true test of emergency system operation.

2. Emergency Operations Simulation (EOS)

Moderate time, cost, and resource commitment. More realism than a tabletop exercise. Tests integrated response of entire emergency management system. Good method of testing command and control before "going public."	Heavily dependent on written scenario. Realism/effectiveness can suffer when not well planned and executed.

3. Drill

Moderate time, cost, and resource commitment. Easiest exercise to design. May be very realistic. Provides good hands-on training. Allows a single component of the system to be isolated and practiced in depth.	Does not test entire system. Player safety is critical.

4. Field Exercise

Good realism. Good test of integrated communications. Provides means to evaluate mobilization of resources and first responder capability. Good opportunity to increase public awareness of program.	Major time, cost, and resource commitment. Scenario development is very critical. Player/public safety is critical. Can create problems when poorly planned.

2.4   REFERENCES

1.   U.S. EPA Chemical Emergency Preparedness Program, Interim Guidance, Revision 1, 9223.0-1A, November 1985.

2.   Cathcart, Christoper. U.S. Chemical Industry Emergency Response Initiatives, Avoiding and Managing Environmental Damage from Major Industrial Accidents. In: Proceedings from Air Pollution Control Association International Conference, Vancouver, British Columbia, Canada, November 1985.

3.   Chemical Manufacturers Association, CAER Handbook, Washington, D.C., 1985.

4.   Lees, F.P. Loss Prevention in the Process Industries, Vol. 2, Butterworth's, London, England, 1980.

3. Facility Siting and Layout

The consequences of an accidental release can be reduced by incorporating mitigative features into the design of a facility's layout and siting. Siting is the location of the process facility within a community or region, while layout is the positioning of equipment within the process facility. Although it is important to consider siting and layout when designing a new facility, the guidelines discussed in Subsection 3.2 apply equally to facility expansions. Siting could also be an important factor when deciding whether to expand a particular facility.

3.1 BACKGROUND

Siting and layout design principles that can reduce the effects of an accidental release include the following:

- Locating the facility within the community or region in a way that reduces the number of people that could be affected; and/or

- Arranging equipment within the facility in a way that reduces the effect that a potential release from one location within the facility would have on the rest of the facility, or on the surrounding community.

One study of the contribution of different hazard factors to accidental releases found that poor facility siting played a role in 5.8% of the cases and that poor facility layout was a factor in 3.9% of the cases (1). The study did not distinguish between factors that could have prevented the

incident and those that could have mitigated its effects. Although this
section focuses on the mitigation of accidental releases, the information
presented in Subsections 3.2 and 3.3 may also help prevent the occurrence of
an accidental release. Preventive measures for facility layout and siting are
discussed in the Control Technologies Manual of this Prevention Reference
Manual series (2).

3.2 SITING

Siting refers to the location of the process facility within a community
or region. Often, siting is carefully examined only for new facilities;
however, the expansion of an existing facility may require a reevaluation of
the site to assess the suitability of the expansion. This will be especially
true if the expansion involves chemicals or processes that pose more hazard
than presented by the original facility. One author discusses siting from a
process hazard perspective and cites a number of references (3).

A study of factors that have contributed to losses in several hundred
large fires and explosions identified the following inadequacies in facility
siting: poor utility service; poor emergency response and fire protection;
off-site traffic congestion hindering response by emergency vehicles; and poor
drainage (1). The effect of these inadequacies can be magnified by meteorologic factors and population densities around the facility, two factors that
will determine the effects of an accidental release on off-site populations
(3,4).

Literature on chemical facility siting has generally been limited to
discussions of traditional siting criteria, i.e., distance from raw material
resources and markets and suitability of the transportation system and
utilities (5,6,7). However, the siting of facilities such as nuclear power
plants and hazardous waste disposal units has been more problematic historic-
ally, and specific siting criteria have been developed to address public and

environmental health issues in these industries. Some of the issues
identified in the related literature could be relevant to the siting of a
chemical facility that uses or produces toxic chemicals.

3.2.1 Population Density Near Facility Site

Clearly, the fewer people that live near an industrial facility, the
fewer that would be affected by an accidental release. One way to minimize
the number of people affected by a potential accidental release is to maintain
a buffer zone immediately around or downwind of the facility. The buffer zone
for a nuclear power facility, for instance, is a circle with a 0.3 to 1.3 mile
radius (6). A distance of five miles downwind was required during the siting
of a spill test facility for liquefied fuels, ammonia, and chlorine (4).
Setting aside a controlled buffer zone, however, has not been standard
practice in industrial facility siting. Although industrial facilities may be
sited in what is initially a rural area, the facility itself may become the
focus of later development; therefore, setting aside a buffer area is one way
for a hazardous industrial facility to protect itself against the problems of
future urbanization.

Another way to reduce the population density around a facility is to
maximize the distance to population centers. This issue has not been explicitly addressed in the siting of hazardous industrial facilities. The site of
the Bhopal Union Carbide facility, where a MIC release resulted in over 2,000
deaths, was found in 1975 to be "not suitable for hazardous industries" because of the population density around the facility (8). Two hundred thousand
people lived within a 4-mile radius (9). The facility did, however, receive
approval to operate despite this evaluation. Contrast this with siting
requirements for commercial nuclear power reactors or hazardous waste disposal
facilities. The New Jersey hazardous waste facility siting act requires a
minimum distance of 2,000 feet between the facility and the closest residence

or school. For some types of hazardous waste disposal facilities, a half mile distance from residences or schools is specified (10).

The regulations governing commercial nuclear reactors take a different approach to specifying distances from population centers. The regulations specify two zones, calculations of which are based on maximum exposure from a major accident and a population center distance. The population center distance is the distance to the nearest densely populated center with more than 25,000 residents. The population center distance must be at least one and a third times the distance from the reactor to the outer boundary of the second zone (11). Thus, the distance to the population center will be a function of the design of the facility and the definition of a major accident.

Increased distance from population centers also allows more time to respond to an emergency and evacuate or inform the local population. This relationship is addressed indirectly by the nuclear plant siting regulations discussed above. However, the literature does not examine the relationship between time available for emergency response and facility distance from population centers. Clearly, the distance could not be fixed arbitrarily, but would be a function of factors such as emergency preparedness and population mobility (i.e., access to public transportation, cars per household, road network).

An example of a risk-based approach to facility siting can be found in the Canvey study, an investigation of a heavily industrialized area in England (12). The study examined existing and proposed facilities in the Canvey Island/Thurrock area and calculated the individual and total risk posed by the petrochemical and shipping facilities to area residents. The Commission recommended that the proposed facilities be approved on the condition that risk reduction measures be undertaken by the existing facilities. The Canvey study illustrates how quantitative risk analysis can be used during the facility siting process. This type of approach would be especially suited to facility expansion decisions.

The literature on the siting of hazardous industrial facilities does not provide guidelines as to what constitutes a safe distance or an effective buffer zone. Examples of buffer zones considered during the facility siting process are given in the literature about other fields, but these examples may not directly apply to chemical processing facilities.

3.2.2 Meteorology and Climate

Meteorologic considerations involve identifying the prevailing wind direction, wind speed, and atmospheric stability. Meteorologic data can be used to rule out inappropriate facility sites or to locate units within the facility boundaries. Hazardous units should be located to minimize the effects of a release on off-site and on-site populations and should be downwind (for the prevailing wind direction) from major population concentrations when possible.

Wind patterns can be characterized in three ways. One is the large-scale movement of major air pressure systems. An intermediate scale involves air movements caused by regional topograhic features and differential heating. A more local effect can be caused by the presence of ridges and valleys. Data on all three types of patterns may be used to predict areas affected by accidental releases (4). However, of most interest will be site-specific wind characteristics. For example, ridges and valleys may channel surface winds that would carry heavier-than-air releases. Ridges serve as barriers in this instance; whereas valleys are possible conduits for the accidental release. The reader is referred to Section 5 on dispersion modeling and Section 6 on meteorological instrumentation for a discussion of the use of meteorological data and the modeling of release impacts.

Temporal variability in wind direction may occur daily or seasonally. Temporal variability may be an important characteristic when there is a basis

for determining the most likely time for a release; for example, if the
chemical is only stored or processed on site for a specific period.
Otherwise, it could be assumed that the most frequently occurring wind
direction is the one of greatest interest in modeling impacts. However, it
will often be necessary to investigate the effects of other than prevailing
wind conditions.

Examples of accidental release incidents in which the meteorological conditions aggravated the release's impact can be found in the literature. At
Bhopal, for instance, the area was experiencing an atmospheric inversion at
the time of the release. An inversion increases the time needed to disperse a
chemical cloud to safe levels. A region that experiences inversions
frequently may not be an appropriate site for a hazardous industrial facility.
Other incidents in the literature illustrate how unstable or neutral
atmospheric conditions are beneficial, since the wind movement disperses
dangerous concentrations quickly; see case history A-24, Lees (3). Another
example of basing the selection of a site on good dispersion characteristics
(reliable wind direction and wind speed) can be found in Miller, (4).

3.2.3 Emergency Response

A rapid response by facility or local emergency response personnel can
mitigate the effects of a release. If a facility is to be located in an area
that does not have an adequate emergency response system or if the addition of
the facility will strain the area's emergency response capabilities, the
facility owners might consider giving support to local emergency response
services. New Jersey now requires this of major hazardous waste facilities.
The New Jersey hazardous waste facilities siting act specifies that five
percent of a facility's gross receipts will be paid to the host municipality
for the cost of additional emergency response training and equipment (10).
Several chemical companies also report supporting the emergency response
capabilities of host communities (6).

3.2.4 Topography and Drainage

A fairly level site is necessary to prevent flammable liquids from flowing down a sloped terrain to other equipment or buildings. Also, the facility site should not drain highway runoff, since flammable liquids from a highway accident may spread a fire. Hills may be used as natural barricades, separating operations with potentially explosive chemicals (13). The effect of the terrain on wind patterns should be considered also, since local features such as valleys and ridges will channel winds. Hills or ridges can also block, divert, and help disperse an air release of a toxic chemical.

3.2.5 Accessibility to Off-Site Emergency Response Vehicles

If the facility is to be served by off-site emergency response personnel, it must be easily accessible. Facility entrances should be free of traffic congestion or blockage by trains (moving, stopped, or derailed); the facility should have multiple entrances/exits, ideally onto different streets or roads so that one accident cannot block all escape routes (13). It should also be considered whether the facility site is accessible year round. In areas with snow and ice, reliance on measures in addition to or instead of outside emergency response services might be warranted.

3.2.6 Utility Service

Water, gas, power, and any other utilities should be reliable. An adequate supply of water for fire fighting is essential. Failure of several utilities services simultaneously can aggravate a release incident (13). Water and electricity are the most important utilities, since they are essential to fire fighting, cooling systems, operation of pumps, lighting, and instruments, etc. Back-up electrical power and looped water mains are typical safeguards. A looped water main originates at a source, is routed through the facility and then returns to the source. This arrangement allows water to flow from the source in either direction through the loop. Thus, a break in the line at any point will not cut off the rest of the loop from the water source.

3.3 LAYOUT

Layout concerns the placement and spacing of the components of a process facility, including the individual equipment in each component. A properly designed facility will minimize the consequences of an accidental release by facilitating process operability and by segregating hazardous areas. Both the Chemical Manufacturers Association (CMA) and the National Fire Protection Association (NFPA) have issued standards and guidelines for facility layout (14). Key features to be considered during the layout of a facility site are its boundaries, the work boundaries, through railway lines, ignition sources, control rooms, buildings that concentrate personnel, production units, storage, loading and unloading, pipebridges, roadways and waste disposal areas.

A study of factors that have contributed to losses in several hundred large fires and explosions identified hazard factors and grouped these into nine categories. Problems with facility layout and spacing were characterized in the following way: congested process and storage areas; lack of isolation of extra-hazardous operations, lack of proper emergency exit facilities; sources of ignition too close to hazardous critical facility areas; and inadequate hazard classification of facility areas (1). These problems can be addressed by applying the general principles discussed below.

3.3.1 Adequate Spacing Between Units

It is important to increase the distance between process units to reduce the impact of an accidental release. An explosion in one process unit can result in an accidental release in another unit because of damage from the blast wave. The blast wave pressure front travels rapidly, but its intensity decreases rapidly with distance. A vapor cloud associated with an accidental release travels much more slowly than does a blast wave but usually affects a larger area before it is diluted enough to present no danger. Explosions of vapor clouds can be particularly destructive in congested plants.

A quantitative method may be used to specify the spacing between facility units. First, a relative hazard ranking is developed of the individual process units, including manufacturing units, unloading/loading operations for tankers or drums, major pipebridge sections and major cross-country pipelines. One common method of quantifying the potential fire and explosion hazard of a given process unit is the Dow Index (15). The Dow Index has been modified to address the potential hazards of toxic chemical releases. This Mond Index is one of several methods developed for quantifying spacing requirements between facility units (15,16). Other methods for quantifying spacing requirements have also been developed (15,16).

Safe spacing requirements can be developed for a particular site, as described above, or reference may be made to published sources, a number of which recommend minimum safe separation distances for process and storage units. Reference 3 lists several other sources that discuss safe separation distances. Reference 3 notes that most of the published guidelines for spacing requirements give little explanation of the basis for the distances and that many sources incorporate recommendations from each other.

The two principal factors that have been traditionally used for determining safe separation distances are: 1) the heat produced by the burning liquid and 2) the potential to ignite the accidentally released gas or vapor. Separation of units from identified ignition sources, such as furnaces, electrical switchgear, flarestacks or transport vehicles is one of the most common recommendations encountered in the literature. A review of case studies in Reference 3 shows that the presence of an ignition source near an accidental release source can turn an initial release incident into a fire that can cause a greater loss of life and property. For instance, in the Feyzin refinery case cited in Reference 3, a source 160 meters distant from the plant unit is suspected of igniting a cloud of propane. Note that the toxic hazard associated with an accidental release may be more severe than the hazards associated with its heat of combustion or flammability. In such cases, traditional methods of determining safe separation distances may not be appropriate.

Some special considerations adapted from the literature for the safe spacing of facility units are:

- In general, units with the most toxic materials should be suitably distant from all facility buildings containing appreciable concentrations of personnel, and also from activities outside the works boundary. This particularly applies to off-site population centers such as schools, hospitals, places of entertainment, etc.

- Units having the greatest potential for explosion should not be located close to a facility boundary; they should be separated by areas occupied by low-risk activities with low population densities (up to 25 people per acre).

- Independent elements should be spaced so that a fire or explosion in one unit has a minimum effect on other units. When units cannot be adequately spaced and segregated, the use of barricades, protective construction, and fixed fire protection is recommended. For further information see Rindner and Wachtell (18) and Lewis (17).

- Ignition sources, including trucks, railroads, boilers, personnel smoking areas and direct-fired equipment should be located away from and upwind of hazard areas processing or storing flammable liquids, gases, or vapors. Some standards call for at least 10 feet of separation between vapor hazard areas and ignition sources, although this distance may be reduced by using vapor barriers or other special protective systems (13). Yet in the Feysin refinery example, the ignition source was thought to be 530 feet away, which illustrates the importance of evaluating each situation individually and identifying high-hazard areas.

- Facility work units should not be located at the facility boundary. Distance to the facility boundary will vary, depending on the hazard of the unit. (17).

Finally, the initial design must allow sufficient space for future facility expansions, and the expansions must also adhere to safe layout principles.

There are two schools of thought on the safest spacing of interconnected process units. One, which emphasizes the danger of individual process units and specifies minimum safe distances between types of units, focuses on the dangers of fire and explosion. Another school of thought emphasizes the danger of releases from long pipe runs and considers specific maximum spacing between the units to minimize such long runs. This approach applies more to non-flammable, acutely-toxic materials. In many cases, both fire and toxic release hazards are present and spacing design will be a compromise between

them. Both hazards can be addressed by planning for enough space (to satisfy
fire/explosion concerns) and by breaking up the resulting long pipe runs with
several emergency excess flow valves (to reduce the size of a release if the
pipe should rupture at any point).

3.3.2 Grouping or Isolation of Facility Units

How units are arranged within the facility is important, since this
partly determines whether the initial accidental release remains localized or
spreads throughout the facility processes. General guidelines for locating
units within a facility are noted below (13,17);

- Whenever possible, high hazard-units should be separated from each
  other by units of mild, low, or medium hazard.

- Control rooms, amenity buildings, workshops, laboratories and offices should be adjacent to units of mild or low hazard, which act as
  a barrier from higher-hazard units. Medium-hazard units are only
  acceptable adjacent to populated buildings (a) if lower-hazard units
  are not available to separate them, and (b) if the hazard level is
  only just inside the "medium" band of overall hazard rating, as
  assigned by the Mond Index.

- Storage units, such as tanks and container storage areas, should be
  adequately separated from operational areas and, as far as possible,
  located away from road and rail traffic routes within the plant
  site.

- Major pipebridges with a medium to high overall hazard rating
  assigned by the Mond Index method should be protected from accidents
  that could happen on tall process units and from vehicle collisions.
  Incidents on tall process units can spread by the collapse of the
  tall unit onto adjacent pipebridges.

3.3.3 Transportation

Planning an adequate transportation system is an integral part of plant layout. An adequate transportation system mitigates the effects of an accidental release in two ways: 1) emergency responsiveness depends on ready access to the affected units to reduce the spread of the emergency in the plant, and 2) a timely evacuation of potentially affected populations reduces the number of people affected. Design principles of an adequate transportation system are described below (13,17).

- Roads should allow entry to the facility from at least two points on the site perimeter, preferably from opposite sides and onto different roads/streets.

- All units in the facility with a moderate or high fire risk should be accessible to for emergency vehicles from at least two directions.

- Control rooms should have access to an easy exit, since control room personnel may be the last plant personnel to leave the facility site.

- Facilities with access from heavily traveled super-highways should also have alternative emergency entrances.

3.3.4 General Layout

A few other recommendations appear below (13,17):

- Control rooms, amenity buildings, workshops, laboratories and offices should be close to the site perimeter and away from hazardous storage and process areas.

- Pipebridges should be laid out to minimize the transfer of accidents from one unit to another. Key pipebridges should be assessed individually for hazard potential, independently of adjacent units.

- As far as possible, units should be laid out for a logical process flow to minimize the pipebridge requirements.

- Pipebridge routes should be chosen so that they are not likely to contribute to the spread of an accident. Alternative routes should be selected to retain as much process control as possible to be retained in the event of an incident occurring.

- A rectangular block layout is often used for chemical and petro-chemical plants so that emergency response vehicles can gain access to all units. Roads can also act as a fire break.

- Water systems should be adequate in all parts of the facility. For a larger facility, the water system may consist of a number of loops, rather than one large loop.

- Sewers and drainage should be laid out to allow rapid and safe removal of spilled chemicals and water used in fire fighting.

- Critically important facility elements (power plants, computer and control rooms, special process units) should be given maximum protection, which might include extra space dividing them from other units, proper location, or barricades.

- Hazardous units in the facility should be accessible to fire stations and other emergency facilities.

- Waste disposal and hazardous units processing, storing, or disposing
  of toxic gases and liquids should be located downwind of concentrations of people on site and off site.

For a detailed discussion of facility layout, refer to Lees (3) and
Mecklenburgh (19).

3.4 COSTS

The cost of incorporating mitigation measures into facility siting and
layout cannot be estimated quantitatively since the costs will be specific to
the site. A general discussion of the types of costs to be encountered is
presented here.

Mitigation measures such as increased spacing between units increase
facility construction costs in two ways. Most directly, the facility might
require additional land to create adequate spacing between all units. Also
considered a mitigation cost is the additional expense of longer piping and
utility runs within the facility. Additional spacing between units will
result in increased pipe length and longer utility lines.

Directly measurable are the costs of including additional land as a
buffer for the facility, which is a function of land prices and of the size of
the area designated as a buffer.

It is difficult to estimate the cost of designing mitigation measures
into the facility siting without considering a particular site and facility.
Sites rated high according to mitigation criteria (i.e, not upwind of a
populated area, for example) would not necessarily cost more to develop than
sites selected according to traditional criteria. However, the initial site
cost is only a portion of the facility costs. If transportation costs
increase because of a greater distance to market, there is a direct cost to
the firm for locating the facility in the site selected for mitigation

purposes. To evaluate the overall effects mitigation will have on costs, an analysis of the facility's construction and operating costs would be needed. Comparing the total costs at two sites, one selected according to mitigation criteria, the other according to traditional siting criteria, is a way to determine how the incorporation of specific mitigation measures affects overall costs.

3.5 REFERENCES

1. American Insurance Association: "Hazard Factors" in Hazard Survey of the Chemical and Allied Industries. Engineering and Safety Service. Technical Survey No. 3, 1979.

2. Davis, D.S., G.B. DeWolf, and J.D. Quass. Prevention Reference Manual: Control Technologies. Vol. 1. Prevention and Protection Technologies for Controlling Accidental Releases of Air Toxics. EPA-600/8-87-039a (NTIS PB87-229656), U.S. Environmental Protection Agency, Research Triangle Park, NC, August 1987. (Part II of this book.)

3. Lees, Frank P. Loss Prevention in the Process Industries. Volumes 1 and 2, Butterworth's, London, England, 1983.

4. Miller, C.F., F.A. Leone, and W. Bryan. Liquified Gaseous Fuels Spill Test Facility: Site Evaluation Study. Lawrence Livermore National Laboratory, February 1985.

5. Urang, Sally. How Experts Pick a Chemical Plant Site. Chemical Business, July 26, 1982. pp. 17-22.

6. McDavid, Catherine. "Prevention of Environmental and Public Health Damage from Major Industrial Accidents through Innovative Siting Approaches" in Avoiding and Managing Environmental Damage from Major Industrial Accidents. Air Pollution Control Association Conference, Vancouver, Canada, November 1985. pp. 237-251.

7. RPC, Inc. Siting Industrial Facilities on the Texas Coast. Texas Coastal Management Program, Texas General Land Office, September 15, 1978.

8. Shrivastava, Paul. The Accident at Union Carbide Plant in Bhopal: A Case Study in Avoiding and Managing Environmental Damage from Major Industrial Accidents. Air Pollution Control Association Conference, Vancouver, Canada, November 1985. pp. 90-95, 68.

9. Pesticide Plant Leak Wreaks Disaster in India. Nature, Vol. 312 (5995): 581, December 13, 1984.

10. New Jersey Statutes Annotated. 13:1E-49.

11. Nuclear Regulatory Commission. Reactor Site Criteria, 10 CFR 100, pp. 860.

12. Health and Safety Executive. Canvey: An Investigation of Potential Hazards from Operations in the Canvey Island/Thurrock Area. London, England, May 1978.

13. American Insurance Association. Hazard Survey of the Chemical and Allied Industries. Engineering and Safety Service, Technical Survey No. 3, 1979.

14. National Fire Protection Association. Flammable and Combustible Liquids Code, NFPA 30. Quincy, MA, 1984.

15. Battelle Columbus Division. Guidelines for Hazard Evaluation Procedures, American Institute of Chemical Engineers, The Center for Chemical Plant Safety, New York, NY, 1985.

16. Kletz, T.A., Plant Layout and Location: Methods for Taking Hazardous Occurrences into Account. Loss Prevention, Volume 13, American Institute of Chemical Engineers, 1980.

17. Lewis, D.J., The Mond Fire, Explosion and Toxicity Index Applied to Plant Layout and Spacing. Loss Prevention, Volume 13, American Institute of Chemical Engineers, 1980.

18. Ridner, R.M. and S. Wachtell. Establishment of Design Criteria for Safe Processing of Hazardous Materials. Loss Prevention, Volume 7, Chemical Engineering Progress, New York, NY, 1973.

19. Mecklenburgh, J.C. Plant Layout. Leonard Hill, London, England, 1973.

4. Detection and Warning Systems

Detection and warning systems are widely used in the chemical process industries to alert plant operators and personnel of process upsets and potentially hazardous situations that could lead to loss of production, property, or in the extreme case, human life. Common to most chemical processing facilities are detection and warning systems built into the process control systems. Such systems monitor process operating conditions such as the temperature, pressure, and flow rate, and trigger audible and visual alarms when these process variables exceed design limits. Other detection systems identify hazards after a release from a process or storage tank has occurred. These systems may detect leakage of the chemical itself, flame, heat, smoke, or gaseous products of combustion (1,2). Plant personnel also play a role in detecting and warning of hazardous releases. However, in the absence of a visible plume or fire, the reliability of plant personnel depends greatly on the particular material being released, ambient conditions, and the individual's physical senses. Reliance on plant personnel for detection and warning can result in dangerous exposure of plant personnel to hazardous materials. Visual observations, however, either directly by personnel in the vicinity of the release or by personnel monitoring closed-circuit television systems, can provide information on the direction and speed of a hazardous release (3).

This section discusses systems used to detect and quantify concentrations of leaks or releases of chemicals to the atmosphere after the release occurs but before an actual fire and/or explosion. Because of the large number of sampling and analytical methods available for collecting and analyzing hazardous gases and vapors, the discussion is limited to direct reading instruments or indicators, which, along with their sensing elements, can yield test results in the short time needed to quickly respond to a hazardous release incident. Post release detection systems are important because the more quickly an airborne release of a hazardous material is detected, the greater the opportunity to minimize the effects on the community by implementing on-site and off-site emergency response procedures.

4.1 BACKGROUND

The purpose of a gas or vapor detection system is to provide a quantifiable indication of the concentration and location of a released chemical that could lead to fire, explosion, or exposure of facility personnel or the general public to hazardous concentrations of an airborne chemical. With adequate warning from an automatic or manually operated warning system, facility personnel and off-site emergency response personnel can take appropriate action to reduce the effects of the chemical release. Specific mitigation procedures taken in response to a leak or release are discussed in other sections of the manual.

The general types of gas detection systems include (1,4,5):

- Multiple devices for sampling the air at various points throughout the facility, using vacuum pumps and tubing which then convey the sample to a common analyzer;

- Portable sampling/analyzing/monitoring devices that are battery or manually operated and carried by plant personnel to determine the concentration of a gas at a particular point of interest in the facility;

- Devices where the sensing element and readout are contained in a single unit mounted in the area of the hazard;

- Multiple sensing elements wired to a single control panel located at a point remote from the area being monitored; and

- Remote optical systems.

These systems often include an audible or visual alarm that activates when it detects a release (2). Audible alarms include bells, whistles, sirens, horns, or easily understood voice messages. In high-noise areas, visual alarms in the form of lights are often used.

Once an alarm signals that a fire or release has occurred, the emergency must be analyzed, and facility field personnel and/or the process operator in the control room must take appropriate action to manually control the release. In some cases, the detection system may activate mitigation devices such as steam curtains or automatically released firefighting agents (1). A fully automated computer system may be able to detect the release, determine whether the release constitutes a hazard, display the location of the release, sound warning alarms, provide instruction concerning mitigation of the release to the operator, or automatically engage control devices, and telephone emergency messages to the surrounding community public service agencies (1,2,3). However, the danger and the damage that could occur if emergency response procedures were falsely activated means that an automated system must be highly reliable.

4.2 DESCRIPTION

Detection instruments use a variety of operating principles, including (6):

- Electrical conductivity, potentiometry, calorimetry, and ionization;

- Radioactivity;

- Thermal conductivity and heat of combustion;

- Electromagnetic methods such as infrared photometry, ultraviolet photometry, and other photometric methods;

- Chemi-electromagnetic such as colorimetry and chemiluminescent methods;

- Magnetic methods such as paramagnetic analysis and mass spectroscopy; and

- Gas chromatography.

In cases where releases of toxic chemicals occur at remote locations, such as during transportation accidents, direct reading colorimetric indicators are sometimes used for qualitative and quantitative measurements of the released chemical. Methods include liquid reagents, chemically treated papers, and glass indicating tubes containing solid chemicals that change color and/or exhibit a length of stain that can be compared with a standard to determine the concentration of the toxic gas (7). These types of manually operated indicators are used in addition to other direct-reading instruments based on the principles previously cited (8).

A detailed description of the numerous detection principles and instruments available for detecting flammable and/or toxic gases is included in References 6, 7, and 8. The following subsections are summary discussions of the types of instruments typically used to detect flammable and hazardous gases associated with an accidental release.

4.2.1 Combustible/Flammable Gas Detectors

Combustible/flammable gas detectors are commonly used to measure the lower explosive limit (LEL) or lower flammable limit (LFL) of flammable gases in air. (According to Reference 1 the terms are used interchangeably.) The

below which the mixture cannot be ignited to yield a self-sustaining flame
(9). Examples of detectors are catalytic combustion sensors, solid-state
electrolytic sensors, and infrared analyzers (10). Table 4-1 presents the
characteristics of these and other combustible/flammable gas detection
systems.

Catalytic combustion sensors use a sensing element such as an electrically heated platinum wire across which a sample of gas is drawn. The
flammable gas burns on the wire, which produces heat in direct proportion to
the concentration of the flammable gas in the air. The temperature of the
wire changes the electrical resistance. Electronic circuitry in the
instrument creates a readout indication as a percent of the LEL or percent
concentration. If the LEL is exceeded, alarm circuits and warning signals may
be activated to alert personnel of dangerous concentrations of gas in the
vicinity.

In a solid-state electrolytic sensor, the gas diffuses into the sensing
element (a semiconductor), resulting in a decrease in the electrical resistance of the cell (10). The current flow resulting from this diffusion is
related to the concentration of the gas in the air, which is shown on a meter
as percent LEL.

Infrared analyzers draw air samples into a cell where an infrared light
shines perpendicular to the flow of the liquid. A detector on the side of the
cell opposite the light source measures the intensity of the light. The gas
of interest absorbs infrared radiation at a certain wavelength ranging from 2
to 15 micrometers (10,11). Thus, the concentration of the gas in air will be
proportional to the amount of infrared light absorbed. Two types of infrared
analyzers are available: the nondispersive type that operates at a specific
wavelength, and the dispersive type that can operate at several different
wavelengths. Infrared analyzers may draw air samples from several locations
or they may be set up to monitor only one location.

TABLE 4-1. COMPARISON OF CHARACTERISTICS OF VARIOUS COMBUSTIBLE/FLAMMABLE VAPOR DETECTION TECHNIQUES

Characteristic	Gas Chromatography	Hydrogen Flame Ionization Detector	Thermal Conductivity	Interferometry	Infrared	Semiconductor[a]	Catalytic Combustion
Sensitivity	Excellent	Best	Poor to Good	Good	Very good	Excellent	Very good
Reliability	Good	Good	Good	Good	Good	Poor	Very good
Selectivity	Excellent	Very Good[b]	Poor to Good	Poor[c,d]	Very good[b]	Poor[d]	Very good
Response time	Poor	Good	Good	Good	Good	Excellent	Excellent
Stability	Good	Very Good	Poor to Good	Good	Very good	Good	Good
Simplicity	Very complex	Medium complexity	Simple	Medium complexity	Medium complexity	Very simple	Very simple
Sample system required	Yes	Yes	Yes	Yes	Yes	No	No
Relative cost	Very high	Medium to high	Medium	Medium	Medium	Low	Low
Rangeability	ppm. to 100%	ppm. to low %	Wide	Wide	Very Wide	Very wide	Below UEL[e]
Maintenance	High	Medium	Medium	Medium	Medium	Low	Low
Auxiliary gas supplies	Yes	Yes	Sometimes	No	No	No	No

[a] Diffusion head-type sensor.
[b] Will not detect hydrogen.
[c] Hydrocarbons and hydrogen of opposite polarity signals.
[d] Responds to products of emission.
[e] Upper explosive limit.

Source: Reference 5.

4.2.2 Toxic Vapor Detection Systems

Detecting toxic vapors requires much greater sensitivity than is required
for detecting flammable vapors. Detection systems for toxic vapors often rely
on detectors that respond to a specific chemical. Less often, a single device
can be used to monitor a variety of hazardous vapors. These devices measure
the concentration of the toxic gas and allow the operator or automated system
to compare the measured value with some set ceiling value. If this
concentration value is exceeded, an alarm system is activated.

The design of detection systems is based on a number of principles and
methods, including optical absorption, photoionization, flame ionization, mass
spectroscopy, gas chromatography, pH, measurement, and infrared
spectrophotometry (12). Detailed descriptions of operating principles and
available instruments appear in References 6, 7, and 8. Before committing to
any type of detection system, discussions should be held with equipment
vendors or manufacturers to determine the correct application of the detection
device to the vapors or gases to be monitored. For vapors and gases with both
a lower explosive limit (LEL) and an employee exposure limit, an instrument
capable of measuring the LEL may not be appropriate for measuring the
exposure limit concentration. The LEL may cover only a single decade of vapor
concentrations, for instance 1 to 10 percent, while the toxicity-based
exposure limit may include concentrations from fractions of a part per million
(ppm) to several thousand ppm (5). The wide range of concentrations may
require different types of detection systems.

For example, Table 4-2 lists the various types of instruments available
from one manufacturer for monitoring several toxic gases. A survey instrument
is typically a portable, battery-powered device used to locate leaks or to
measure ambient concentrations of toxic gases for a short time. Personal
sampling instruments are devices that are typically worn by personnel who may
be exposed to high concentrations of toxic gases. These devices are usually
equipped with alarms to warn the wearer when dangerous concentrations are

TABLE 4-2. EXAMPLES OF INSTRUMENTS AVAILABLE FOR VARIOUS TOXIC CHEMICALS[a]

Substance	Survey Instrument	Personal Sampling Instrument	Area Monitoring Instrument
Ammonia	Autospot	—	PSM-8 Multipoint Monitor, PSM-8e Multipoint Monitor, Series 7000 Continuous Monitor, Series 7100 Continuous Monitor
Chlorine	Autospot	Monitox/Chronotox	PSM-8 Multipoint Monitor, PSM-8e Multipoint Monitor, Series 7000 Continuous Monitor, Series 7100 Continuous Monitor, Statox Multipoint Monitor
Hydrazine	Autospot	Monitox/Chronotox	Series 7000 Continuous Monitor, Series 7100 Continuous Monitor
Hydrogen Cyanide	Autospot	Monitox/Chronotox	PSM-8 Multipoint Monitor, PSM-8e Multipoint Monitor, Series 7100 Continuous Monitor, Statox Multipoint Monitor
Hydrogen Fluoride	—	—	PSM-8e Multipoint Monitor, Series 7100 Continuous Monitor
Hydrogen Sulfide	Autospot	Monitox/Chronotox	PSM-8 Multipoint Monitor, PSM-8e Multipoint Monitor, Series 7000 Continuous Monitor, Series 7100 Continuous Monitor, Statox Multipoint Monitor, Statox E Multipoint Monitor
Phosgene	Autospot	Miniature Continuous Monitor, Monitox/ Chronotox	PSM-8 Multipoint Monitor, PSM-8e Multipoint Monitor, Series 7000 Continuous Monitor, Series 7100 Continuous Monitor, Statox Multipoint Monitor
Sulfur	Autospot	—	PSM-8 Multipoint Monitor, PSM-8e Multipoint Monitor, Series 7000 Continuous Monitor, Series 7100 Continuous Monitor
Toxic Vapors (General)	Vacu-Sampler	Accuhaler, Low Flow, Sampling Pump	—

[a] Note: The instruments listed are manufactured by MDA Scientific, Inc. Their listing does not constitute an endorsement.

reached. Area monitoring instruments are designed for continuous monitoring
at fixed locations. A single area monitor can sample several locations if a
switching valve and a sampling manifold are used.

4.2.3 Remote Monitoring

The previous detection methods are applied at a specific single location
or set of locations within a facility. However, a number of techniques have
been developed for remotely detecting hazardous vapors and gases. Remote
detection uses sensors not physically located at the point of release or in
the area directly covered by the plume or cloud.

Most of these systems use some form of light beam directed through the
plume or cloud. These detectors cover an area rather than a single location.
A variety of remote detection systems are available (4,8). Two of the most
widely used laser-based optical monitoring technique are the Differential
Absorption Light Detection and Ranging (LIDAR, also abbreviated as DIAL) and
the Raman LIDAR remote detection technique. Figure 4-1 is a schematic
representation of these systems.

The DIAL technique uses two lasers at different wavelengths. The first
wavelength coincides with band peak of the target gas and the other coincides
with a band where the gas does not absorb. The lasers are positioned so that
scattered or reflected light from natural or positioned targets is detected by
a receiver, which determines the concentration of the target gas by comparing
the wavelengths of the reflected light (since the concentration is
proportional to the difference between the return signals at the two specified
wavelengths).

The Raman LIDAR technique differs from the DIAL technique in that only
one laser beam is used. The advantage of this technique is that the beam does
not have to be tuned to any specific wavelength. Operation of the Raman LIDAR
technique is based on the phenomenon that light scattered by gas molecules

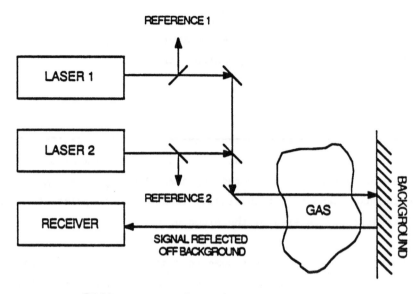

DIAL remote detection system

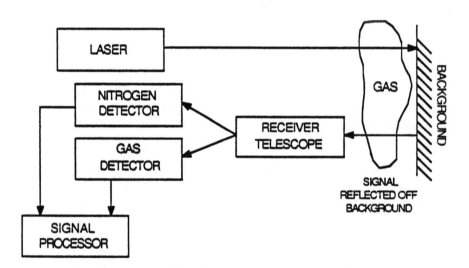

Raman LIDAR remote detection system

Figure 4-1. Schematic representations of typical remote monitoring techniques for hazardous vapors.

produces a shift in wavelength. The wavelength shifts that occur are independent of the incident wavelength and are characteristic of specific compounds. Signals reflected off a background are collected by a receiving telescope and are sent to a detector for concentration determination. The Raman frequency of nitrogen in the air is used as a reference frequency, which allows the system to discriminate between scattering from the target gas and that from other atmospheric compounds.

References 4 and 8 discuss in more detail these and other remote detection systems.

4.3 DESIGN

The design of an effective detection and warning system for hazardous vapor releases must consider the following (5,13,14,15):

- Selection of the proper detection equipment (for the vapor of interest and its particular hazard) which meets the appropriate code or is approved for use within the facility environment;

- Location of sampling points and positioning of detectors;

- Installation of the detection system;

- Type of alarm system; and

- Maintenance, checking, and calibration.

The selection of equipment is based on the likely concentrations of the specific chemicals to be detected. In some situations, combinations of devices might be required. The selection process should include information from equipment manufacturers and require certification that the equipment will perform properly in the given application.

Equipment used should meet the National Electrical Code and have Factory Mutual, Underwriters Laboratory, Canadian Standards, or Bureau of Mines approval for use in Class 1, Division 1, Group A,B,C, and D classified areas (14).

Detector and sampling points should be located either around the perimeter of a potential source or at the source. Sources may include pumps, valves, flanges, compressor seals, rupture discs, vents, pressure relief valves, storage containers, stacks, and ventilation and air conditioning system intakes (5,14). The location of detectors or sampling points should take into consideration the local wind conditions and structures that may affect air flow in the immediate vicinity of the source. Multiple detectors should be located no more than 30 feet apart (13). One reference suggests that sampling points or detectors should be located 1-1/2 feet above ground for heavier-than-air combustible compounds and 6 to 8 feet above ground for lighter-than-air combustible compounds (13). For monitoring measurements of work place exposure detection points should be selected at the height of breathing levels (5).

Although most equipment is designed for simple installation, care must be taken to follow the manufacturer's instructions concerning proper installation and coordination with other equipment. When sampling lines are used, these should be of a material that will neither contaminate nor degrade the sample being routed to the analyzer. Detection sensors and sampling probes should be protected from the weather, including extremes of temperatures (5). Installation of equipment should be located so that maintenance can be performed easily.

Alarm systems can vary greatly in complexity. A simple alarm system may provide visual and/or audible warning signals once a single or multiple preset concentration of vapor or gas is exceeded. Facility personnel can respond appropriately. Complex alarm systems may initiate automatic shutdown of the process or signal automatic mitigation efforts.

Maintenance should be conducted periodically according to the manufacturer's instructions. The system should be checked daily to ensure that the detector or sampling lines are not plugged. Calibration and adjustment of equipment should be performed periodically with vapor or gas standards of known concentrations. Proper operation and maintenance of the equipment requires proper training of facility personnel. This training may be included in the contract for the equipment, where plant personnel can learn the proper calibration and maintenance techniques from the supplier actually performing the maintenance during the equipment's warranty period (13).

4.4 APPLICABILITY AND PERFORMANCE

A variety of equipment is available to monitor flammable and/or toxic vapors from releases within plant environments and to use at remote chemical spill locations. As indicated in Subsection 4.2, these systems operate according to numerous detection principles. References 6 and 7 include detailed descriptions of available equipment. In these references, approximately 150 different direct-reading instruments and colorimetric indicators are described and the operating principles, performance data, and inferences are discussed. In addition, Reference 8 discusses several single-component and multicomponent vapor monitoring devices for monitoring hazardous releases from railroad accidents.

4.5 RELIABILITY

The reliability of detection and warning systems depends primarily on proper system maintenance and on visual and functional tests. For catalytic combustion detector elements, a 3 percent failure per month can be expected (14). For the portion of the instrument where adjustments for the span, zero, and alarm set points can be made, an 0.5 percent failure per month can be expected (14).

The reliability of detection systems may be affected by contaminants. Sulfur compounds may form acids that permanently damage the sensor catalyst, while halides can coat the sensor and cause a loss of sensitivity (14). For other types of detection systems, there may be chemical compounds that inter-fere with the accurate detection of the hazardous vapor being analyzed.

4.6   SECONDARY HAZARDS

Because of potential fire hazards, the use of detection and warning sys-tems should follow all applicable electrical codes. Where necessary, systems should be approved for use in classified hazardous areas.

Personnel maintaining and checking release alarm situations should wear approved safety equipment to prevent exposure to hazardous concentrations of vapors and physical injuries.

An indirect secondary hazard of detection and warning systems is that they may engender complacency. These systems should not be substitutes for proper accidental release prevention measures.

4.7   COSTS

The costs of detection equipment vary widely, depending on whether the device is monitoring a single vapor or gas or several. Some typical estimated costs of different types of detection systems are presented in Table 4-3 (4,8,16).

TABLE 4-3.   EXAMPLES OF ESTIMATED COSTS

Equipment	Capital Cost ($)
Manually operated portable sampling pump for colorimetric indicating tubes	135
Battery powered portable combustible gas analyzer	300 - 7,000
Multicomponent monitoring devices (gas chromatographs and infrared analyzers)	5,000 - 35,000
Remote area-wide detection systems (laser based, infrared, and ultraviolet systems)	15,000 - 175,000
Mobile mass spectrometer	500,000

4.8   REFERENCES

1. Lees, F.P. Loss Prevention in the Process Industries. Butterworth's London, England, 1983.

2. Soden, J.E. Basics of Fire-Safety Design. Fire Protection Manual for Hydrocarbon Processing Plants, Volume 1, 3rd Edition. Gulf Publishing Company, Houston, TX, 1985.

3. Prugh, R.W. Post-Release Mitigation Design for Mitigation of Releases. Presented at International Symposium on Preventing Major Chemical Accidents, American Institute of Chemical Engineers, New York, NY, February 1987.

4. Atallah, S. and E. Guzman. Remote Optical Sensing of Fire and Hazardous Gases. Presented at International Symposium on Preventing Major Chemical Accidents, American Institute of Chemical Engineers, Washington, D.C., February 1987.

5. Dailey, W.V. Area Monitoring for Flammable and Toxic Hazards. Loss Prevention, Volume 10, American Institute of Chemical Engineers, New York, NY, 1976.

6. Nader, J.S., et al. Direct Reading Instruments for Analyzing Airborne Gases and Vapors. Air Sampling Instruments for Evaluation of Atmospheric Contaminants, 6th Edition. American Conference of Governmental Industrial Hygienists, Cincinnati, OH, 1983.

7. Saltzman, B.E. Direct Reading Colorimetric Indicators. Air Sampling Instruments for Evaluation of Atmospheric Contaminants, 6th Edition. American Conference of Governmental Industrial Hygienists, Cincinnati, OH, 1983.

8. Hobbs, J.R. Monitoring Devices for Railroad Emergency Responses Teams. Department of Transportation. Federal Railroad Administration Report Number DOT/FRA/ORD-86/02. Cambridge, MA, February 1986.

9. National Fire Protection Association. Fire Protection Handbook, 15th Edition. Quincy, MA, 1981.

10. Brown, L.E. and L.M. Romine. Flammable Liquid Gases. Hazardous Materials Spill Handbook. McGraw-Hill, New York, NY, 1982.

11. Rodgers, S.J. Commercially Available Monitors of Airborne Hazardous Chemicals. Hazardous Materials Spill Handbook. McGraw-Hill, New York, NY, 1982.

12. Rome, D. Personnel Safety Equipment, Hazardous Materials Spill Handbook. McGraw-Hill, New York, NY, 1982.

13. MDA Scientific, Inc. Toxic Substance Detection and Measurement
     Instrumentation. Catalogue No. 970358, October 1985.

14. Johanson, K.A. Design of a Gas Monitoring System. Loss Prevention,
     Volume 10, American Institute of Chemical Engineers, New York, NY, 1976.

15. St. John, K. Use Combustible Gas Analyzers. Fire Protection Manual for
     Hydrocarbon Processing Plants, Volume 1, Third Edition. Gulf Publishing
     Company, Houston, TX, 1985.

16. Schaeffer, J. Use Flammable Vapor Sensors? Fire Protection Manual for
     Hydrocarbon Processing Plants, Volume 1, Third Edition. Gulf Publishing
     Company, Houston, TX, 1985.

17. Telephone Conversation between R.C. Keeney of Radian Corporation and a
     representative of Mine Safety Appliances Company, Houston, TX, May 1987.

5. Vapor Dispersion Modeling

Mathematical models that accurately simulate the movement of a hazardous vapor cloud from an accidental release are valuable for predicting the effect of various accidental release scenarios and, to a limited extent during an actual release event, for determining whether community evacuation is required. Models are available today that support both of these tasks. However, the usefulness of a model during an actual release event is often limited by a lack of information and by the time required to activate the modeling process. Vapor dispersion models should not be used as the only source of information for decision making during an actual release event, but they can help the emergency response expert choose the appropriate response for the specific situation.

This section of the manual discusses vapor dispersion models for predicting the impact of an accidental release. Types of models are discussed in terms of their applicability, performance capabilities, and reliability.

5.1 BACKGROUND

Mathematical vapor dispersion modeling to predict the potential impact of an accidental release is a relatively new field. Traditional vapor dispersion modeling has focused on the atmospheric dispersion of pollutants from an elevated stack. Such models deal with low concentrations of pollutants in air continuously emitted over a long period of time. Models for predicting the effects of an accidental release, however, must be able to handle short-term releases at both high and low concentrations and at variable release rates. They must be capable of modeling a release/dispersion of heavier-than-air and lighter-than-air materials, and materials that have the

same density as air. Such models should simulate a variety of possible
release forms, such as a release from a boiling pool of liquid, or a release
from a hole in a pressurized vessel.

One purpose of a vapor dispersion model is to predict the area that might
be adversely affected by the vapors from a release. The adverse effects will
depend partly on the properties of the released chemical. For flammable
materials, the presence of vapor concentrations within the flammable limits of
the material are of concern because such concentrations can be ignited. For
toxic materials, the presence of vapor concentrations greater than the
compounds IDLH (immediately dangerous to life and health) will be of concern
(1).

One modeling method used to predict vapor cloud movement is experimental
physical modeling, such as in wind tunnels or water channels. In these
experiments, a scale model is used to simulate the release, using air and the
actual chemical of interest or liquid substitutes. Some success has been
achieved using these methods (2,3,4); however, the rest of this section
focuses on mathematical models because of their convenience and intense
development in recent years.

Mathematical vapor dispersion models may be used for two purposes: to
assess hazards and plan an emergency response, and to provide emergency
response guidance information during an actual accidental release. Modeling
may be used to predict the effects of various accidental release scenarios and
to estimate which scenarios present the greatest threat to plant personnel and
to the community. Modeling may be used to estimate what portions of the
community might be affected during various accidental release scenarios. This
information can be used to develop community emergency response plans. Models
may also be used during an actual release event to help the emergency response
coordinator make decisions about what action should be taken if a release
occurs.

A distinction should be made at this point between individual dispersion models and what could be called emergency response modeling packages. Many of the models sold commercially would come under the category of emergency response modeling packages. Such a package will typically contain a set of source models, a dispersion model with a number of submodels to handle non-idealities, a database with chemical property data for a variety of potential release candidates, the ability to use real-time meteorological data if an actual release occurs, and a visual display that shows the physical motion of the cloud. Such packages are often customized for specific sites with a map of the area shown under the released vapor cloud. Some packages can be set up to communicate directly with detection and emergency response notification systems.

Some mention of "real-time" response models should be also be made. For a model to be "real time" it must react or change its simulation as actual events occur. In most accidental release situations, a model is activated some time after the release has begun (from minutes to hours). Information describing the actual event, necessary for model input, may also not be available for some time after the release has begun (if at all). A true "real-time" model would have to be monitoring sensors at possible event sites. When it appears that a release may be imminent, the calculations are activated. The necessary data would be obtained by the model itself. Such a system would be quite complex, and no such system is known to exist. The existing "real-time" response models usually only use real-time meteorological data in a predefined release scenario.

5.2 DESCRIPTIONS

Two key components of models must be available to fully describe an accidental chemical release: a source model component and a vapor dispersion model component.

5.2.1 Source Models

A source model is required to calculate the effect of the initial release conditions on vapor dispersion. For example, a release from a refrigerated storage vessel containing a liquefied gas will behave differently than a release of the same material from a pressurized, non-refrigerated storage vessel. The difference between these two releases is not a difference in the material released, but in its physical state. Source models must be used to calculate the effects of these differences. Examples of the types of information generated from a source model which becomes information that a dispersion model can use include the following: the rate of vapor generation, the height of the release, the initial velocity and temperature of the vapor, and the initial concentration of the released vapor.

Each type of release will require a different set of mathematical equations to define the nature of the release. The physical state of the released material and the way in which the loss of containment occurs will both affect the calculations required to define the release. Examples of possible physical states of released materials are:

- Release of a pressurized liquefied gas;

- Release of a chilled liquefied gas, stored at atmospheric pressure;

- Release of a pressurized gas;

- Release of a liquid stored at ambient or elevated temperature and pressure; and

- Release of a two-phase slurry.

Some examples of different ways a release may occur are:

- Release from a ruptured pipe;

- Release from a hole in a vessel;

- Release from an activated pressure relief device;

- Sudden release of the entire contents of a vessel;

- Opening a valve that results in a release from a non-blanked pipe; and

- Any of the above releases within a diked area, or any of the above released within an enclosure.

A wide variety of release scenarios can be modelled by combining the physical state of the released material with the conditions under which the release occurs. Figure 5-1 illustrates a few possible release situations. Ideally, a source model will be able to take these combinations and ambient weather conditions and determine whether a boiling pool of liquid is formed, whether the release is entirely gaseous, whether an aerosol is formed or if the release is some combination of each. This information is then used to define the initial conditions of the vapor from the release.

Source models can be obtained either as part of a model package that contains both source and dispersion models, or as separate packages. Table 5-1 gives the names of some source models, although this list is not exhaustive. Examples of modeling packages that contain both source and dispersion models are presented in Subsection 5.2.2 on dispersion models. Additional information about these models will be discussed in Subsections 5.3 and 5.4.

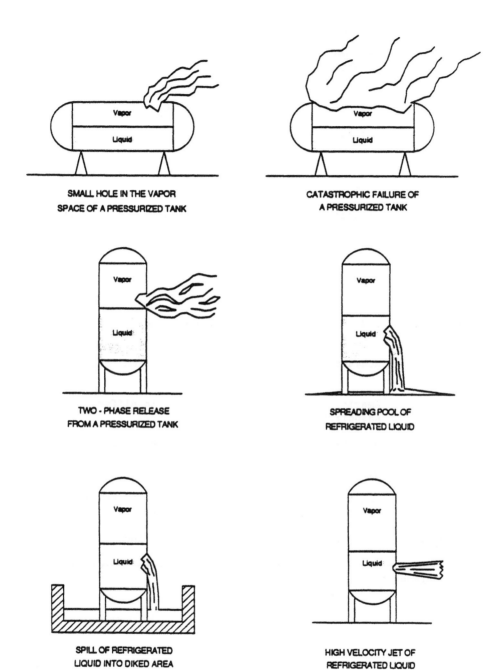

Figure 5-1. Potential release scenarios.

TABLE 5-1.   EXAMPLES OF SOURCE MODELS

Authors	Sponsor	Application
Tang et al (1980)	U.S. Coast Guard	Oleum/H_2SO_4 spills on water
Clewell (1983)	U.S. Air Force (ESL)	Parameters presented for approximately 25 chemicals
Shen (1982)	New York State Dept. of Env. Conservation	Volatilization from exposed water surface
Whittaker (1977)	Alberta Environment	H_2S risk assessment modeling
Kahler et al (1978)	U.S. Air Force	Specific to N_2O_4 (pool evaporation)
Whitacre and Myriski (1982)	U.S. Army	Toxic liquid spill (pool evaporation)
Ille and Springer (1978)	U.S. Air Force	Dispersion modeling of propellant spills (pool evaporation)
Wu and Schroy (1979)	Monsanto Company	Pool evaporation, chemical spills
Reid and Wang (1978)	American Gas Association	LNG spills in different substrates
Shan and Brisiac (1978)	U.K. Safety and Reliability directorate	LNG applications with suggestions other liquids
Reid and Smith (1978)	California Public Utilities Commission	Estimates of LNG release rates for subsurface spills
Georgakis et al (1979)	Distrigas Corp.	LNG and Gasoline

Source: Reference 5

5.2.2   Dispersion Models

Of the two extreme approaches for modeling vapor dispersion, attempts to solve the problem exactly by using differential equations that describe each process affecting the motion and dispersion of the vapor in air. The equations are integrated with time over three-dimensional space. This type of model is called a three-dimensional numerical model.

The second extreme is one that makes simplifying assumptions for each differential term used in the three-dimensional numerical model until the equations can be combined and simplified to form a single equation.

A large number of models can be formed with characteristics of both these extremes. By making a limited set of simplifying assumptions, models can be created whose solutions cover a specific range of accidental release cases. In this manual, such models are called similarity models.

Similarity models --
Similarity models avoid the complexity of specifying the details of how the physical properties of the cloud vary from point to point and, instead, are written in terms of the bulk or integrated properties of the cloud. Instead of calculating how the velocity, concentration, and other characteristics change along a stream line, they determine how these bulk properties change as the cloud moves downwind. These models generally assume that the material behaves as a massless quantity that in no way influences the turbulent atmosphere in which it disperses. One of the best known examples of this type of model is the gaussian plume model.

When the cloud formation time is very small compared with the time of cloud travel, the cloud may be approximated by the formation of an instantaneous cloud or "puff", as it is typically referred to. If the release is continuous, some similarity models will approximate the release by modeling a series of instantaneous "puffs." The concentration at any point may be determined by summing the contributions at the point from each of the puffs.

By making the assumption of continuous source strength and a constant wind speed and direction, this summing can lead to a single expression describing the concentration anywhere in the plume. This expression is known as the continuous Gaussian plume model. By using puffs, noncontinuous time-dependent source strengths and meteorology can be simulated (5,6).

In practice, similarity models often solve differential equations with simplifying assumptions. For example, a similarity model may account for the change in the volume of a puff caused by entrainment by using differential equations of motion and entrainment, but make the simplifying assumption that the concentration in the puff is Gaussian in nature.

A wide number of similarity models are available both publicly and from commercial sources. The differences, or lack of difference, between these models was summarized by one author as follows (7);

The current generation of public domain and commercially available source or dispersion models is not notably innovative in formulation. This is a result of the current state of the science of dispersion meteorology. Differences among models do not represent innovation, but are related to the selection and combination of different theoretical expressions for efficient programming or the use of different frameworks for data handling and presentation.

Table 5-2 presents examples of similarity dispersion models. Table 5-3 presents examples of modeling packages that contain source and similarity dispersion models. Table 5-4 shows examples of commercial modeling packages that use similarity dispersion models. These lists are not exhaustive. Additional information is presented in Subsections 5.3 and 5.4.

Three-dimensional numerical models --
    In three-dimensional models, the distributions of velocity and concentration are determined by the solutions of the conservation equations of fluid flow and the prescribed boundary conditions. These models take a much more

TABLE 5-2. DISPERSION MODELS

Model Name	Sponsoring Organization
British Gas	Cremer and Warner
(Chlorine Institute)	Chlorine Institute (CI)
DENZ	U.K. Safety & Reliability Directorate (SRD)
DEGADIS	Dense Gas Dispersion Model U.S. Coast Guard
INPUFF	EPA Integrated Puff Model
(IEPA)	Illinois EPA
HEGADIS-II	Heavy Gas Dispersion from Area Sources Koninklijke/ Shell Lab
OB/DG	Ocean Breeze/Dry Gulch (U.S.A.F.)
EIDSVIK	Norwegian Institute for Air Research
SLUMPING	Royal Netherland Meteorological Institute
SLAB	Lawrence Livermore National (LLNL)
SRI PUFF	SRI International
WINDS	Savannah River Lab (SRL)

Source: Reference 5

TABLE 5-3.   SOURCE AND DISPERSION MODEL PACKAGES

Model Name	Sponsoring Organization
AGA (LNG)	American Gas Association (AGA)
AGA	Germeles & Drake
DDESB	Department of Defense Explosives Safety Board (DDESB)
ICARIS	Industrial Chemical Accident Response Information System of the Association of American Railroads
HACS	Hazard Assessment Computer System U.S. Coast Guard
FUMING ACID	Proctor & Gamble
(Standard Oil Co.)	Standard Oil Co. (Indiana)
TOXCOP	Toxic Corridor Prediction U.S. Army Atmospheric Sciences Lab (ASL)
SPILLS	Shell Development Co.

Source: Reference 5

TABLE 5-4.   COMMERCIAL MODELING PACKAGES

Model Name	Sponsoring Organization
Dispersion Models	
DAISY	Dispersion Analysis Information System, Dow Chemical
TRPUF	Trinity Consultants
Source and Dispersion Model Packages	
CEES/ERADAS	Energy Impact Associates
CHARM	Complex Hazardous Air Release Model Radian Corporation
COBRA	Concord Scientific Corporation Dames & Moore
CARE	Environmental Systems Corporation (ESC)
AIRTOX/HASTE	Environmental Research and Technology (ERT)
ETA	ETA, Incorporated
MIDAS	Pickard, Lowe, and Garrick, Inc.
MESOCHEM	IMPELL Corporation
SAFER	Safer Emergency Systems

Source: Reference 5

rigorous and theoretical approach to the diffusion problem than do similarity models. A primary use of these models is to predict the dispersion of a vapor cloud around obstructions and irregular terrain.

Several three-dimensional numerical models have been developed for use on toxic releases. All are experimental at this point. Additional information may be found elsewhere (6,8).

5.3   APPLICABILITY AND PERFORMANCE

The models presented in Tables 5-1 thru 5-4 are not all representative of the state of the art; some have been superseded by further developments. Also, some of these models were developed for very narrow applications and will be of limited use for general accidental release modeling. Some models were developed for a purpose that did not require the accuracy necessary for modeling accidental releases of toxic materials. These models incorporate simplifications that are not appropriate for accidental release modeling. The potential users must determine their requirements and select a model accordingly. One source has summarized the important questions in model selection as follows (7);

- What is the time frame in which results are required?

- Who is the intended user?

- What is the source configuration and what are likely accident scenarios?

- What are the chemical and physical characteristics of the material released?

- Does local terrain influence dispersive patterns and what is the local dispersion climate?

The present generation of models is most valuable as an emergency response planning tool. Models should not be used as the sole decision-making tool during an accidental release, and with the current limitations of "real-time" modeling they may be of little use during an actual event. Instead, model results should be used to help the emergency response expert during an accidental release situation.

Ideally, before a release occurs, single page reports containing response information for likely releases would be created by using a model. These single pages could then be used by nontechnical persons in an actual event until more information was available for a refined estimate of impact.

Only those models that use an empirical or simplified approach to model vapor dispersion (similarity models) are useful in the emergency response model because of the calculation time involved. These simplifications restrict the accuracy of the models. However, these same simplifications allow the models to fairly rapidly produce a result accurate enough for the purposes mentioned above. At this time, models that take a more rigorous approach (three-dimensional models) cannot produce rapid results. Also, these models require large quantities of data to achieve accurate results and such data may not be available.

Several potentially desirable modeling features are either not available, or are not fully developed at this time. Some of these features are discussed below:

- The effects of post-release mitigation measures such as foams applied to the spill surface, or water sprays;

- Adequate complex terrain models;

- Modeling of atmospheric turbulence;

- Models for certain types of sources, such as high-pressure jets, or two phase releases;

- The ability to model multi-component releases;

- Accounting for chemical reactions, such as the effect of atmospheric moisture on water reactive releases; and

- The effect of fire on the release and the dispersion of the result-ing fire decomposition products.

5.3.1 Source Models

Most source models used in conjunction with air models are designed to calculate the evaporation rate from pools of spilled liquid. These models calculate an evaporation rate based on mass and heat transfer equations. All evaporation source models apply some simplifying assumptions. Some models assume that the pool forms instantaneously and is of a fixed radius defined by the model user. Other models calculate the size of the pool as a function of time. Most models have restrictions on the range of materials that can be modeled. Additional simplifications made by most vaporization models include (5);

- Models are for single component systems;

- Ice formation in the soil is not considered;

- Percolation into the soil is specified (as opposed to calculated), or not considered; and

- Runoff from the spill area is specified or not considered.

For models that calculate the size of a spill pool, calculation of the rate of spilled material into the pool is based on user-defined spill parameters. Most often these parameters are the size and location of a hole in a vessel. The models will typically use an orifice flow equation to calculate the flow into the pool. Source models that handle vapor releases will also use some type of orifice flow equation to calculate the rate of vapor release.

The modeling of more complicated sources has not been incorporated into vapor dispersion modeling. Examples of complicated sources would be the release of two-phased flow or the release from a high-pressure jet. There are two reasons why these types of flow have not been incorporated into atmospheric dispersion models. First, the behavior of such releases is not well understood and the mathematics of the sources have not been developed. Second, the complexity of these sources would make it difficult to get mean- ingful results when used with similarity models. A complex source would probably need to be used with the more complex three-dimensional models, and three-dimensional models themselves are not yet practical for modeling acci- dental releases.

Another issue of concern when considering source model reliability is the selection of an appropriate source model for a given situation. Each poten- tial type of release will require a unique combination of calculation proce- dures to model the result of the release. The issue of source model selection is of particular concern when a release is being modeled as it occurs. In this event, the modeling package used must not only have a source model for the type of release that has actually occurred, but must be able to select that model and set all of the model parameters to be consistent with what has actually occurred. This shows one of the limitations of using dispersion models for modeling "real-time" events.

5.3.2 Dispersion Models

Similarity Models --

As was discussed above, similarity models use a simplified approach for solving the problem of the dispersion of gases from an accidental release. Even with these simplifications, similarity models offer the best available option for modeling such releases. The simplifying assumptions allow the models to operate rapidly enough on a computer to be used to predict the movement of a release even after the release has begun. All of the emergency response modeling packages sold commercially incorporate similarity models. As will be discussed below, modifications to these models have allowed some of them to handle some of the nonideal conditions encountered in real release situations.

A state-of-the-art similarity model, interfaced with a state-of-the-art source modeling package will produce accurate enough results to be used for the following purposes;

- By modeling a series of credible worst case releases, plant designers can determine whether the siting and layout of a potential plant is appropriate.

- Plant designers can also use these modeled events to evaluate what prevention, protection, or mitigation technologies might be appropriate at various locations within the plant.

- A series of credible worst-case releases can be modeled for an existing plant. Plant management and engineers can use the results to evaluate the same features mentioned above for a new plant. A heavier emphasis would be placed on prevention, protection, and mitigation than on siting and layout. The effect of some of these control measures could be tested using the model. For example, many models allow the effect of diking to be tested.

- An emergency response system that incorporates a modeling package could be set up to help the emergency response coordinator decide what actions would be appropriate in the event of an accidental release.

As has been discussed, similarity models are not accurate enough to serve as the sole basis for decision making during an actual accidental release event.

The assumptions and simplifications incorporated into similarity models impose basic restrictions on prediction accuracy. Individual models will incorporate varying degrees of complexity and varying degrees of accuracy. However, certain simplifying assumptions will be common among most similarity models. One source has summarized these assumptions as follows (5);

- Meteorological conditions are assumed to be constant during the simulation period, which is typically one hour. The effects of any systematic change or trend in wind speed, wind direction, or stability conditions during an hour would not be described by the model.

- Winds and turbulence are assumed to be the same at all locations throughout the boundary layer. The effects of wind speed, or wind direction shear, or of changes in turbulence with height are not considered within the Gaussian formulation.

- The basic model averaging time is assumed to be long, compared with the time scale of turbulent atmospheric motion and with the transport time from source to receptor.

- Pollutant mass is conserved within the Gaussian formulation. Processes that add or remove mass, such as deposition, decay, or chemical transformations, are assumed to be of secondary importance.

Many models have incorporated modifications in an attempt to improve the performance for non-ideal situations. Some incorporate mathematical expressions to account for the movement of gases that are heavier than air (many toxic materials fall into this category) (5,9,10). Methods have been developed to account for simple wind shear and for occasional changes in wind speed and direction (5). Some work has been done to allow for the deposition of materials from the vapor cloud, or for the time-dependent decay of the material. A few similarity models have incorporated methods for simulating the effect of certain types of complex terrain. All of these modifications are made using simplified approaches to complex phenomena. However, most of these phenomena have not been modeled using more rigorous approaches, and for the present time, the simplified approaches offer the only practical solution.

Naturally, the accuracy of predictions made by a similarity model will depend somewhat on the accuracy of the source model used. As a result, predictions made using an adequate source model can be misleading. Additionally, the way the source model interacts with the dispersion model may affect the overall accuracy.

Three-Dimensional Models --

Three-dimensional models can, in principle, simulate the dispersion of vapors as a function of space and time without the simplifications incorporated into the similarity models. These models should be able to account for the effects of terrain and wake turbulence. Although they appear to be theoretically correct, several factors limit their usefulness at this time.

First, the model theories and assumptions have been developed from an inadequate experimental data base. A variety of methods have been developed for handling the effects of turbulence on diffusion, yet the accuracy of the assumptions in each case are difficult to evaluate because of the present lack of experimental data (6).

Second, these models require large quantities of data, large computer storage and computational capacity, and long calculation times, features that make it difficult to run multiple comparative runs for planning purposes, and that make their use during actual release events impractical.

Finally, most of these models have yet to develop computational routines that give highly accurate results. For example, most of these models compute diffusion by dividing the physical vapor cloud and time into many pieces or elements and performing momentum, mass, and energy balances around each element. An error will be introduced into the calculation that is inversely proportional to the number of elements that the cloud is divided into and the size of the time increments chosen. An infinite number of elements will result in no error, while dividing the cloud into one element will result in a very large error. Also, there are rounding errors produced by carrying a finite number of significant figures in each calculation. These errors accumulate and can become quite significant. The size of the accumulated error will be inversely proportional to the amount of computer computation time required. This compounds the problem mentioned in the previous paragraph.

Because of their potential accuracy and ability to model complex situations, three-dimensional models may eventually replace similarity models as the preferred method for modeling accidental releases; however, the areas of dispersion theory, numerical methods, and computer technology need improvements before this will occur.

5.4 RELIABILITY

One way to measure model reliability is to compare actual test data with model prediction. A limited number of such tests have been performed, and comparisons of the data with several models have been published. A major comparison of models was done for the Chemical Manufacturers Association (5).

The list of models mentioned in the report is quite large, but actual data comparisons were only performed for nine of the models. The material released was Freon.

The U.S. Air Force has compared other models with field data. The results are shown in References 11 and 12. Comparisons were for nitrogen tetroxide spills and an inert tracer. In 1986, Amoco Chemical Company performed a series of full-scale hydrogen fluoride releases at the Department of Energy's spill test center in Nevada. Liquid hydrogen fluoride, stored under pressure and at ambient temperature, was released to determine if such a release could be modeled accurately by existing source and dispersion models. Test results were presented in late 1987 (13).

There are plans to continue test releases of toxic substances to gather data for model comparison/development. For the most part, these tests are carried out at the Department of Energy's Nevada Test Site.

5.5 COSTS

Models in the public domain are available for a copying fee that may range up to several hundred dollars. The main expense associated with most of the public domain models is the time the user needs to become familiar with the model and adapt it to specific requirements. Most of the models in the public domain have limited applicability; therefore, several models may be required to meet all the needs of the user.

Most of the commercial modeling packages are assembled so that comparatively little time is needed to become familiar with the model. Also, many companies that market the commercial modeling packages will customize the model for the specific needs of the user. Several man-months may be required for such an effort. Such features are reflected in the higher cost of commercial modeling packages.

Commercial modeling packages may be divided into two cost groups. The first is composed of modeling packages that sell for less than $50,000 (includes hardware and software). At the bottom end of the cost range is one company that has packaged the EPA puff model (with no heavy gas dispersion capability) for use on a personal computer. At the top end of this cost range are packages that include multiple source models, data bases with the physical and chemical properties of many different chemicals, and a meteorological tower that sends data to the system to be used during an actual accidental release. These higher-priced packages incorporate a degree of flexibility so that the user can adapt the package for specific needs or use the model for a number of different release situations.

The second group of commercial modeling packages are those that sell for over $100,000 (including hardware and software). These systems have the capabilities of the higher-priced models from the first group, while adding extensive customization for the specific location where they will be applied. The packages can also include a number of features that aid the emergency response effort but that are not part of the modeling system. Such features may include recorded phone messages that are automatically sent to local fire and police in the event of an emergency.

5.6   REFERENCES

1. Mackison, F.W., Stricoff, R.S. (eds.). NIOSH/OSHA Pocket Guide to Chemical Hazards. U.S. Government Printing Office, DHHS (NIOSH) Publication No. 78-210, Washington, D.C., 1980.

2. Wighus, R. Simulation of Propane Dispersion from Spills in a Petrochemical Plant Using a Water Tunnel Technique. Presented at Fourth International Symposium on Loss Prevention and Safety Promotion in the Process Industries. The Institution of Chemical Engineers, Symposium Series No. 80, London, England, 1983.

3. Cheah, S.C., J.W., Cleaver, and A. Millward. The Physical Modeling of Heavy Gas Plumes in a Water Channel. Presented at Fourth International Symposium on Loss Prevention and Safety Promotion in the Process Industries. The Institution of Chemical Engineers, Symposium Series No. 80, London, England, 1983.

4. Bardley, C.I., and R.J. Carpenter. The Simulation of Dense Vapour Cloud Dispersion Using Wind Tunnels and Water Flumes. Presented at Fourth International Symposium on Loss Prevention and Safety Promotion in the Process Industries. The Institution of Chemical Engineers, Symposium Series No. 80, London, England, 1983.

5. McNaughton, D.J., A.A. Marshall, P.M. Bodner, and G.G. Worley. Evaluation and Assessment of Models for Emergency Response Planning. Prepared for Chemical Manufacturers Association by TRC Environmental Consultants, Inc., 1986.

6. Havens, J.A. Mathematical Models for Atmospheric Dispersion of Hazardous Chemical Gas Releases: An Overview. Presented at International Symposium on Preventing Major Chemical Accidents. American Institute of Chemical Engineers, New York, NY, 1987.

7. McNaughton, D.J., G.G. Worley, and P.M. Bodner. Evaluating Emergency Response Models for the Chemical Industry. Chemical Engineering Progress, January 1987.

8. Havens, J.A. Evaluation of 3-D Hydrodynamic Computer Models for Prediction of LNG Vapor Dispersion in the Atmosphere. Gas Research Institute. Contract No. 5083-252-0788, Annual Report, March 1984 - February 1985.

9. Van Olden, A.P. On the Spreading of a Heavy Gas Released Near the Ground. Presented at First International Loss Symposium, The Hague, Netherlands, 1974.

10. Van Olden, A.P. A New Bulk Model for Dense Gas Dispersion: Two-Dimensional Spread in Still Air. I.U.T.A.M. Symposium on Atmospheric Dispersion of Heavy Gases and Small Particles. Delft University of Technology, The Netherlands, 1983.

11. Carney, T.A., and M.M. Lukes. A Comparative Study and Evaluation of Four Atmospheric Dispersion Models with Present or Potential Utility in Air Force Operations. Air Force Office of Scientific Research (AFOSR), 1987.

12. McRae, T.G., et al. Eagle Series Data Report: 1983 Nitrogen Tetroxide Spills. Lawrence Livermore National Laboratories. Report Number UCID-20063, June 1984.

13. Results presented at International Conference on Vapor Cloud Modeling. American Institute of Chemical Engineers (Center for Chemical Process Safety) and U.S. Environmental Protection Agency. Cambridge, MA, November 2-4, 1987.

6. Meteorological Instrumentation

Meteorological data can serve several roles in the mitigation of accidental releases. First, historical data can be used to run vapor/gas dispersion studies for planning emergency response procedures. Second, real-time meteorological data are essential for choosing correct mitigation and emergency response actions at the time of an actual release.

This section presents guidelines for gathering and using meteorological data. General specifications for instrumentation are recommended, and basic considerations for the siting of meteorological systems are discussed.

The information presented in this section is general in nature; meteorological data needs and the design of meteorological systems for specific applications must be assessed case by case.

6.1 BACKGROUND

When applied to facility design and emergency response planning, meteorological data are useful for selecting mitigation measures for accidental releases. When an actual release occurs, these data can facilitate emergency response by real-time estimation of toxic cloud transport and of the potential effects on sensitive receptors. Meteorological data can also be used to analyze past events and to predict the consequences of various accidental release scenarios of possible future events. The following subsections discuss the utility of historical and real-time meteorological data to a facility or the surrounding community for accidental release applications.

6.1.1 Historical Data

Historical meteorological data can be used in facility design, emergency response planning, and the historical analyses of hazardous releases.

Facility Design--
Meteorological data can play a major role in the design of a facility to protect sensitive receptors from the effects of accidental releases. The facility designer can obtain one or more years of representative wind direction data to determine the frequency distribution of wind direction and speed in the vicinity of the facility. With this information, the facility designer can locate potential toxic release points to minimize the probability of their being upwind of sensitive receptors.

Using a more rigorous approach, the facility designer can use hazardous release dispersion models (see the dispersion modeling section of this manual) to determine the optimal locations for potential toxic release points. The model user is allowed to select various meteorological and release scenarios. As noted in the dispersion model section of this manual, wind speed, wind direction, air temperature, humidity, and atmospheric stability class are parameters commonly used by these models. The criticality of the various meteorological parameters depends on the specific chemical and other characteristics of the source during a release.

Examples of the use of meteorological data for facility design purposes have been provided by Boykin and Kazarians (1).

Emergency Response Planning--
Historical meteorological data can be used to help determine where the significant effects of toxic releases would most likely be felt; this information can be used to help assess the degree and type of emergency response planning necessary. Historical meteorological data should be used for emergency response planning during the facility design stage because

careful consideration of meteorological data at this point can reduce the emergency response planning effort if potential toxic release points are located. In the case of an existing facility, meteorological data can still be used to help determine the potential effects on sensitive receptors, which makes possible an assessment of the type and degree of emergency response planning needed. The methodology for using meteorological data for emergency response planning is similar to that described in the previous subsection for facility design.

Historical Analysis of Accidental Releases--

After an actual release, meteorological data from the release period can be combined with other data and observations of the actual release to better understand what has occurred. Meteorological data used for this purpose can provide comparative information on observed effects versus predicted effects. This information may be useful for mitigating future hazardous releases. For example, the placement of potential toxic release points in future facility designs or emergency response planning might be influenced by the analysis of an actual release.

A simplified approach to using meteorological data for historical analysis involves using wind direction and wind speed data for arriving at a crude estimate of the transport path of the released cloud. A more in-depth analysis would use hazardous release dispersion models for predicting the path of transport and the specific effects on selected receptors along the transport path.

On-Site versus Off-Site Historical Data--

Historical data used for the purposes described in the previous subsections should be obtained from on-site instrumentation if the meteorological instruments are sited properly (see Subsection 6.3.4, Siting of Meteorological Instrumentation). In many cases, however, historical data gathered on site may not be available or may be insufficient. In such cases, off-site data may

have to suffice. If off-site data are used, the adequacy of the data must be
carefully assessed. Off-site meteorological data collected near the plant
would be more representative of conditions at the plant than off-site data
collected far from the plant. Data obtained from a distant source or from an
area with significantly different topography may not be useful.

6.1.2 Real-Time Meteorological Data

Real-time meteorological data at the time a release is occurring can be
used by the emergency response team in appropriate models to predict the
concentration levels and areas affected by a cloud or plume, thus enabling the
response team to make more effective decisions about mitigation action.

The ability of the emergency response team to use real time data and a
dispersion model depends on the conditions of the emergency (e.g., available
reaction time, character of the release). In some cases, the usefulness of
the model may be limited to a subsequent analysis of the event, as described
in the previous section.

6.2 DESCRIPTION

This section describes the meteorological equipment that may be used for
accidental release applications. Stability measurement techniques (inferred
from measurement of other variables) and meteorological towers and data
acquisition systems are also included. Much of the discussion of
meteorological instrumentation is extracted from guidelines previously
published by the U.S. EPA (2).

The intended use for the data determines which meteorological parameters
should be monitored. When no dispersion model will be used, wind direction
and wind speed data may be sufficient. When a dispersion model is used,
additional variables such as temperature and humidity will probably be
necessary. The variables needed depend on the model used (see Section 5).

Relative humidity and barometric pressure data are used in some hazardous release dispersion models; however, the sensitivity of the models to variations in these two parameters is minimal, and climatological averages can often be used. Thus, it is generally not necessary to obtain actual measurements of these parameters.

6.2.1 Wind Direction

Wind direction must be known in order to determine which direction the vapors from an accidental release will travel. Wind direction is measured with a wind vane connected to the necessary electronic circuitry to provide a remote readout and to record, if necessary, wind direction. Though rarely used for remote sensing, an elevated wind sock should also be used as a visual indicator of wind speed and direction. An elevated wind sock allows plant personnel to get an instant visual indication of wind direction from many locations at the facility, and it can also be seen off site, depending on its placement.

6.2.2 Wind Speed

Wind speed determines how quickly the plume or cloud from an accidental release will reach off-site receptors. A number of wind speed measurement systems that operate on different principles are available. The most commonly used sensors are the rotational cup and propeller anemometers. Cup anemometers operate on the principle that net torque (lift greater than drag) causes a rate of rotation roughly proportional to wind speed. The three conical cup design performs best. With propeller anemometers, the propeller turns at a rate almost directly proportional to the wind component parallel to its axis. Electronic circuitry converts the rotary motion of the cups or propeller to an electrical signal for remote readout equipment.

6.2.3 Ambient Temperature

Ambient temperature data are used in dispersion modeling to calculate cloud buoyancy and the heat available for evaporating volatile liquids. The most appropriate temperature measurement devices for meteorological applications are linear thermistors and resistance temperature detectors.

Thermistors are electronic semiconductors made from metallic oxides. The resistance of a thermistor varies inversely with absolute temperature. The system configuration is typically designed with the thermistor connected to a bridge circuit that provides an output voltage that varies directly with the temperature of the thermistor.

Resistance temperature detectors (RTDs) operate on the principle that the electrical resistance of a pure metal increases with temperature. An RTD operates connected to a bridge circuit that provides an output voltage that varies directly with the temperature of the RTD.

6.2.4 Stability

When accidental release dispersion modeling is conducted, data for atmospheric stability are essential for estimating the spread of the plume. Stability data are derived indirectly from measurements of other parameters. There are several methods for computing atmospheric stability. For example, one method calculates stability by evaluating the cloud cover, ceiling height and wind speed (Turner's 1964 method). Another method uses the standard deviation of the vertical wind direction; one uses the standard deviation of the horizontal wind direction; another uses vertical temperature differences. These methods are discussed extensively in the literature and in U.S. EPA guidelines (3).

The method used for calculating stability will depend on the method used in the specific vapor dispersion model. Some models allow a choice of methods.

6.2.5 Meteorological Towers

Several types of meteorological towers are commercially available for mounting wind and temperature instrumentation. Free-standing, guyed towers, as well as towers that can be attached to fixed structures, are available from a number of vendors. In some cases, meteorological instruments can be mounted on an existing elevated structure. Lightning protection kits, available from tower vendors, are recommended for areas where cloud-to-ground lightning is common.

6.2.6 Data Acquisition Systems

The most practical type of data acquisition system for the meteorological applications discussed in this manual is a digital system that uses magnetic media such as diskettes, cassettes, bubble memory, and semiconductor memory. Such systems allow immediate retrieval of data during actual emergency situations and provide the convenience of data retrieval and processing at any time. Many types of digital data acquisition systems with a wide range of memory capability are commercially available. System needs will have to be determined case by case.

6.3 APPLICABILITY AND PERFORMANCE

The utility of on-site meteorological data for hazardous release applications depends on the careful design of the meteorological data acquisition systems. Important considerations, discussed in the following sections, are variable selection, instrumentation selection, and instrumentation siting.

6.3.1   <u>Instrument Selection</u>

A number of wind vane manufacturers offer wind sensors with a rated
accuracy of ±3 degrees, which meets EPA recommendations for prevention of
significant deterioration (PSD) monitoring in standard air pollution applications (4), and which should be adequate for hazardous release applications.
As with all meteorological sensors, the accuracy is actually the accuracy of
the entire measurement system, including the readout device. Additional
information on important wind vane performance criteria may be found in
References 5, 6, and 7.

A wind sock is often used as a visual indication of wind direction.
Although wind socks are rarely used for remotely sensing wind direction, they
are extremely valuable to facility personnel as quick visual indicators. They
should be used wherever an accidental release could present a hazard to
facility personnel or to the surrounding community.

Many manufacturers offer wind speed sensors with a rated accuracy of
±0.82 feet per second or better, which meets EPA guidelines. As in the case
of wind direction, the accuracy guidelines apply to the entire measurement
system. Additional information on wind speed performance criteria may be
found in Reference 8.

A wind sock may be used as a visual guide for estimating wind speed.
Most manufacturers offer guidelines as to how the deflection of the wind sock
correlates with wind speed. As mentioned in the previous discussion, wind
socks should be used wherever an accidental release could present a hazard to
the facility personnel or the surrounding community. Even when more sophisticated instruments are available, the wind sock should be used as a backup
visual indicator.

Temperature sensor accuracies of 0.9°F or better are easily achievable (9). A number of manufacturers offer sensors with much better accuracy, although an accuracy of 0.9°F should be adequate for most accidental release applications (an exception is the case in which delta temperature measurements are needed for stability calculations; in this case, an accuracy of 0.18°F or better would be appropriate). It is important to note that temperature sensor accuracy is often limited by poor sensor exposure, improper coupling, and signal interference, rather than by the accuracy of the sensor itself. A serious problem in temperature measurements is radiation error, which can amount to several degrees at midday. Aspirated temperature radiation shields, readily available from manufacturers, can minimize radiation error.

6.3.2 Siting of Meteorological Instrumentation

The objective of proper meteorological instrumentation siting is to place the instruments at locations where measured data represent the atmospheric conditions in the area of interest. Proper instrumentation siting is of special importance for accidental release applications, since the data may be used for making critical decisions that affect human lives and property.

This section discusses instrumentation siting criteria, including the number and location of measurement sites, distances from obstructions, measurement heights and topographical considerations. The discussion focuses on the variables of wind direction, wind speed, and ambient temperature. Siting considerations for wind direction and speed receive the greatest attention, since data for these parameters are most sensitive to siting factors.

It is often difficult to locate a site that meets all the desired siting criteria. In any instance, site accessibility and security should never be overriding considerations; the integrity of the data may be significantly compromised if factors such as obstructions and topography are given too low a priority.

Siting for Wind Direction and Speed--

For accidental release applications, it is of utmost importance that reliable wind data be measured, since wind data (including stability calculated from wind data) determine the path and spread of the toxic cloud. The reliability of the wind data depends on the appropriateness of wind sensor siting.

Usually, the more wind data sites available, the more reliable the modeled estimates of accidental release transport. Some vapor dispersion models take advantage of multiple sites by weighting wind values according to the locations of the wind sites relative to the position of the plume at a given time.

The number and location of wind data measurement sites should be based on consideration of probable paths of cloud transport, the demographics of the facility vicinity, and expected variations in wind speed and direction caused by obstructions and non-uniform topography.

Using historical wind frequency distributions, the meteorological network designer should determine the relative probability that populations or properties in various directions from the facility will be located in the path of a hazardous cloud release so that the possible number of wind data sites and their locations can be determined. Wind data sites should be located as close as possible to the facility property boundary line, since data from a property line location should provide the best estimate of cloud transport into adjacent areas once the cloud has cleared the obstructions created by facility structures.

Once the number and general locations of meteorological sites have been determined, the network designer must analyze obstructions and topography to find specific sites that will produce data representative of areas potentially affected by a hazardous release.

It is standard meteorological practice to locate wind direction and wind speed sensors where the distance between the instruments and any obstruction is at least ten times the height of the obstruction (2). Examples of common obstructions are industrial facility structures, buildings, and large trees. Sites that are too close to obstructions will normally record wind speeds that are too low, wind directions that are unrepresentative of the area, and wind direction fluctuations that are too great (as indicated by the standard deviation of the wind direction).

While the adverse effects of obstructions on wind instrumentation siting can be determined and avoided rather easily in many cases, it is a significant challenge, at least in complex terrain, to assess the effects of topography on wind flow and to locate wind sensors where measured data are truly representative of the areas of interest. The meteorological network designer should carefully review topographical maps and determine the potential of the local topography (hills, creek beds, valleys, canyons) to cause channeling and upslope and downslope flows. The network designer must then locate the wind instrumentation sites so that the measured data are representative of the areas that may be affected by a hazardous release. When there are no significant obstructions to the flow of wind and the topography is level, wind measurements at any one site in the facility area should provide representative wind data.

The standard height for surface wind measurements in scientific applications is 33 feet (2). An appropriate wind measurement height for accidental release applications, however, will depend on the expected height of potential releases. For example, in the case of a non-buoyant toxic cloud that may flow near the ground, a wind measurement height in the range of 7 to 17 feet may be more appropriate than one of 33 feet. If a toxic cloud could be released at an elevated point, wind measurements at that height would be appropriate, especially during night-time periods when wind directions can vary significantly with height. When a facility contains multiple potential toxic release points at various heights, selection of an optimal wind measurement height may be difficult. In such a case, a measurement height that reflects the average of the release heights may be used.

Siting for Ambient Temperature--

The calculations made by hazardous release dispersion models of the dispersion of the released cloud are based on the ambient temperature at the time of release, and they assume that the temperature remains constant along the path of transport. It is usually sufficient to measure the ambient temperature at one location in the site vicinity.

The standard ambient temperature measurement height is 3 to 7 feet (2). For accidental release applications, however, the measurement height should be adjusted to correspond with the height of potential toxic releases. Such a height adjustment is most important during nighttime periods when variations in temperature with height can be significant. In case there are multiple potential toxic release points located at various heights, it is most practical to select a single height that best represents the heights of the release points.

Ambient temperature sensors should be sited so that they are not unduly influenced by heat sources such as exhausts. Locations directly over concrete or asphalt should also be avoided (2).

6.4 RELIABILITY

Meteorological instrumentation can be highly reliable if proper preventive maintenance procedures, as prescribed by the manufacturers, are followed. However, because of the importance of meteorological measurements for accidental release applications, spare parts and sensors should always be readily available.

For a meteorological system to produce valid and useful data, a well-designed program of quality assurance/quality control must be implemented and maintained. A stringent quality assurance/quality control program is particularly important for accidental release applications, since decisions affecting human health and the environment will frequently be based on meteorological data.

Examples of quality control measures include calibration checks, preventive maintenance, and spot data checks by personnel directly involved in operating the instrumentation. Quality assurance activities include equipment audits and detailed screening of the data sets by parties who are independent of routine system operations. A comprehensive set of quality assurance/quality control procedures has been developed by the U.S. EPA (2). These procedures are appropriate for meteorological systems deployed for accidental release applications.

6.5   COSTS

The cost of equipment and accessories for a station measuring wind speed, wind direction, and temperature typically ranges from $10,000 to $20,000. A number of vendors offer a variety of equipment models and configurations.

Components that should be considered when projecting costs for meteorological systems include the sensors; translator modules, racks, and power supply; junction boxes; cabling; wind sensor mounting crossarms; surge and lightning protection; data acquisition system; telemetry devices; calibration equipment; spare parts; and equipment shelter.

6.6 REFERENCES

1. Boykin, Raymond F., and Mardyros Kazarians. Quantitative Risk Assessment for Chemical Operations. Presented at International Symposium on Preventing Major Chemical Accidents. Pickard, Lowe, and Garrick, Inc., Washington, D.C., 1987.

2. U.S. Environmental Protection Agency. Quality Assurance Handbook for Air Pollution Measurement Systems, Volume IV: Meteorological Measurements. EPA-600/4-82-060, February 1983.

3. U.S. Environmental Protection Agency. Guideline on Air Quality Models (Revised). EPA-450/2-78-027R (NTIS PB86-245248), July 1986.

4. U.S. Environmental Protection Agency. Ambient Monitoring Guidelines for Prevention of Significant Deterioration (PSD). EPA-450/4-80-012 (NTIS PB81-153231), November 1980.

5. Finkelstein, Peter L. Measuring the Dynamic Performance of Wind Vanes. Journal of Applied Meteorology, 20, pp. 588-594, 1981.

6. MacCready, P.B. Dynamic Response Characteristics of Meteorological Sensors. Bulletin of American Meteorological Society, 46 (9), pp. 533-538, 1965.

7. Wierinza, J. Evaluation and Design of Wind Vanes. Journal of Applied Meteorology, 6(G), pp. 1114-1122, 1967.

8. Lockhart, T.J. Bivanes and Direct Turbulence Sensors. MRI 70 Pa-928, EPA Institute for Air Pollution Training, 1970.

9. U.S. Environmental Protection Agency. Strimaitis, David, G. Doffngole, and A. Bass. On-Site Meteorological Instrumentation Requirements to Characterize Diffusion from Point Sources. EPA-600/9-81-020 (NTIS PB81-247223), C-1-C-5, 1981.

7. Secondary Containment

In the event of the accidental release of a volatile, toxic and/or flammable liquid, the spread of the liquid must be controlled to reduce the rate of evaporation and the resulting generation of a vapor cloud. Physical barriers provided by judicious siting of storage and process facilities can contain and thus reduce the effects of a hazardous release. However, such barriers often do not exist or do not provide sufficient containment. In such cases, other methods of secondary containment are needed. Stopping or reducing the flow of a released chemical at the source will also help contain an accidental release. Sometimes the flow of material can be stopped by closing a valve of stopping a pump. At other times stopping the flow of released material requires the application of a leak plugging procedure.

This section of the manual discusses several methods for plugging leaks and stopping the flow of material upstream of a leak. A similar discussion and evaluation of several secondary or partial containment systems is also included. A brief discussion of costs is presented.

7.1 BACKGROUND

Post-release mitigation measures attempt to reduce the effects of an accidental release. The most effective method for reducing the effect of a release is to stop the release before large quantities of material have escaped. Using strategically placed, remotely operated emergency isolation valves is the most effective way of stopping the flow of material from an accidental release; however, in some cases, it is not possible to use valves to isolate a leaking portion of a process. An example would be a release from a hole in a large vessel. Such a release would continue even if the tank were isolated, as long as the liquid level was above the leak point or the internal pressure was sufficient to force vapors out. In such cases, leak plugging techniques might stop the flow of material.

Leak plugging specifically applied to accidental releases is not a common practice, although the Chlorine Institute has designed a number of leak-plugging devices for chlorine cylinders, one ton containers, and tank cars and tank trucks used to transport chlorine (1,2,3). In the chemical process industries, the desire to perform repairs without shutting down a process has led to the development of several plugging methods. Although these methods would have limited applicability to accidental release mitigation, the principles may be useful. The use of various chemical sealing materials for plugging chemical leaks has been investigated experimentally; none is in wide-spread commercial use. Such methods could be potentially applied to a number of accidental release situations. The following section discusses some of the leak plugging methods that may apply to mitigating releases.

In addition to efforts to stop the flow of the release at the source, the spread of any released liquid must be controlled. Limiting the spread of released liquids will limit the ground surface area affected by the release and will allow emergency response personnel to work close to the release source. In addition, containing the release will limit the surface area of the liquid, which will limit the rate of vapor generation, which can substantially reduce downwind vapor concentrations of volatile liquids. Finally, containment can be used to divert released liquids into a temporary containment vessel, thus substantially reducing the impact of the release.

Secondary containment systems in the form of curbing, trenches, basins and dikes have been successfully used for many years in the chemical industry. Many improvements have been made over the years. The following section discusses some of the secondary containment systems available at this time.

7.2   DESCRIPTION

7.2.1   Leak-Plugging Techniques

Leak-plugging techniques may be divided into three groups: chemical patches and plugs, physical patches and plugs, and methods for stopping the flow upstream of a leak. Examples of each of these categories are discussed below.

Chemical Patches and Plugs--

A chemical patch or plug is made by filling a hole or crack with a material that solidifies in place after application. No chemical patching system has been developed for widespread commercial use. The concept has been experimentally tested and requires further developmental work. Three categories of materials have been tested for this application (4). Urethane foams conform to an irregularly shaped hole and expand in place, resulting in a tight seal. Nonexpanding materials react in place to achieve crosslinked structures which are relatively resistant to chemical attack. Examples of crosslinked materials include: various epoxy systems, quick-curing silicone rubber, and catalyzed polysulfide rubbers. The third category of materials that have been tested are thermosetting nonexpanding plastics (5,6,7).

Several different methods for applying chemical plugs and patches have been tested. In one method, the patching material is forced into the leak and held in place until it sets and forms a plug (4). A second technique has been tested for improving the ability of nonexpanding materials to plug a hole. In this method, the chemical patching material is used in conjunction with a supporting or backing material such as nylon cloth (4). Another application technique uses an elastomer "balloon" or cloth sack attached to the end of a lance. The lance is inserted into the leak and the balloon or sack is filled with a plugging chemical, thereby plugging the leak (4,8,9). This type of device is illustrated in Figure 7-1. A variation of this last type of device is a cloth bag that is wrapped loosely around a leaking valve or fitting with

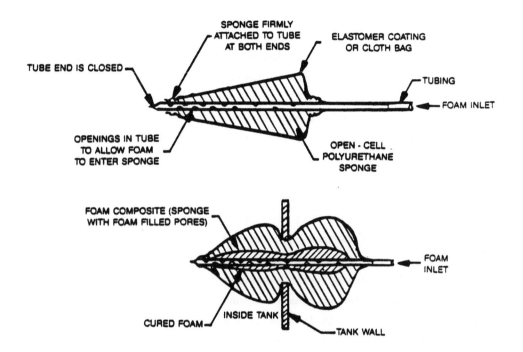

Figure 7-1. Foam-injected leak plugging device.

Source: Adapted from Reference 10.

a leak-plugging chemical injected into the annular space between the cloth and the leak. The cloth holds the chemical in place around the leaking fitting until it can harden in place.

Physical Patches and Plugs--

A physical patch or plug is any prefabricated device used to fit into or around a leak; a multitude of such devices are available. The Chlorine Institute has assembled three emergency kits used to temporarily stop leaks from valves and fittings on cylinders, one-ton containers, and tank cars and tank trucks (1,2,3). Figure 7-2 illustrates some of these patching and plugging devices. The kits have patching devices for small punctures of cylinders and one-ton containers. This type of leak repair equipment is available from numerous other sources.

Several physical patching devices have been developed for leaking pipes, flanges or other fittings, so that pipe repairs can be made without shutting down the process. These devices, illustrated in Figure 7-3, are usually used to repair leaks of nontoxic materials such as steam, but they may be useful for preventing the release of toxic materials (6). This type of leak-plugging equipment is usually bolted or clamped around the leaking pipe or fitting, forming a leak-tight case. It is usually sized for a specific type of fitting or pipe. These devices are often used in conjunction with a chemical sealing material. For example, a metal jacket is bolted around a leaking pipe or fitting and a chemical sealing compound is injected into the annular space to more fully seal the leak (11). This type of arrangement is illustrated in Figure 7-4.

Methods for Stopping Flow Upstream of a Leak--

Analogous to leak-plugging methods are techniques developed for stopping the flow of material upstream from a leak. The most practical method for stopping flow would be shutting an emergency isolation valve. However, such valves are not always available. A few methods have been developed for stopping the flow of material in a pipe so that repairs can be made without

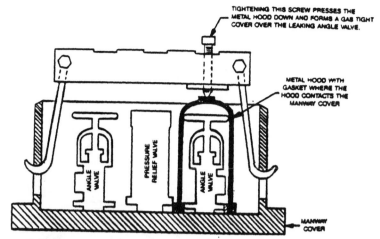

TIGHTENING THIS SCREW PRESSES THE
METAL HOOD DOWN AND FORMS A GAS TIGHT
COVER OVER THE LEAKING ANGLE VALVE.

METAL HOOD WITH
GASKET WHERE THE
HOOD CONTACTS THE
MANWAY COVER

ANGLE VALVE

PRESSURE RELIEF VALVE

ANGLE VALVE

MANWAY COVER

Mechanical Leak Repair Device for Leaking Angle Valve
on Manway Cover on Chlorine Tank Car

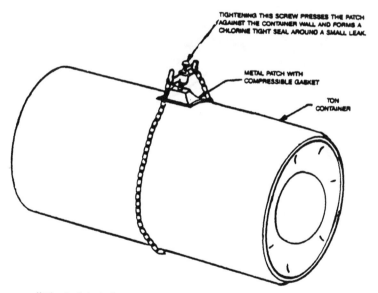

TIGHTENING THIS SCREW PRESSES THE PATCH
AGAINST THE CONTAINER WALL AND FORMS A
CHLORINE TIGHT SEAL AROUND A SMALL LEAK.

METAL PATCH WITH
COMPRESSIBLE GASKET

TON CONTAINER

Mechanical Leak Repair Device for One Ton Chlorine Container
With Small Side Wall Leak

Figure 7-2.   Examples of mechanical leak repairing devices offered
by the Chlorine Institute.

Source: Adapted from References 2 and 3.

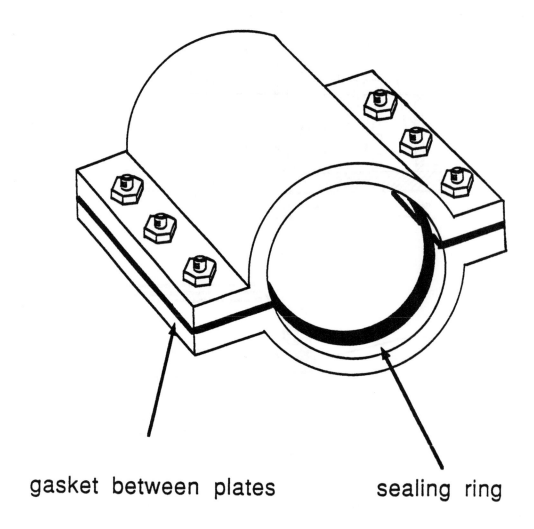

gasket between plates   sealing ring

Figure 7-3.   Jacket that bolts around a leaking pipe,
              forming a leak-tight seal.

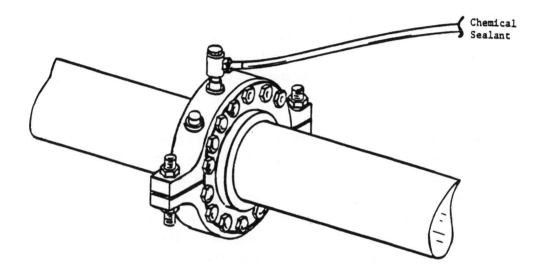

Figure 7-4.   Mechanical sealing ring with injected chemical
               sealant to fit around and seal a leaking flange.

Source: Adapted from Reference 12.

shutting down a process. These techniques are probably not useful for most accidental release situations. However, they are discussed here to illustrate what the state of the art is at this time.

Pipe freezing, one method for stopping the flow in a pipe, has been developed primarily as a maintenance tool for making repairs in a pipe without shutting down an entire process unit (13). A temporary jacket is placed around the pipe where the freeze is to be made. The cooling liquid, such as liquid carbon dioxide or liquid nitrogen, is injected into the jacket and replenished until a solid, frozen plug is formed in the pipe. The plug is maintained by monitoring the temperature in the jacket and replenishing the coolant as required.

One company has developed a device for high-pressure steam applications capable of shearing a pipe and sealing the shear in place, thereby stopping the flow of material through the pipe (12). A similar method, called "stoppling," has been used for stopping flow in a line. In this technique, a reinforcing saddle is first welded or bolted around the line. A hole is drilled into the side of the line and a plug is inserted to stop flow in the line. The entire operation is kept under sufficient pressure to prevent any liquid in the pipe from escaping.

7.2.2 Physical Barriers and Containment Systems

Secondary containment systems typically consist of the following: curbing, trenches, excavated basins, natural basins, and earth, steel, or concrete dikes. The type of containment system best suited for a particular facility depends on the consequences associated with an accidental release from that location. The inventory of toxic material and its proximity to other portions of the plant and to the community are primary considerations in the selection. The secondary containment system should be able to contain spills with a minimum of damage to the facility and its surroundings and with minimum potential for escalation of the event.

In general, secondary containment systems:

- Prevent the spread of hazardous liquid from the immediate spill area to nearby areas;

- Limit the size of the pool of spilled liquid;

- Reduce the rate and quantity of vapor generated;

- Prevent hazardous vapor concentrations from spreading to areas outside the facility boundary; and

- Prevent the material from reaching a source of ignition, if it is flammable.

The most common type of secondary containment for storage and process vessels is a low-wall dike (14). A dike, which is a physical barrier around the perimeter of process equipment or storage vessels, is designed to confine the spread of liquid spills. A dike can be little more than a curb or a high wall rising to the top of a storage tank. It can be constructed of earth, steel, or concrete. Figure 7-5 shows a typical use of a dike.

The area contained by the dike is usually sufficient to contain the largest tank or process unit served. Generally, a limited number of process units or storage tanks are enclosed within one diked area; as the number of units increases, so also does the risk of an accidental release.

For most applications, dike heights range from three feet to twelve feet, depending on the area available to achieve the required volumetric capacity (15). For a specific capacity, a low-wall dike requires the greatest amount of land area but is usually the least expensive to construct. On the other hand, higher wall dikes may cost more but they may also result in some vapor holdup and contribute to a reduction in the size of the vapor cloud (5).

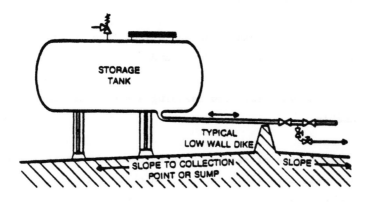

Figure 7-5.   Conceptual diagram of a typical low-wall dike.

Dike walls are designed to be liquid tight and to withstand the hydrostatic pressure and temperature of the spill. Dikes are typically located a sufficient distance from the tank to allow access to the vessel and to equipment near the vessel. Piping is routed over dike walls; penetrations through walls are avoided if possible. Vapor fences may be situated on the top of the dikes to give additional vapor retention capacity. In addition, if there is more than one tank or process vessel in the diked area, the vessels are often elevated above the maximum level attainable in the impoundment if more than one tank or vessel loses its entire contents. Common diking of incompatible materials should be avoided.

Provision should be made for draining rainwater from diked areas. This will involve the use of sumps and separate drainage pumps, since direct drainage to stormwater sewers would allow any spilled material to follow the same route. Alternatively, a sloped rain hood may be used over the diked area that could also serve to direct the rising vapors to a single release point.

In areas where it is critical to minimize vapor generation, surface insulation may be used in the diked area to further reduce heat transfer from the floor and walls or the dike to the spilled liquid. In addition, earthwork dikes may require a layer of clay, asphalt, plastic film, or similar material to prevent the contained material from seeping into the ground (4).

The ground enclosed within a dike should be graded to cause the spilled liquid to accumulate at one side or in one corner. This will help minimize the area to which the liquid is exposed and from which it may gain heat.

High-wall dikes may be an appropriate secondary containment choice for selected systems (8). High-wall dikes may be constructed of low-temperature steel, reinforced concrete, or prestressed concrete. A weather shield may be provided between the tank and wall, with the annular space remaining open to the atmosphere. The available area surrounding a storage tank will dictate the minimum height of the wall. High-wall dikes may be designed with a volumetric capacity greater than that of the tank or vessel to provide vapor

containment. Increasing the height of the wall also raises the elevation of
any released vapor.

As with low-wall dikes, piping should be routed over the wall if possible. The closeness of the wall to the tank may necessitate placement of the
pump outside the wall, in which case the outlet (pump suction) line will have
to pass through the wall. In such a situation, a low dike encompassing the
pipe penetration and pump may be provided, or a low dike may be placed around
the entire wall.

An alternative to simply accumulating the material in one corner of the
diked area is to provide a system such as the one depicted in Figure 7-6,
which consists of a concrete or steel basin located at the lower edge of the
dike. A one-way valve allows the toxic material to flow into but not out of
the basin. A submerged pump is used to remove the material from the basin,
and it is either sent to storage or to an appropriate treatment and disposal
system. In addition, the vapor generated from the spilled liquid could be
sent to a scrubber or flare. Such basins may be required to conform to
regulations for underground storage of toxic chemicals.

A remote basin is another type of secondary containment. It is especially well suited to storage systems where more than one tank or process
vessel is served and a relatively large site is available. The flow of a
liquid spill is directed to the basin by dikes and channels under the storage
tanks. These channels are designed to minimize exposure of the liquid to
other tanks and surrounding facilities. Often the channels that lead to the
remote basin, as well as to the basin itself, are covered to reduce the rate
of evaporation of the spilled liquid. For nonflammable materials, the
impounding basin is usually located near the tank to minimize the amount of
toxic material that evaporates as it travels to the basin. For flammable
materials, two conflicting spacing requirements must be considered. First,
the impounding basin must be close to the tank to minimize evaporation.
Second, the impounding basin must be far enough from the tank that a fire in
the basin would not expose the tank to excessive heat.

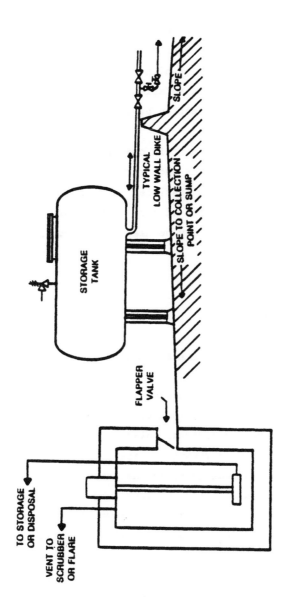

Figure 7-6.   Conceptual diked storage facility with integrated containment sump.

A neutralization basin is an alternative to an impounding basin. This
system is essentially the same as the impounding basin, except that the
spilled material is mixed with a neutralizing material to form a mixture that
can then be disposed of safely (9). For acidic materials, an appropriate
material might be lime, which could be stored in the basin as a premixed
slurry or as solid limestone. However, a high heat of reaction may cause an
increased rate of evaporation with some materials. For example, Figure 7-7
illustrates two layouts for a neutralization basin.

A final type of secondary containment is one structurally integrated with
the primary system to form a vapor-tight enclosure around the primary
container (15). Many arrangements are possible. A double-walled tank, as
illustrated in Figure 7-8, is an example of such an enclosure. A low-wall
dike placed around the entire systems may be used as a backup, in the event of
wall or foundation failure.

7.3   APPLICABILITY AND PERFORMANCE

7.3.1   Applicability and Performance of Leak-Plugging Devices

The leak plugging and patching techniques listed in this section cannot
be used for catastrophic failures of equipment. They perform best on small
leaks and most will work better on circular holes than on cracks. Four other
limitations common to most of these techniques are: first, personnel must be
able to get close enough to the leak to apply the patch or plug; second, it is
difficult to prevent the person applying the leak from being sprayed with the
leaking material; third, all of these techniques have limited capabilities:
no appropriate leak-plugging technique may be available at the time of a
specific release situation; finally, all of these techniques require time to
set up and implement, and a large amount of hazardous material may escape
before the leak can be stopped. It could be very hazardous to attempt to plug
the flow of a flammable material. Plugging such a leak would probably require
personnel to enter a cloud of material within the flammable range.

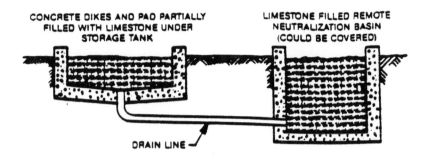

CONCRETE DIKES AND PAD PARTIALLY
FILLED WITH LIMESTONE UNDER
STORAGE TANK

LIMESTONE FILLED REMOTE
NEUTRALIZATION BASIN
(COULD BE COVERED)

DRAIN LINE

LIMESTONE NEUTRALIZATION BASIN

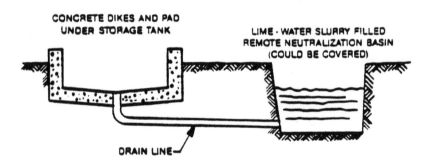

CONCRETE DIKES AND PAD
UNDER STORAGE TANK

LIME - WATER SLURRY FILLED
REMOTE NEUTRALIZATION BASIN
(COULD BE COVERED)

DRAIN LINE

LIME WATER NEUTRALIZATION SYSTEM

Figure 7-7.   Conceptual diagram of potential layouts for a remote
              neutralization basin system.

Source: Adapted from Reference 9.

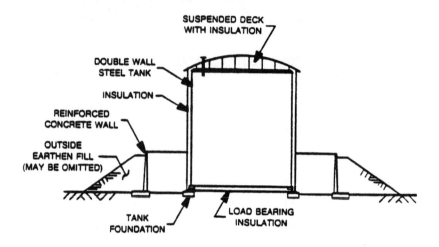

Figure 7-8.   Conceptual diagram of a double-walled tank system.

Source: Adapted from Reference 15.

Where hazardous materials are involved, a patch or plug should never be regarded as a permanent form of repair. The purpose of leak plugging is to stop the leak just long enough to shut down the process, drain the storage tank or effect whatever changes are necessary to stop the flow of material upstream of the release.

Chemical Patches and Plugs--

A limited amount of experimental testing has been done to measure the performance of chemical patches and plugs. One report prepared by the Environmental Protection Agency summarized the results of leak-plugging tests on a variety of expanding foams and nonexpanding sealants (5). In these tests, the plugging chemical was pushed into the leak and held in place until it hardened. Table 7-1 shows some of the test results. The EPA report also calculated a practical applicability range where chemical plugs might be used; Figure 7-9 shows this range.

Chemical patches and plugs have several advantages over alternate plugging techniques: the plug can conform to irregularly shaped holes and cracks; the plug can be applied to holes in irregularly shaped surfaces; and no elaborate application devices are required. One disadvantage of chemical plugs is that the chemical plugging material must be resistant to attack by the chemical that is escaping; a facility may need to have several different plugging materials on hand. In addition, a chemical plug will be subject to failure by chemical attack or physical or thermal shock.

Several studies have tested the performance of applicators that use a foam-filled elastomer or fabric bag for leak plugging (6,7). Using such a device increases the physical strength of the plug. Also, since a narrow lance is first inserted into the hole, less spray of leaking material may occur than with other methods, thus reducing the risk of exposure to personnel. The disadvantages of using this type of applicator are the same as those associated with the use of chemical patches without an applicator. The

TABLE 7-1. HAZARDOUS CHEMICALS TESTED AND SMALL-SCALE SEALING TEST RESULTS

Hazardous Material	Starfoam® Urethane Foam	Sea-Goin® Epoxy Putty	Other Successful Sealants
Phenol	Sealed (A)[+][*]	Failed (A)	Butyl Rubber (A)
Methyl Alcohol	Sealed (A,W)	Sealed (A,W)	Epoxy Putty (A) MSA Urethane (A)
Insecticides, Rodenticides: DDT 95% Solu./Water Dieldrin	Sealed (A)	Sealed (W) Sealed (A)	
Acrylonitrile	Failed (A)	Failed (A)	Polysulfide Rubber (A)
Chlorosulfonic Acid		Failed (A)	
Benzene	Sealed (A,W)	Sealed (A,W)	MSA Urethane (A,W)
Phosphorus Pentasulfide		Failed (A)	
Styrene	Sealed (A)	Sealed (A)	
Acetone Cyanohydrin		Sealed (A)	
Nonyl Phenol	Sealed (A,W)	Sealed (A,W)	
Isoprene	Sealed (A)	Sealed (A)	
Xylene		Sealed (A)	
Nitrophenol	Sealed (W)	Sealed (W)	

[+]Sealed leaks of aqueous solutions (50 to 90 percent) for 18 hours, then failed.
[*]"A" and "W" indicate tests performed in air and submerged in water, respectively.
Source: Reference 5

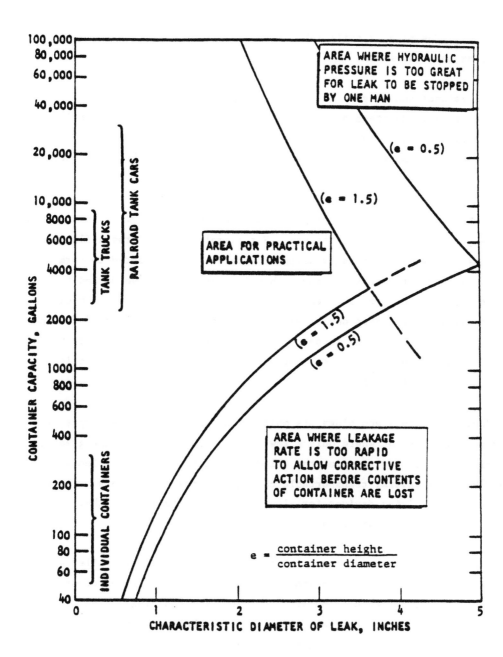

Figure 7-9.   Envelope of practical applications for sealing
leaks in nonsubmerged containers.

Source: Reference 5.

applicator. The plug is still subject to failure by chemical attack or
thermal or physical shock.

Physical Patches and Plugs--

The primary advantage of physical patches and plugs is their ability to
form a very tight seal nearly as strong as the original container. They can
be constructed of material more resistant to chemical attack than a chemical
plug.

The primary disadvantage of physical leak-repairing devices is that most
are designed for a very specific use. For example, a leak from a flange can
be stopped by using a specially designed encasement constructed to form a
tight seal around the entire flange; this is a very effective method for
stopping leaks. However, such an encasement will work only for flanges and
often a different encasing device will be required for each size of flange. A
facility would need a large inventory of physical plugging devices to be able
to handle any accidental release that might occur; this is an impractical
solution for most situations.

Physical patches and plugs are most useful where only a few types of
leaks are possible, or where only a few types of leaks present the greatest
potential hazard. For example, the Chlorine Institute has assembled leak
repair kits for several transportable chlorine storage containers. The kits
contain leak-repairing devices for all the types of leaks most likely to occur
in each of the containers (5 to 9 different types of leaks).

Leak-plugging methods where a physical jacket or shell is filled with a
chemical plugging material have some advantages over other types of devices.
The outer shell provides high physical strength for the plug. Also, the use
of the chemical sealing material, means that the shell does not have to form a
tight seal around the leaking equipment; therefore, one type of shell could be
used for plugging leaks from several line sizes or for several types of
fittings.

One disadvantage of this method is that the chemical plugging material must be chemically resistant to the leaking material and must be thermally resistant to the temperature of the leaking material. These types of devices are used for plugging leaks from pipes or other small diameter fittings by enclosing the entire line. They are not used for filling a leak, and thus are not practical for leaks from large diameter vessels.

Methods for Stopping Flow Upstream of a Leak--

The main advantage of any method that stops the flow of material upstream of a leak is that it reduces the potential of personnel exposure. However, all of these methods have severe limitations that restrict their usefulness in the event of an accidental release.

Some technical literature addressing the use of pipe freezing for plugging lines has examined the utility of pipe freezing for maintenance purposes (15). Many organic materials cannot form a frozen plug with the strength sufficient to stop flow. Many materials will require extremely low temperatures to create a solid plug. Even if a material can form a physically strong plug at an achievable temperature, the method would probably require too much time to implement in the event of an accidental release.

Stoppling and line shearing and sealing techniques have the advantage of forming a mechanically tight seal. These are proven methods that have a low potential for exposing personnel to the leaking material. The disadvantage of the methods is that they require elaborate and expensive equipment. Usually, a company that specializes in this work is contracted to plug a line. Such an arrangement would not be practical in the event of an accidental release because of the time required for the plugging equipment to be brought to the site.

7.3.2   <u>Applicability and Performance of Physical Barriers and Containment</u>
         <u>Systems</u>

Secondary containment systems can minimize the spread of a liquid spill
and/or the rate of toxic vapor generated from an accidental release. However,
the type of containment system best suited for a particular storage or process
unit will depend on the material being contained and on the risk associated
with an accidental release from that location. This subsection discusses the
factors that must be examined when selecting a secondary containment system.

In general, it is desirable to use some form of secondary containment
around storage facilities to prevent the unconfined spread of spilled liquid.
It is also beneficial to slope the surfaces away from storage tanks to a catch
basin so that liquid does not accumulate in the dike but is still confined to
an area near the dike. This is especially important if the material is toxic
and flammable, since an ignited pool of liquid beneath a vessel could lead to
a BLEVE (boiling liquid expanding vapor explosion).

In process areas, curbing around accumulators, reactors, or other vessels
containing large amounts of liquid serves to prevent the unconfined spread of
spilled liquid. For piping systems, the question of whether or not to provide
secondary containment depends partly on the pressure and temperature of the
material in the pipes and on the location of the piping system. Curbing or
trenches may be used along long pipe runs if a break could result in a large
accumulation of liquid. However, in such situations the curb or dike should
be designed to drain liquid away from the pipe. For a long pipe carrying
flammable materials this will reduce the potential for small leaks to result
in the accumulation of flammable materials around the pipe. Transfer areas
represent high-hazard areas because of the failure of temporary connections
used during transfer. Spills may occur frequently and a method of secondary
containment may be warranted.

Each situation must be specifically examined. The secondary containment system should at least be able to contain spills with a minimum of damage to the facility and its surroundings and with a minimum potential for escalating the event.

The applicability of diking to spills of volatile liquids is readily apparent. By containing the liquid, the dike reduces the surface area available for evaporation, at the same time allowing a liquid to be cooled by evaporation so that the vapor release is diminished. In this way, diking can reduce the rate at which a toxic material is released to the air. The material can be allowed to evaporate at a manageable rate, collected into alternate containers, or neutralized in place. A dike with a vapor fence will give extra protection from the wind and will be even more effective at reducing the rate of evaporation.

A dike will not reduce the impact of a gas or vapor release. A dike also creates the potential for toxic material and trapped water to mix, and, depending on the material, this may accelerate the rate of evaporation or the formation of highly corrosive solutions. If materials that would react violently with the released toxic material are stored within the same diked area, the dike will increase the potential for mixing the materials in the event of a simultaneous leak. A dike also limits access to the tank during a spill.

High-wall diking, on the other hand, can be used to minimize vapor generation rates and to protect the tank or vessel from external hazards such as missiles from nearby explosions. Maximum vapor generation rates will generally be lower for high-wall dikes than for low-wall dikes or remote impoundments because of the reduced surface contact area. Insulation on the wall and floor within the annular space can further reduce these rates.

One disadvantage of high-wall impoundments is that the high walls around a tank or vessel may hinder routine external observation. Furthermore, the closer the wall is to the tank or vessel, the more difficult it becomes to reach the tank or vessel for inspection and maintenance.

A remote basin, on the other hand, allows for removal of the spilled liquid from the immediate tank or process area, allows access to the tank during the spill, and reduces the probability that the spilled liquid will damage the tank, vessel, piping, electrical equipment, pumps, or other equipment. In addition, a covered impoundment will reduce the rate of evaporation from the spill by protecting it from wind or from heating by sunlight.

A limitation of a remote basin is that the potential still exists for water or other incompatible materials to be trapped in the impoundment and mix with the spilled material. Additionally, remote basins do not reduce the effects of a gaseous toxic release.

The effectiveness of a secondary containment system will depend on the specific material being contained. For some liquefied gases, such as chlorine, a spill from a pressurized storage or process vessel results in a significant amount of flashing and vapor generation and little liquid accumulation (14). In such cases, conventional low-wall diking may not be effective. All flashing gases "self-refrigerate" to some extent, and there will likely be some liquid accumulation. Low-wall diking can be constructed around a liquified gas tank in an area too small to contain the whole tank volume, since some will flash off, and a reduction in vapor evolution should be achieved. On the other hand, during releases of pressurized liquefied gases with higher boiling points or during releases of refrigerated liquefied gases, liquid will accumulate near the release point. In these cases, diking will be effective. Each situation must be examined in detail to determine the most appropriate applicable secondary containment system.

A limited amount of data is available concerning the performance of the various types of secondary containment systems. However, dispersion modeling can be used to illustrate the effects of diking and vapor holdup.

For example, a leak from a vertical 10,000-gallon refrigerated storage tank containing chlorine was modeled using the Complex Hazardous Release Model (CHARM®) of Radian Corporation. The chlorine was stored at -40°F and one atmosphere of pressure with a surrounding ambient temperature of 78°F. Vapor dispersion patterns for the following scenarios were investigated:

- Spill of chlorine directly onto soil from a 2-inch hole and allowed to spread unconfined; and

- Spill of chlorine directly onto soil from a 2-inch hole, confined in a 25-foot diameter.

The results of the modeling are presented in Figures 7-10 and 7-11. The results are based on a westerly wind of 10 mph and atmospheric conditions not conducive to mixing and dispersing gas or vapor clouds. Similar results were obtained by Eidsrick (16) for a ruptured pipe containing refrigerated propane.

Figures 7-10 and 7-11 show that confining the spill with a dike dramatically reduces the size of the vapor cloud by limiting the spread of the spill and reducing the surface area available for evaporation.

This example shows the potential effectiveness of a secondary containment system. In general, the effectiveness of any system depends on the properties of the particular chemical being contained and on the properties of the system itself. The secondary containment system should be able to contain spills with as little damage to the facility and its surroundings as possible and with a minimum potential for escalation of the event.

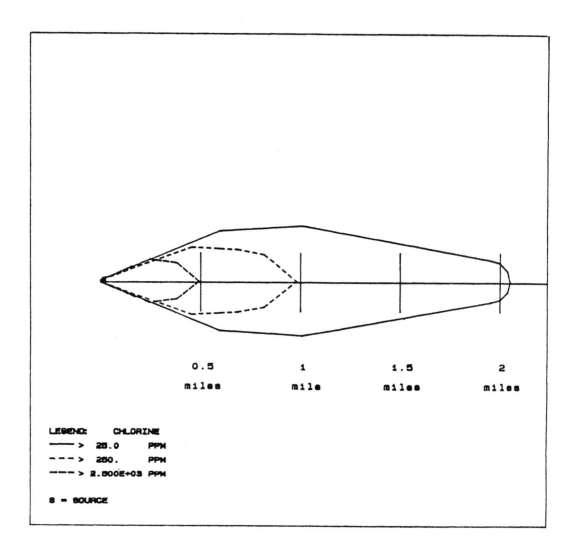

Figure 7-10.   Vapor dispersion patterns for a release of chlorine
from a 2-inch hole in a 10,000 gallon refrigerated
storage tank using CHARM® model.

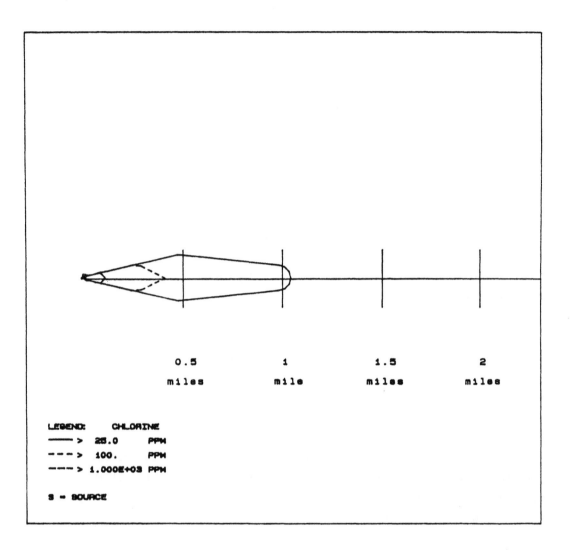

Figure 7-11. Vapor dispersion patterns for a release of chlorine from a 2-inch hole in a 10,000 gallon refrigerated storage tank surrounded by a 25-ft diameter dike using Charm® model.

7.4   RELIABILITY

7.4.1   Reliability of Leak-Plugging Techniques

A qualitative evaluation of the reliability of a leak-plugging device can be based on the ability of the method to securely stop the flow of material from a leak. Reliability could also be assessed in terms of how quickly a leak can be stopped in the event of an accidental release.

Chemical patches and plugs can be applied to a leak relatively quickly. In most cases, some equipment will be required to generate the plugging compound. Since this equipment may sit idle for very long periods of time, routine inspection and maintenance of the equipment will be required to maintain a quick response time.

Chemical patches and plugs are mechanically and chemically weak when compared with other leak repair methods. Vibration, shock, or high pressure can all result in the failure of a chemical plug. These plugs are also subject to failure when exposed to high temperatures or to chemically incompatible materials.

Mechanical patches and plugs can be applied to a leak relatively quickly. They form seals that can sometimes have the structural strength of the original process equipment; however, mechanical vibration, shock or improper application can cause a mechanical leak plugging device to fail.

Methods for stopping the flow of material upstream of a leak are characterized by long preparation and implementation times. Once in place, however, these methods are reliable ways of stopping flow.

7.4.2   Reliability of Physical Barriers and Containment Systems

The reliability of a secondary containment system is directly related to how it is designed and constructed. In particular, the reliability will be affected by the following:

- Ability to withstand the hydrostatic pressure and temperature of a spill;

- Ability to withstand the dynamic stresses caused by fire or explosion, in the case of flammable materials; and

- Ability to withstand damage caused by earthquakes, floods, high winds, frost, or similar events.

Systems must be inspected and maintained regularly. Undetected corrosion and cracks in dike walls, for example, could lead to the failure of a secondary containment system during an accidental release.

7.5   SECONDARY HAZARDS

7.5.1   Secondary Hazards of Leak-Plugging Techniques

All leak-plugging techniques can potentially expose plant personnel to the leaking material. Often, material flowing from a leak will spray out around the plugging device as it is being applied. Also, where flammable vapors are involved, these personnel will probably have to enter a flammable gas cloud when applying the plug.

Where hazardous materials are involved, a temporary leak repair should never be treated as a permanent fix. Once the leak has been plugged, actions should be taken to shut down the portion of the process involved with the release so that permanent repairs can be made.

7.5.2   <u>Secondary Hazards of Physical Barriers and Containment Systems</u>

Mixing incompatible materials is one potential secondary hazard associated with most types of secondary containment. This is a particular problem
where water-reactive materials are stored within the containment area.
Rainwater can collect in diked areas or in containment basins. Spilled
materials could mix with the water in the event of an accidental release.
Though not so likely, two incompatible chemicals could mix if they were both
stored within the same secondary containment area, and if both were released
simultaneously.

A spill from a vessel surrounded by a dike can fill up the dike to a
level that prevents access to the vessel, which could prevent repair or
emergency response crews from stopping the release at its source. Also, a
fire in this situation would expose the vessel to extreme heat, which could
result in an explosion. For both of these reasons, dikes should be designed
so that material accumulates away from vessels within the dike.

A covered collection basin can trap flammable vapors when a liquid is
involved. Fire could spread to such a basin and result in an explosion.

7.6   COSTS

Table 7-2 presents some order-of-magnitude costs for a few leak-plugging
devices. These costs are for equipment only and do not include installation.

Although the variety of types and sizes of secondary containment systems
makes it difficult to estimate costs, costs for dikes built around storage
facilities can be estimated easily. The cost of a rectangular dike can be
expressed as (17):

$$COST = (aD^2 - \pi r^2)F + 2(1 + a)DHW$$

TABLE 7-2. COSTS OF VARIOUS LEAK PLUGGING DEVICES (1986 Dollars)

Equipment Item	Cost ($)	Reference
Small transportable polyurethane foam system	5,000 - 8,000	18
Chlorine Institute Emergency Repair Kit for Chlorine Cylinders	800	1
Chlorine Institute Emergency Repair Kit for Chlorine Ton Containers	900	2
Chlorine Institute Emergency Repair Kit for Chlorine Tank Cars and Tank Trucks	1,000	3
Stainless steel band clamp, for leaking 4" pipe, working pressure less than 150 psi	100	12
Carbon steel bolted sleeve, for leaking 4" pipe, working pressure less than 1000 psi	500	18
Clamp for sealing pinhole leaks on a 4" pipe, working pressure less than 1000 psi	70	18

where:

a - ratio of dike length to dike width

D - dike width

r - tank radius

H - dike height

F - cost of dike floor per unit area

W - cost of dike wall per unit area.

In general, the cost of secondary containment systems will be small compared with the costs associated with the vessels that they protect. Besides the initial investment, additional costs are associated with the necessary cleanup equipment needed such as pumps, for example, when a release occurs within the containment systems.

7.7 REFERENCES

1. The Chlorine Institute, Inc. Chlorine Institute Emergency Kit "A" for 100-lb and 150-lb. Chlorine Cylinders. Washington, D.C., 1986.

2. The Chlorine Institute, Inc. Chlorine Institute Emergency Kit "B" for Chlorine Ton Containers. Washington, D.C., 1981.

3. The Chlorine Institute, Inc. Chlorine Institute Emergency Kit "C" for Chlorine Tank Cars and Tank Trucks. Washington, D.C., 1981.

4. Feind, K. Reducing Vapor Loss in Ammonia Tank Spills. CEP Technical Manual, Ammonia Plant Safety and Related Facilities. Volume 17. American Institute of Chemical Engineers, New York, NY, 1975.

5. Mitchell, R.C., Hamermesh, C.L., Lecce, J.V. Feasibility of Plastic Foam Plugs for Sealing Leaking Chemical Containers. EPA-R2-73-251 (NTIS PB 222627), U.S. Environmental Protection Agency, May 1973.

6. Vrolyk, J.J., Mitchell, R.C., Melvold, R.W. Prototype System for Plugging Leaks in Ruptured Containers. EPA-600/2-76-300 (NTIS PB 267245), U.S. Environmental Protection Agency, 1976.

7. Cook, R.L., Melvold, R.W. Development of a Foam Plugging Device for Hazardous Chemical Discharges. In: Proceedings of the 1980 National Conference on Control of Hazardous Material Spills, Vanderbilt University, Nashville, TN, 1980.

8. Hendriks, N.A. Safety Wall Systems for Ammonia Storage Protection. CEP Technical Manual, Ammonia Plant Safety and Related Facilities. Volume 21. American Institute of Chemical Engineers, New York, NY, 1979.

9. Benson, R. Hydrogen Fluoride Exposure - Prevention in the Operation of HF Alkylation Plants. Industrial Medicine 13(1): 113-117, 1944.

10. Zajic, J.E., Himmelman, W.A. Highly Hazardous Materials Spills and Emergency Planning. Marcel Dekker, Inc., New York, NY, 1978.

11. Bond, J. Leak Sealing Using Compound Injection. Loss Prevention Bulletin. Issue #069. 1986. (No publisher or other identification provided in the publication.)

12. The Pipe Line Development Company, PLIDCO. Piping Repair and Maintenance Products. Cleveland, OH, 1987.

13. Schelling, Carter. Reducing Repair Downtime with Pipe Freezing. Plant Engineering, July 24, 1986.

14. Lees, F.P. Loss Prevention in the Process Industries - Hazard Identification, Assessment, and Control. Volumes 1 & 2. Butterworth's London, England, 1983.

15. Aarts, J.J. and D.M. Morrison. Refrigerated Storage Tank Retainment Walls. CEP Technical Manual, Ammonia Plant Safety and Related Facilities. Volume 23. American Institute of Chemical Engineers, New York, NY, 1983.

16. Eidsvik, K.J. A Model for Heavy Gas Dispersion in the Atmosphere. Atmospheric Environment, Volume 14, 1980.

17. Johnson, D.W., and J.R. Welker. Diked-In Storage Areas, Revisited. Chemical Engineering, July 31, 1978.

18. Batch Air Equipment, San Antonio, TX: vendor.

8. Spray, Dilution, and Dispersion Systems

Water spray systems are routinely used in the chemical process industries for a variety of fire protection purposes. However, they can also be used to reduce the effects of toxic and/or flammable gas or vapor releases. Theoretical studies and experimental research have shown that such sprays can be effective in aiding the dispersion and dilution of gas or vapor clouds resulting from an accidental release (1,2,3,4,5,6,7,8). However, many of the results are based on specific systems operating under a specific set of conditions. Details concerning the overall effectiveness of these systems are limited in scope. Additional research in this area is needed to refine the present data.

Steam curtains are another technique for reducing the effects of toxic and/or flammable gas/vapor releases. Steam curtains act similarly to water sprays in that the primary dispersing mechanism is the dilution of the gas or vapor with air. However, steam curtains provide enhanced buoyancy to the toxic and/or flammable cloud by heating the gas or vapor passing through the steam curtain.

This section of the manual presents a detailed discussion of the use of water spray systems and steam curtains for the mitigation of toxic and/or flammable gas or vapor releases. Both mobile and fixed installations are discussed in terms of their applicability, performance capabilities, reliability, secondary hazards, and costs. In addition, reactive spray systems are briefly discussed.

8.1   BACKGROUND

8.1.1 Spray Systems

A spray is defined as a dispersion of liquid droplets in a gas. Sprays can achieve a variety of objectives, including, but not limited to, the forced dispersion and dilution of a gas or vapor, absorption of water-soluble materials, confinement of released vapor or gas in a particular area, or diversion of the vapor or gas away from a particular area.

In mitigating toxic and/or flammable gas or vapor releases, the primary purpose of sprays is to dilute the gas or vapor with air. This is brought about by the entrainment action of the sprays. The momentum and energy of the spray causes air to be pulled into the cloud and creates momentum and turbulence in the gas or vapor, thus improving mixing and enhancing dilution. The performance of a spray depends in part on the drop size distribution. Fine droplets lose their momentum quickly and move large volumes of air relatively slowly. Coarse droplets entrain less air but induce greater velocities.

Dilution is also achieved to a lesser degree by absorption of the gas in the liquid drops. Gases that are highly soluble in water (e.g., ammonia) can be partially absorbed by a water spray. However, the mass transfer characteristics of the system may limit the efficiency of absorption; it is difficult to create a sufficiently high liquid-to-gas ratio using sprays. A "fog" consisting of fine water droplets often improves the mass transfer efficiency. Adding a chemical to the spray solution that reacts with the vapor can improve the absorption efficiency. An example would be using an alkaline spray solution to absorb an acidic vapor release.

Finally, spray-induced warming of cold vapor clouds that form from liquefied gas releases can enhance the dilution of a heavier-than-air cloud.   The individual water drops transfer heat to the cloud, thereby increasing both the buoyancy of the gas and its dispersal.

In addition to dilution, liquid sprays are also useful for containment and diversion. In some situations, the movement of vapor clouds, especially if visible because of aerosol formation or condensation of moisture, can be controlled by hand-held nozzles and fire monitors. A released gas cloud may also be contained in a particular area by a series of spray nozzles around the perimeter.

8.1.2 Steam Curtains

As with water sprays, the primary purpose of a steam curtain used to mitigate toxic and/or flammable gas/vapor releases is to dilute the gas or vapor with air. The energy of the steam causes air to be pulled into the cloud and creates turbulence in the gas or vapor, thus improving mixing and enhancing dilution.

Steam curtains will also heat the gas cloud and enhance its buoyancy. Heating the cloud will decrease its density and help the cloud to rise from the ground, which can result in decreased ground level concentrations of the toxic vapors downwind of the release.

The following subsection describes spray systems and steam curtain systems that may assist in the mitigation of accidental vapor releases.

8.2 DESCRIPTION

In recent years, increased attention has been focused on the use of aqueous sprays and steam curtains for the control and dispersion of toxic and/or flammable vapor releases. Research studies have revealed that such systems are effective; however, the technology is still in the early stages of development.

8.2.1   Spray Systems

A spray system may consist of a single nozzle, such as a fire monitor, or a series of specially designed nozzles that discharge in a predetermined spray pattern. The type, design, and method of application of spray systems vary, depending on the specific situation. Two systems are most common: fixed water sprays and mobile water sprays. These systems may be used individually or in combination. A reactive water solution is sometimes used with either a fixed or mobile spray system.

Fixed Water Spray Systems--

Fixed water spray installations are similar to fire protection fixed pipe deluge systems. They may be located in the open air or in enclosures where there is a high probability of an accidental release of a toxic and/or flammable gas/vapor or where there is a need to protect a certain area (i.e., a control room). These spray systems are used where there is a need for quick application of water spray.

Fixed installations may consist of a series of spray nozzles elevated off the ground with the spray directed downwards or located at ground level with the spray directed upwards. Figure 8-1 is a conceptual drawing of a typical system using downward spraying nozzles. Two facilities in England have installed such fixed water spray barriers (9); however, they are limited to the control of small scale leaks.

Figure 8-2 shows a system where the spray nozzles completely surround a desired area. The advantage of a completely-surrounding system is that it also provides a degree of containment. However, total spray barrier systems may use excessive quantities of water since upwind sprays will have little effect on the gas cloud. Spray barriers that totally surround a system often are designed so that only downwind sections are activated in the event of a

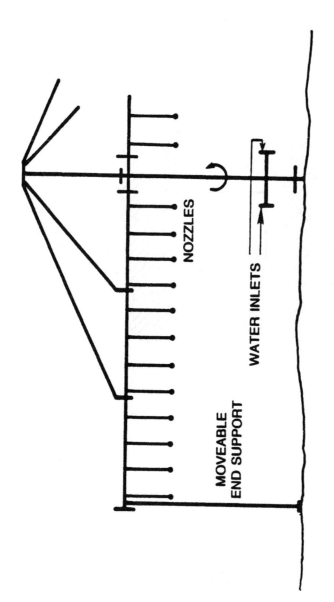

Figure 8-1.   Conceptual diagram of water spray barrier using downward spraying nozzles.

Source: Adapted from Reference 8.

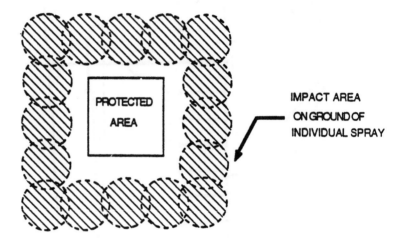

IMPACT AREA
ON GROUND OF
INDIVIDUAL SPRAY

PROTECTED
AREA

Figure 8-2. Conceptual diagram of fixed-spray barrier
surrounding protected area.

release, as shown in Figure 8-3. Activation of fixed systems can be manual or automatic (by using gas detectors and switches).

Fixed water sprays incorporated into semi-enclosed structures have also been proposed as a possible mitigation technique, although this is still an experimental concept (10). Figure 8-4 shows such a system. These systems offer several potential advantages over other systems. First, it can provide a degree of direct containment by partially enclosing the released vapors. Second, release of the material occurs at a known location, rate, and concentration.

Thus, if sources with a high-release probability can be identified and the exposed area and release direction can be defined, a properly designed fixed water spray system can be effective in reducing the hazards of an accidental toxic release.

Mobile Water Spray Systems--

A mobile water spray system, illustrated in Figure 8-5, consists of hand-held firefighting nozzles and/or fire monitors. It can be used as an alternative to a fixed water spray system or when a release has moved beyond the reach of a fixed installation.

Mobile water sprays in the form of a "fog" have been used for absorbing and diluting of water-soluble materials. A "fog" is defined as a water spray having a mass median drop diameter of 0.03 in. or less (11). The fine spray or "fog" allows for more effective mass transfer. A technique for controlling ammonia releases has been developed from several hundred outdoor ammonia workshops for firefighters (11). As shown in Figure 8-5, fire hoses and fog nozzles are used to create a "capture" zone downwind of a release by positioning the nozzles in designated locations. This technique was developed for a specific type of release (i.e., an ammonia tank rupture) and thus may not apply in all situations.

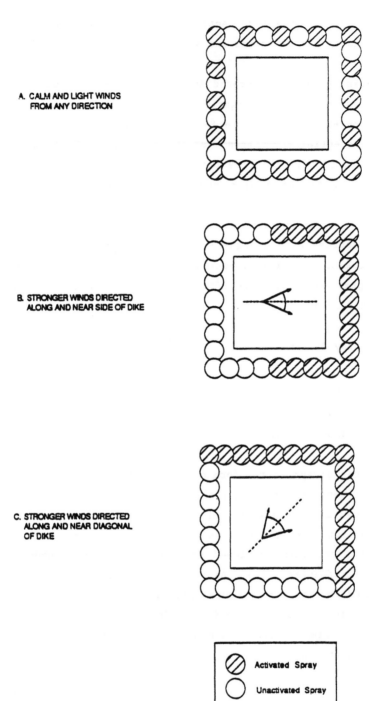

A. CALM AND LIGHT WINDS
FROM ANY DIRECTION

B. STRONGER WINDS DIRECTED
ALONG AND NEAR SIDE OF DIKE

C. STRONGER WINDS DIRECTED
ALONG AND NEAR DIAGONAL
OF DIKE

Activated Spray

Unactivated Spray

Figure 8-3.   Conceptual diagram of fixed-spray barrier with activation
according to wind direction.

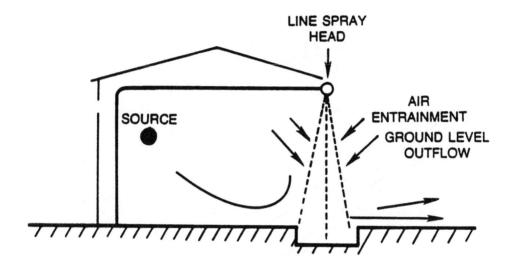

Figure 8-4.   Conceptual diagram of water sprays incorporated
into semi-enclosure.

Source: Adapted from Reference 10.

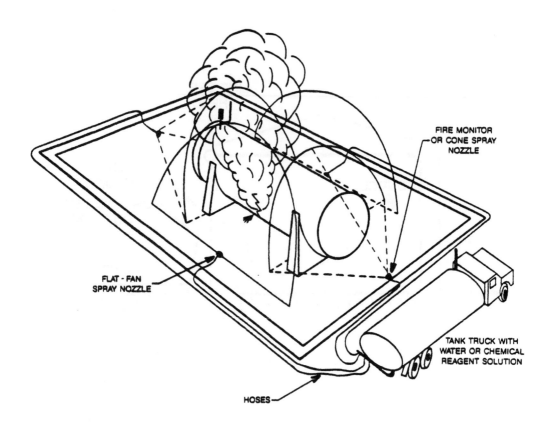

FIRE MONITOR
OR CONE SPRAY
NOZZLE

FLAT - FAN
SPRAY NOZZLE

TANK TRUCK WITH
WATER OR CHEMICAL
REAGENT SOLUTION

HOSES

Figure 8-5. Typical mobile water spray system.

Source: Adapted from Reference 11.

Mobile systems have also been used as effective barriers. In particular,
Beresford (12) has developed a system whereby coarse water spray discharging
from flat fan sprays and wide-angled spray monitors in an upward direction
completely surround a flammable and/or toxic gas leak in the form of a "chimney." This technique is shown in Figure 8-6. Large amounts of air are
induced at ground level, and dilution of the gas is achieved as the sprays
push the gas out the top of the "chimney" allowing it to disperse safely.

A mobile water spray system must be capable of rapid deployment to the
source of a release. In many releases, the worst may be over by the time such
a system is operational. However, the cost of mobile systems are usually less
than fixed installations since much of the same equipment can also be used for
fire protection. In fixed installations, additional piping and nozzles may
have to be added to existing fire protection systems. Thus, for very large
installations, a mobile system may be the most cost-effective.

Reactive Spray Systems--

An alternative to a water spray system is the use of a mild aqueous
alkaline spray system (e.g., reactive spray system). Often the use of water
in absorbing and diluting a toxic gas or vapor is limited by the mass transfer
characteristics of the system. In addition, dilution of some materials with
water sprays results in the formation of highly corrosive mists. A mild
alkaline spray would both enhance absorption and act as a neutralizing agent
in the case of acid formation.

Limited data are currently available on the use of reactive spray
systems. Several major users of phosgene have experimented with ammonia-
injected water spray systems at their facilities (13). These systems consist
of an ammonia cylinder connected to a water feed line of a water spray system
via an injector system. When the spray system is activated, a valve opens
allowing ammonia, under pressure, to enter the water line and create the
alkaline solution.

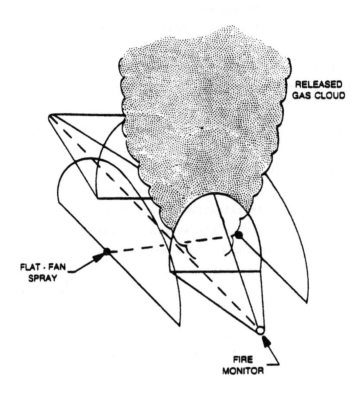

RELEASED
GAS CLOUD

FLAT - FAN
SPRAY

FIRE
MONITOR

Figure 8-6. Conceptual diagram of mobile water spray
barrier and associated "chimney" effect.

Other alkaline solutions could also be used. Several examples are sodium carbonate, calcium hydroxide derived from slaked lime, or a weak solution of sodium hydroxide. However, adequate supplies of these solutions would have to be stored or a quick means of production would be required in an emergency situation.

Additionally, the hazards associated with spraying an alkaline solution into the air must be considered. Unless the spray is applied so that the alkaline material remains in a confined area (e.g., a diked storage facility), reactive sprays present an additional health hazard to plant personnel. Also, such sprays may cause the corrosion of machinery and equipment.

In conclusion, reactive spray systems may be more efficient than water sprays in applicable situations. However, more research is needed to define potential applications, performance capabilities, and system design.

8.2.2 Steam Curtains

Steam curtains are fixed-pipe systems used to contain and disperse releases of toxic and/or flammable gas or vapor. Cairney and Cude (14) and Simpson (15,16) have described such systems in detail. A steam curtain consists of a horizontal steam pipe with a row of small holes in the top and mounted near the top of a wall. Figure 8-7 illustrates a typical system. When operated, a wall of steam approximately 15 to 20 feet high is produced. The steam pipe is designed so that all the individual jets combine to form a continuous curtain of steam that entrains sufficient air to dilute the gas or vapor concentration to below its toxic and/or flammable limit. Additionally, the steam pipe is usually divided into sections that are individually supplied with steam from a distribution main, allowing plant operators or an automated activation system to select which sections of the steam curtain will be activated in the event of a release.

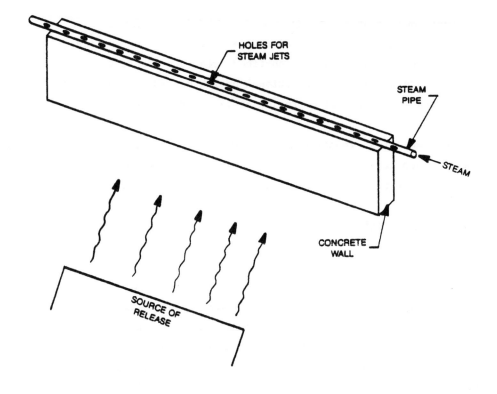

Figure 8-7.   Conceptual diagram of a steam curtain.

The steam curtain can be activated automatically or manually. However, in practice, steam curtains are typically controlled manually by remotely operated valves (17,18).

8.3 DESIGN

8.3.1 Spray Systems

In recent years, much research and a number of experimental studies have contributed to the development of several experimental design methods for water spray systems (6,7,12,19). However, the optimum set of parameters for a particular set of circumstances (i.e., nozzle type, size, location, number, orientation) are not always clear, and, indeed, each situation requires a unique set of parameters. This section of the manual focuses primarily on the circumstances considered in designing water spray systems for mitigating toxic and/or flammable gases and vapors. A brief discussion of several of the proposed design methods mentioned above are also included.

Detailed procedures for the design of fixed water spray barriers have been proposed. All of these methods use different approaches. The procedures have been developed from limited experimental data and may not apply in all situations, and since many factors, such as large obstructions (building, process units, other structures), and shifting wind direction will modify the dispersion characteristics of a gas cloud, an appropriate application and design must be evaluated case by case. Therefore, the following discussion merely indicates the types of procedures currently available for designing water spray systems.

McQuaid (19) developed a method using an experimental correlation whereby a water spray barrier can be sized by equating the total air entrained by a barrier (a summation of the quantity of air entrained by each nozzle) to the amount of air required to dilute the gas or vapor cloud below a certain level. The method is based on the assumption that adequate mixing requires an induced air velocity of approximately 20 feet/second. The air flow rate and the water flow rate at the nozzles are required. This method can give the type, number, and position and direction of nozzles required for a given installation. This method was derived from work carried out on the entrainment properties of water sprays used in coal mines for auxiliary ventilation. Predicted results were compared with test data on large-scale sprays as reported in the literature (19).

In contrast, Moodie (7) developed a design technique based on the assumption that a water spray barrier behaves as a jet in a crossflow. From this, a characteristic length can be defined in terms of the square root of the momentum flow rate of the barrier and a characteristic wind speed. For a specific wind speed and specified degree of dilution, the momentum flow rate can be determined from an experimental correlation. Using this information and a procedure developed by Moodie (7) for nozzle selection, the barrier can be sized in terms of its width, nozzle type, and water consumption. This method was developed from extensive test data generated using water sprays to mitigate carbon dioxide releases; therefore, it is based on substantial field data.

Apart from these proposed design methods, the only other technique that has been developed to aid in the design of fixed systems is the use of computer models to predict the performance of various spray systems (20,21,22,23). In these models, water sprays are treated as sources of momentum imparted to the air at the same location as the actual water spray. The quantity of momentum imparted is related to the flux of the air entrained by the water spray. It is assumed that the air mixes instantaneously with the gas/air mixture in the plume at the spray location, resulting in a sudden change in geometry and composition of the plume. The accuracy of such models depend on the accuracy of the dispersion model used. At best, they indicate how a spray barrier can influence the downwind development of a gas plume. However, they can help the planner decide how the dispersing effects of a water spray system can best be achieved.

Whether the system is mobile or fixed, designed as a barrier or as a fog, the following are important design characteristics:

- Water pressure,
- Water flow rate,
- Nozzle type,
- Nozzle spacing, and
- Nozzle orientation.

Research has shown that the efficiency of water spray systems increases as the water pressure increases (8,24). One source suggests that a minimum nozzle water pressure of approximately 145 psi be used for fixed systems (8), which is consistent with experience in gas-cleaning systems where scrubbing efficiency increases with energy input.

Research has also shown that the most efficient spray systems are designed so that the adjacent sprays just impinge on each other (6,24). This helps prevent the passage of the cloud between the sprays. Significant overlap is inefficient. Thus, the spacing will depend on the nozzle type, size, and spray angle selected.

Water spray systems perform as deluge type systems with all nozzles open. Based on the number of nozzles, nozzle spacing, and flow to each nozzle, up to several thousand gallons per minute of water may be required. The water supply system must be properly sized to permit operation at the designed pressure and flow rate for spray system. A municipal water system may not be able to deliver the necessary flowrate. A second water supply may be required for the spray system. Many facilities will already have this type of water system in place for fire protection. It may be appropriate to use this system to supply the spray system as well. A backup water system will often be

composed of a large water storage tank that is automatically kept full. The
spray system designer must decide which accidental release events are most
likely to occur and whether the waste supply contains a sufficient quantity of
water to mitigate these releases. In addition, drainage must be adequate to
prevent flood damage.

Finally, the nozzle type and size must be chosen to provide a density of
water spray at a velocity and droplet size that maximizes the dilution and/or
absorption of the gas cloud. There are many basic types of nozzles; however,
based on recent test data, the most useful are those producing hollow-cone and
fan-tail type sprays, as illustrated in Figures 8-8 and 8-9 (6,8,24). Moodie
(7) has developed a scheme where the choice of nozzle, its size, spray angle,
and separation distance can be determined in a systematic fashion, based on
the specific momentum flow rate for a spray.

Additional consideration must be given to the piping system when fixed
systems are designed. Two general types are used: wet systems and dry systems. In a wet system, the pipes are kept full of water under pressure. In
dry systems, the pipes are empty until a master valve is opened when the
system is activated. If there is a danger of freezing, a dry system is required to ensure that the system is not rendered ineffective by freezing.
Consideration should also be given to other hazards, such as potentially
explosive processes nearby, when designing such systems.

8.3.2   **Steam Curtains**

Only limited information is currently available on the effective design
of steam curtains. Seifert (25) has developed a mathematical model that can
be used as a preliminary basis for designing such systems, and experimental
evidence verifies this model (25).

The model is based on the assumption that the toxic and/or flammable gas
cloud moves toward the curtain on one side while the steam jets entrain

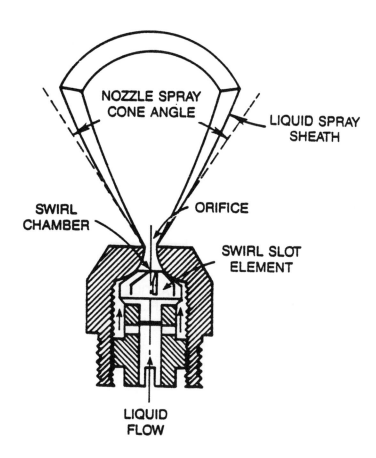

Figure 8-8. Typical hollow-cone spray nozzle.

Source: Adapted from Reference 26.

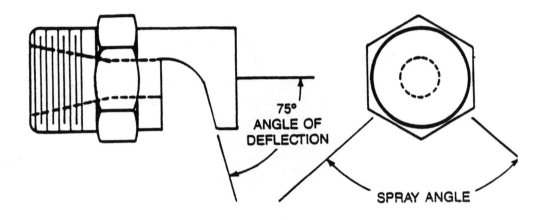

75°
ANGLE OF
DEFLECTION

SPRAY ANGLE

Figure 8-9.   Typical fan-tail spray nozzle.

Source: Adapted from Reference 5.

surrounding air from the other side, as illustrated in Figure 8-10. When the released cloud reaches the curtain, the steam jets mix the gas or vapor with air and heat the vapors which provides buoyancy, causing the mixture to rise, dilute, and disperse.

The design method requires knowledge of the mass flow rate and density of the released gas or vapor. With these data, a subsequent system design can be determined for a given reduction in concentration.

8.4 APPLICABILITY AND PERFORMANCE

8.4.1 Spray Systems

Water sprays can be used to help mitigate accidental toxic and/or flammable gas or vapor releases. However, such systems are often limited in their scope and may not be capable of dealing with catastrophic releases. Although much theoretical and experimental work has been done on the use of these spray systems, the use of water spray systems is not a proven mitigation technology for all types of toxic and/or flammable gas or vapor releases.

Hazard analysis can be used to determine where a spray system might be appropriate. Such an analysis can identify the locations within a chemical facility with the highest potential for an accidental release of a hazardous vapor. The analysis can also indicate the potential impact that each release would have on the surrounding community. Sometimes there will be only a few locations within a facility with a high accidental release risk. In these situations, a fixed spray system at each potential release site may be appropriate. However, most facilities will have a number of possible release sites.

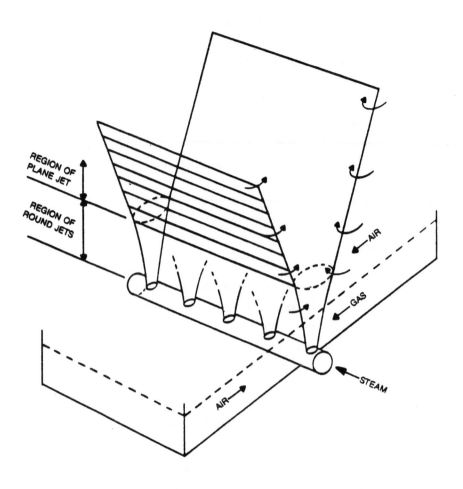

Figure 8-10.   Conceptual diagram of steam jets emerging
                from steam curtain.

Source: Adapted from Reference 25.

Since mitigative spray system technology is still in a developmental stage, there are no proven rules for designing an appropriate spray system when there are multiple release sites.

If a fixed system is installed, the potential exists that a release will occur in an area not covered by the system. Even if a release occurs in a location with a fixed spray system, the individual spray nozzles may not be directed directly at the vapor plume. The sprays from a mobile system can be directed to optimize dilution of the plume. However, a mobile system cannot be activated as quickly as a fixed system. A significant quantity of vapors can be released before the mobile system is in place.

For some facilities it may be appropriate to install fixed spray systems in the highest risk locations and have a mobile spray system on hand as a backup.

For flammable vapors, a spray system can help dilute the vapors below the flammable limit. However, the vapors can still ignite if they are present in flammable vapor phase concentrations. Mobile spray systems should only be used in situations where the personnel operating the sprays can be located away from the flammable gas cloud (this is sometimes impossible to do). A fixed spray system is safer to use for flammable vapors.

The results of the theoretical and experimental studies in recent years show that water sprays can be used effectively under certain circumstances. However, limited and often contradictory performance characteristics, optimal system design, and reliability data have been reported. Table 8-1 lists some of these studies, along with major results and performance characteristics.

Several investigations have attempted to quantify the performance of water spray systems. The most comprehensive method is that developed by Rees (6) and Moodie (7). It consists of ranking the spray system performance in

TABLE 8-1.   RESULTS OF SEVERAL EXPERIMENTAL STUDIES ON THE EFFECTIVENESS OF WATER SPRAYS

Description of Study	Summary of Results	Reference
Clouds of ethylene and vinyl chloride vapor directed from ground level towards 20-foot spray barrier, with sprays directed downwards	If velocity of entrained air exceeds that of vapor cloud, spray acts as barrier. Vapor clouds with velocities of 3 and 9 mph were stopped by sprays with 100 psig nozzle pressure. Only lower velocity cloud stopped with 40 psig pressure. Where cloud passed through barrier, there was still an appreciable dilution effect. Water spray did not prevent ignition of flammable cloud and flame speed actually increased.	1 (1976)
Clouds of propane vapor directed towards spray barrier, with downward directed spray nozzles at 1, 5, 9, 15, and 20 feet above the ground, 55-feet long with nozzles 5 feet apart.	A properly designed and installed water spray system can dilute a flammable mixture below the LEL. However, the system does not act as a flame arrester. Ignition of vapor cloud depends on the spacing of nozzles and water pressure.	2 (1976)
Effectiveness of water sprays in suppressing combustion in mists of heat transfer media, specifically Dowtherm A.	Combustion suppressed by a scrubbing mechanism that reduced concentration of droplets in the mist.	3 (1976)
A water curtain generated by mobile monitors was used on a chlorine gas cloud.	The curtain successfully protected a small area against the gas cloud.	4 (1975)
A series of wide-angle, flat spray nozzles was used to disperse LNG vapors.	Two mechanisms are responsible for dilution: heating the vapor and increased turbulence. Such sprays are effective for small, confined area LNG spills. Large spills may require systems too large to be cost effective.	5 (1977)

(Continued)

TABLE 8-1.   (Continued)

Description of Study	Performance Characteritics	Reference
A series of eight experiments ranging up to full scale were conducted to evaluate the performance and effectiveness of water sprays. Propane was used as gas source.	Upward water sprays are significantly more effective than downward. The best type of nozzle is a conical, narrow angle, high velocity.	6 (1981)
A series of full-scale tests using hollow-cone spray nozzles inclined downwards were conducted using carbon dioxide.	Water spray barriers can be effective in dispersing a heavy gas cloud. Nozzles pointing in a vertical direction or angled into the gas cloud at a 45° angle are most effective. Effectiveness can be optimized in terms of the water pressure.	7 (1985)
Two water spray barriers, one using downward-directed nozzles, the other using upward-directed nozzles were used to develop performance data for full-scale barriers using carbon dioxide.	High momentum discharges from flat-fan nozzles with spray angles of 90° or less are most effective. At a wind speed of 5 mph, a vertical system can produce a ratio of gas concentration with and without sprays of approximately 4. At a wind speed of 20 mph, this ratio decreased to about 1.5.	8 (1981)

terms of a concentration reduction (CR) ratio (Moodie (7)) or a forced dispersion (FD) factor (Rees (6)); both are a ratio of gas concentrations with and without a water spray observed at a point downstream.

Moodie (7) demonstrated that, in general, high momentum discharges from flat-fan spray nozzles with spray angles of 90° or less are the most effective. Experimental results using carbon dioxide indicated that at a wind speed of 5 mph, a properly designed vertical water spray system directed upwards produced a CR ratio of approximately 4. At a wind speed of 20 mph, the effectiveness was reduced to 1.5.

Prugh (27) developed a systematic method in which the maximum theoretical effectiveness of a water spray barrier can be calculated. Material balances are used to determine the mole-fraction of vapor in the air exiting the barrier, which is then compared with the average crosswind concentrations of the vapor in the air entering the barrier, resulting in a reduction factor for the system. However, this method is limited to water-soluble vapors.

In general, water sprays can be an effective means of absorbing water-soluble gas or vapors (e.g., ammonia). However, for many released chemicals the efficiency of absorption may be low because of mass transfer limitations. As a result, a second water spray barrier may also be required further downstream.

Likewise, water spray barriers appear to be limited in their effectiveness. It has been reported that these barriers are only marginally effective in reducing the hazards of gases and vapors not soluble in water (3,19,21,28,29,30). Others have gone as far as regarding them as being limited to water-soluble, low-density, and/or non-flammable vapors (7).

8.4.2 Steam Curtains

Steam curtains were initially designed to dilute and contain heavy flammable vapors and are incorporated into several installations to prevent highly flammable materials from reaching sources of ignition in the event of an accidental release (17,18,31). In principle, steam curtains can also be helpful in mitigating heaver-than-air releases of toxic and/or flammable vapors. At the time this manual was prepared, no commercial steam curtain installations were known to exist for mitigating toxic releases.

In general, the characteristics of water spray systems described in Section 4 are similar to those of steam curtains. However, steam curtains require large quantities of steam, typically 0.15 ton/hr per foot of curtain. Therefore, they must often be limited to small-scale uses. Also, where steam is not available or where the supply of steam is not reliable, water spray systems may be more applicable.

The effectiveness of steam curtains has been investigated experimentally by Rees (32), Seifert (26), and Rulkens (33). These investigations found that steam curtains can reduce the concentration of a vapor cloud from an accidental release. Table 8-2 shows the effectiveness of steam curtains for diluting a flammable gas to levels below its lower flammability limit (9). The primary advantage of steam curtains over water sprays is that steam curtains enhance the buoyancy behavior of the cloud by heating the vapors reaching the curtain.

However, steam curtains are reported to be less effective at mixing than water sprays at the same supply pressure (32). More steam is required for comparable mixing. This can be more important for toxic than flammable vapors or gases since the former are hazardous at lower concentrations. The buoyancy effect may still be beneficial, however. For example, for materials like ammonia, whose density is near that of air, heating from the steam curtain may

TABLE 8-2.   EFFECTIVENESS OF STEAM CURTAINS[a]

Source Strength[b] (lb/s)		44			132			264		
Wind Velocity (ft/s)		3.3	6.6	13.2	3.3	6.6	13.2	3.3	6.6	13.2
Distance from Curtain to Source (ft)	Steam Flow (ton/hr)									
165	25	+	−	0	+	−	−	−	−	−
	50	+	+	0	+	−	−	+	−	−
	100	+	+	0	+	+	−	+	+	−
330	25	+	+	0	+	−	0	−	−	−
	50	+	+	0	+	+	0	+	−	−
	100	+	+	0	+	+	0	+	+	−
500	25	+	0	0	+	−	0	+	−	0
	50	+	0	0	+	+	0	+	+	0
	100	+	0	0	+	+	0	+	+	0

Note: + effective
       − not effective
       0 sufficient dilution
         by wind alone

[a] Based on propylene release diluted below 2 percent.
[b] Amount of propylene released at the source.

Source: Adapted from Reference 9.

be sufficient to cause the plume to become buoyant and airborne, which would result in a significant reduction in downwind ground level concentrations.

8.5 RELIABILITY

8.5.1 Spray Systems

Reliability data for water spray release mitigation systems do not appear to be readily available. In general, the reliability of water spray systems should depend on the design of the system and on the reliability of the individual components that make up the system. However, reliability is more than the degree to which a piece of hardware is free of basic defects that may prevent it from operating properly. It also includes human factors, interactions, and controls that determine whether a systems is effective or not.

System unreliability can often result from events other than simple mechanical failure, including:

- Failure to properly define the hazard;
- Inadequate quality control;
- Poor maintenance;
- Incorrect operations of the system; and
- Changes made to the surrounding systems that were not accounted for in the original spray system design.

Systems must be regularly inspected and maintained. Of particular importance are the spray nozzles, which must be inspected and cleaned as necessary, or provided with some type of nozzle protection to ensure that their small water passages remain clear.

8.5.2 Steam Curtains

A steam curtain system can be no more reliable than the steam source. Steam pressure must be available. Since the system is dormant much of the time, the mechanism that activates it could become inoperable if neglected. Reliability depends on regular inspection, testing, and maintenance.

8.6 SECONDARY HAZARDS

8.6.1 Spray Systems

Water spray systems pose some potential hazards if they are used in certain circumstances. This subsection briefly discusses some of the potential hazards.

For some materials (e.g., hydrogen chloride, hydrogen fluoride, and chlorine), applying water directly at the leakage point will result in an acceleration of the corrosive effect of the chemical. This can result in an enlarged release hole and an increase in the rate of release. Similarly, spraying water on a spill of liquified gas such as chlorine would add heat to the subcooled pool of liquified gas, resulting in increased vaporization.

For some chemicals, water sprays can lead to the formation of highly toxic, acidic, or alkaline liquid droplets that may be hazardous or potentially more hazardous than the original release.

Finally, where mobile systems are used, personnel can be injured if appropriate protective clothing and breathing apparatus are not worn.

8.6.2 <u>Steam Curtains</u>

Steam curtains, particularly those with jets containing wet steam, can create static electricity. The generation of static electricity in steam curtains has been investigated by several individuals (26,33). Static electricity could ignite a flammable vapor or gas. Static discharges from steam jets can be prevented by grounding all pipework and other metal in contact with the steam.

Additional secondary hazards are similar to those presented for spray systems.

8.7 COSTS

8.7.1 <u>Spray Systems</u>

The cost of a water spray system depends on the type of system used (i.e., fixed or mobile) and will be similar to costs for fire protection systems and equipment. Table 8-3 lists the costs of typical components for both a fixed and a mobile system. Design bases for these costs are presented in Table 8-4.

For fixed systems, the major portion of the actual system cost is that associated with the piping and water supply system. If an existing supply of water for fire protection systems can also be used for mitigation purposes, the only major additional cost is that of the extra pipework and nozzles needed for the water spray system. These systems may, however, be limited in their application size. Thus, for very large installations, mobile systems may be more cost-effective.

TABLE 8-3.   TYPICAL COSTS ASSOCIATED WITH FIXED AND MOBILE WATER
             SPRAY SYSTEMS

Item	Capital Cost Range (1987 $)	Annual Cost Range (1987 $/yr)
Mobile Water Spray System:		
Fire Hose	4 - 5 per linear foot	0.35 - 0.90 per linear foot
Fog Nozzle	250 - 300	20 - 25
Fire Hydrant	400 - 500	35 - 45
Fire Monitor	750 - 1,000	65 - 85
Fixed Water Spray System:		
Deluge type, water spray system	300 - 350 per nozzle	25 - 30 per nozzle
Spray nozzle	60 - 80	5 - 7
Deluge valve	1,400 - 1,500	120 - 130
Fire pump	25,000 - 30,000	2,100 - 2,600

Source: See Table 8-4 for references.

TABLE 8-4. EQUIPMENT SPECIFICATIONS FOR MOBILE AND FIXED WATER
SPRAY SYSTEMS USED IN COST ESTIMATES

Item	Specification	Reference
Mobile Systems:		
Fire Hose	High strength, 500 lb. test, 2-1/2 in. diameter.	36
Fog Nozzle	Adjustable nozzle, 1-in. booster inlet	36
Fire Hydrant	Standard 4-in. hydrant with two 2-1/2 in. hose connections.	36,37
Fire Monitor	Standard fire monitor with adjustable nozzle, 500 gpm capacity.	36,37
Fixed Systems:		
Water Spray System	Deluge type, fixed-pipe system capable of delivering 3,000 gpm of water to source. Includes piping and supports.	38,39
Spray Nozzle	Fan-tail or hollow-cone type, carbon steel construction.	36
Deluge Valve	2-inch deluge valve, including trim, pressure-operated relief, emergency release, and pressure gauge.	36
Fire Pump	5-inch centrifugal pump, 99 hp, 100 psi, 1,000 gpm, diesel operated.	36

8.7.2   <u>Steam Curtain</u>

The costs of typical components associated with steam curtain systems are shown in Table 8-5. Since there are many individual variations of components, the system chosen is for illustrative purposes only. These costs provide an order-of-magnitude basis for estimating the total cost of a steam curtain system.

TABLE 8-5.   ESTIMATED COSTS FOR A TYPICAL STEAM CURTAIN SYSTEM[a]

	Capital Cost (1987 $)	Annual Cost (1987 $/yr)	References
Piping	7,000	1,200	30
Concrete wall	2,500	220	39
Globe valve	1,000	90	40
Remote shutoff valve	3,000	450	40
Steam supply[b]	$4.50/hr	$68/hr	40

[a] Basis: 6-inch Schedule 40 carbon steel piping, 100 feet long with 3/16-inch holes, 1 foot apart
250 psig steam supply
Concrete walls, 4 feet high, 100 feet long, 6 inches thick.

[b] Based on a steam requirement of 0.15 ton/hr per foot of curtain and a steam cost of $15 per 1000 pounds.

8.8 REFERENCES

1. Eggleston, L.A., W.R. Herrera, and M.D. Pish. Water Spray to Reduce Vapor Cloud Spray. Loss Prevention, Volume 10. American Institute of Chemical Engineers, New York, NY, 1976.

2. Watts, J.W. Effects of Water Spray on Unconfined Flammable Gas. Loss Prevention, Volume 10. American Institute of Chemical Engineers, New York, NY, 1976.

3. Vincent, G.C., et al. Hydrocarbon Mist Explosions - Part II, Prevention by Water Fog. Loss Prevention, Volume 10, American Institute of Chemical Engineers, New York, NY, 1976.

4. Experiments With Chlorine. Ministry of Social Affairs, Voorburg, Netherlands, 1975.

5. Martinsen, W.E., and S.P. Muhlenkanp. Disperse LNG Vapors With Water. Hydrocarbon Processing, July 1977.

6. Moore, P.A.C., and W.D. Rees. Forced Dispersion of Gases by Water and Steam. In: I. Chem. E. Symposium Proceeding. The Containment and Dispersion of Gases by Water Sprays. Manchester, England, 1981.

7. Moodie, K. The Use of Water Spray Barriers to Disperse Spills of Heavy Gases. Plant/Operations Progress, October 1985.

8. Moodie, K. Experimental Assessment of Full-Scale Water Spray Barriers for Dispersing Dense Gases. In: I. Chem. E. Symposium Proceedings. The Containment and Dispersion of Gases by Water Sprays. Manchester, England, 1981.

9. Health and Safety Executive (U.K.). Canvey: An Investigation of Potential Hazards. London, England, 1978.

10. Smith, J.M., and M. van Doorn. Water Sprays in Confined Applications: Mixing and Release from Enclosed Spaces. In: I. Chem. E. Symposium Proceedings. The Containment and Dispersion of Gases by Water Sprays. Manchester, England, 1981.

11. Greiner, M.L. Emergency Response Procedures for Anhydrous Ammonia Vapor Release. Ammonia Plant Safety and Related Facilities. Volume 24. CEP Technical Manual, AIChE, 1984.

12. Beresford, T.C. The Use of Water Spray Monitors and Fan Sprays for Dispersing Gas Leakages. In: I. Chem. E. Symposium Proceedings. The Containment and Dispersion of Gases by Water Sprays. Manchester, England, 1981.

13. Personal Communication with industry representative. Name withheld by request.

14. Cairney, E.M., and A.L. Cude. The Safe Dispersal of Large Clouds of Flammable Heavy Vapours. Institute of Chemical Engineers Symposium, Series No. 34, London, England, 1971.

15. Simpson. H.G. The ICI-Vapour Barrier - A Means of Containing and Dispersing Leakages of Flammable Vapour. Power and Works Engineering, May 8, 1974.

16. Simpson, H.G., and A.L. Cude. U.S. Patent 3,882,943. May 13, 1975.

17. Lees, F.P. Loss Prevention in the Chemical Process Industries, Volumes 1 and 2. Buttersworth's, London, England, 1983.

18. Barker, G.F., T.A. Kletz, and H.A. Knight. Olefin Plant Safety During the Last 15 Years. Loss Prevention, Volume II. American Institute of Chemical Engineers, 1977.

19. McQuaid, J. The Design of Water-Spray Barriers for Chemical Plants. Second International Loss Prevention Symposium. Heidelberg, West Germany, September 1977.

20. Bucklin, S.M. Aerodynamic Behavior of Liquid Spray Design Method. VKI Report No. 171, 1980.

21. Deaves, D.M. Experimental and Computational Assessment of Full-Scale Water Spray Barriers for Dispersing Dense Gases. Fourth International Loss Prevention Symposium. Harrogate, England, September 1983.

22. McQuaid, J. and R.D. Fitzpatrick. The Uses and Limitations of Water-Spray Barriers. In: I. Chem. E. Symposium Proceedings. The Containment and Dispersion of Gases by Water Sprays. Manchester, England, 1981.

23. Zalosh, R.G. Dispersal of LNG Vapor Clouds with Water Spray Curtains. In: I. Chem. E. Symposium Proceedings. The Containment and Dispersion of Gases by Water Sprays. Manchester, England, 1981.

24. Harris, N.C. The Design of Effective Water Sprays - What We Need to Know. I. Chem. E. Symposium Proceedings. The Containment and Dispersion of Gases by Water Sprays. Manchester, England, 1981.

25. Seifert, H.B. Maurer, and H. Giesbrecht. Steam Curtains - Effectiveness and Electrostatic Hazards. Fourth International Symposium on Loss Prevention. Harrogate, England, 1983.

26. Kirk, R.E., and D.F. Othmer. Encyclopedia of Chemical Technology, 3rd Edition. John Wiley & Sons, Inc., 1980.

27. Prugh, R.W. Mitigation of Vapor Cloud Hazards. Part II. Limiting the Quantity Released and Countermeasures for Releases. Plant/Operations Progress, July 1986.

28. Lees, F.P. Loss Prevention in the Process Industries - Hazard Identification, Assessment and Control. Volumes 1 and 2, Butterworth's, London, England, 1983.

29. Emblem, K., and O.K. Madsen. Full-Scale Test of a Water Curtain in Operation. 5th Loss Prevention Symposium. Cannes, France, 1986.

30. Harris, N.C. The Control of Vapor Emissions from Liquified Gas Spillages. 3rd Loss Prevention Symposium. Basle, Switzerland, 1980.

31. Bockman, T., G.H. Ingebrigtsen, and T. Hakstad. Safety Design of the Ethylene Plant. Loss Prevention, Volume 14. American Institute of Chemical Engineers, 1981.

32. Rees, W.D., and P.A.C. Moore. Forced Dispersion of Gases by Water and Steam. In: I. Chem. E. Symposium Proceedings. The Containment and Dispersion of Gases by Water Sprays. Manchester, England, 1981.

33. Rulkens, P.F.M., et al. The Application of Gas Curtains for Diluting Flammable Gas Clouds to Prevent Their Ignition. Fourth International Symposium on Loss Prevention. Harrogate, England, 1983.

34. R.S. Means Company, Inc. Building Construction Cost Data 1986, 44th Edition, Kingston, MA.

35. American Valve and Hydrant. Arlington, TX: vendor.

36. Spraying Systems Company. Wheaton, IL: vendor.

37. Richardson Engineering Services, Inc. The Richardson Rapid Construction Cost Estimating System. Volumes 1-4, San Marcos, CA, 1986.

38. Yamartino, J. Installed Cost of Corrosion-Resistant Piping--1978. Chemical Engineering, November 20, 1978.

39. R.S. Means Company, Inc. Building Construction Cost Data 1986, 44th Edition, Kingston, MA, 1986.

40. Peters, M.S. and K.D. Timmerhaus. Plant Design and Economics for Chemical Engineers. McGraw-Hill Book Company, New York, NY, 1980.

9. Foam Systems

Foams, which are used in the chemical process industries to control and extinguish certain types of hydrocarbon fires involving spilled liquids, are used when the fire may not be effectively controlled by water spray application. Although originally developed for firefighting purposes, some of the same properties that make foams effective for controlling fires also make them useful for controlling the release of vapors from volatile chemical spills. The application of a foam blanket to a liquid spill may prevent the release of a flammable gas or vapor from reaching an ignition source in concentrations that could result in an explosion or fire. In the case of a nonflammable toxic liquid spill, foam may help prevent personnel or general public exposure to dangerous concentrations of a hazardous gas or vapor being emitted from the surface of the liquid.

This section of the manual discusses the use of firefighting and specialty foams to control vapor hazards from spilled volatile flammable and/or toxic chemicals. The different types of foams and basic design considerations are described, as well as their applicability, performance capabilities, and reliability. Secondary hazards that may limit the use of foams for controlling vapor hazards from certain hazardous liquid spills are also described.

9.1 BACKGROUND

Foams used for firefighting consist of a mass of gas-filled bubbles which, when applied to a liquid hydrocarbon fire, control and extinguish the flames. The properties of foam that make it effective for fighting fires are (1,2):

- Ability to blanket the spilled liquid surface with a material that has a lower density than the liquid, thereby extinguishing an existing fire by cutting off the source of combustion air;

- Suppression of potentially flammable vapors from being emitted to the atmosphere and mixing with combustion-supporting air;

- Prevention of nearby flames from heating the spilled liquid covered by the foams; and

- Cooling of the spilled liquid with water draining from the foam and surrounding surfaces to help prevent reignition.

These properties are also helpful in mitigating the release of flammable and/ or toxic vapors before actual ignition or exposure of personnel.

Two varieties of foams, chemical and mechanical, are used for firefighting. Chemical foam is produced by a chemical reaction, such as the reaction that occurs when an aqueous solution of sodium bicarbonate is mixed with aluminum sulfate and sulfuric acid (3,4). With the addition of other chemicals acting as foam stabilizers, the reaction produces a foam consisting of carbon dioxide-filled bubbles (1). Originally introduced to control coal fires in the 19th century (2), chemical foam has been largely replaced by mechanical foam, which consists of air-filled bubbles (2,3). This "air" foam is produced by mechanical aeration of a mixture of foam concentrate and water using a foam-making device. Mechanical foams developed for fire control include regular protein foam, fluoroprotein foam, surfactant foam, aqueous film-forming foam (AFFF), and alcohol-type foam (ATF) (5). A discussion of each of these conventional firefighting foams and other special foams developed specifically to control vapors is included in Subsection 9.2. The various types of equipment used to generate and apply these foams to spills are also described in Subsection 9.2.

An important quality of firefighting foam is the expansion ratio, which is the ratio of the volume of foam produced to the volume of solution fed to the foam-making device. Foams are classified as low-, medium-, and high-expansion foams (6). Low-expansion foams have expansion ratios of from 2:1 to 15:1. Medium-expansion foams have expansion ratios from 15:1 to 100:1, while high-expansion foams are those with expansion ratios of from 100:1 to 1000:1. While these particular ranges of expansion ratios have been presented in the literature for firefighting foams (6), the ranges are rather subjective and can be different for the low-, medium-, and high-expansion foams used for vapor control. These different range are discussed in Subsection 9.2.

Most firefighting foams, including regular protein, fluoroprotein, aqueous film-forming foam (AFFF), and alcohol-type foam (ATF), are available in 3 percent and 6 percent concentrations. This percent value indicates the amount of foam concentrate that should be mixed with water to form the foam solution, which is then aerated to create the actual foam. For example, to produce 100 gallons of foam solution, 3 gallons of 3 percent foam concentrate would be mixed with 97 gallons of water (1). If the particular foam had an expansion ratio of 10:1, 1000 gallons of foam could be produced for application to the fire or spill.

Other important qualities of firefighting foams are fluidity and drainage rate (4). The foam should flow easily around obstructions in the spill area and quickly cover the liquid surface without breaking up the flame smothering "blanket." A good firefighting foam possesses a shear stress value of 150 to 250 dynes/cm^2, as measured on a torsional vane viscometer (4). The drainage rate of a foam is a critical property for firefighting, as well as for vapor control, as discussed in Subsection 9.2. The drainage rate is referred to as the "25 percent" or "quarter drainage time." The "quarter drainage time" is the time, in minutes, it takes for a foam to loose 25 percent of the liquid used to make the foam. For a good firefighting foam, the drainage rate is 2 to 5 minutes (4).

9.2   DESCRIPTION

The ability of foams to mitigate vapor released from volatile liquid chemical spills has grown out of their ability to extinguish fires and prevent reignition by isolating the spilled material and volatile vapors from ignition sources.   The foam blanket insulates the liquid from ignition sources and radiant heat sources that cause vaporization.   The foam acts as a physical barrier, because of its limited permeability, and suppresses vapor loss to the atmosphere.   The foam may also absorb vapors being emitted from the spill surface (3,7).   For refrigerated liquid spills, the foam can help warm the emitted vapor, which passes through the applied foam layer, thus allowing the vapors to rise and disperse (8).   Foams can also provide water for diluting certain water-reactive chemical spills (3).   On the other hand, the accelerated boil-off of regrigerated liquids or violent reactions with water-reactive chemical spills can occur.   These secondary hazards are discussed in more detail in Subsection 9.6.

The effectiveness of a foam depends on the type of chemical spilled, the type of foam selected to control the spill, the general qualities of foam that help control the release of vapors, and foam generation equipment.

9.2.1   Types of Foams

Six types of foams are used to control vapors from chemical spills: regular protein foams, fluoroprotein foams, surfactant foams, aqueous film-forming foams (AFFF), alcohol-type (ATF) or polar solvent foams, and special foams (9).   The first five were developed for firefighting, while the special foams have been developed specifically to control vapor releases from spills.

Regular Protein Foams--

Protein foams, derived from hydrolyzed protein, are the oldest of the mechanical firefighting foams.   The use of protein foam in controlling vapor release grew out of its use as an extinguishing agent for aromatic hydrocarbon fuel fires and its effectiveness in preventing the emission of ignitable vapors. Some of the important characteristics of regular protein foam are listed below (2,3,6,10):

- Demonstrated spill control capability for water-immiscible, non-polar hydrocarbon fuels such as petroleum products, gasoline, fuel oil, jet fuel, etc.;

- Designed for low expansion foam-making equipment;

- Desirable physical characteristics (good adhesion, stability, water retention, elasticity, and mechanical strength);

- Good heat resistance, resistance to burnback (burning back of the foam blanket if the blanket is broken) and resistance to reignition;

- Non-toxic and biodegradable after dilution;

- Among standard types of foam in use by fire service departments;

- Fresh or sea water can be used to make foam solution;

- Limited shelf life for foam concentrates;

- Poorer flowing ability and slower fire knockdown ability than other foam types; and

- Can be used in temperature range of 20°F to 120°F.

Fluoroprotein Foams--

Fluoroprotein foams are similar to regular protein foams, but also contain fluorinated surfactants, which allow the foam to shed fuel from its surface if the foam becomes coated with hydrocarbon material (3,10). This fuel-shedding characteristic is important when fighting hydrocarbon tank fires

or in cases where the foam is injected below the surface of the fuel fire
(10). With the following exceptions, the characteristics of fluoroprotein
foams are similar to those of regular protein foam. These exceptions are
noted below (2,10):

- Better suppression of hydrocarbon fuel vapors than regular protein
  and AFFF foam;

- Better resistance to burnback than protein foam;

- Better compatability than regular protein foam with dry chemical
  agents; and

- Faster fire knockdown ability than regular protein foam, but slower
  than AFFF foam.

Surfactant Foams--

Surfactant foams are produced for application as low-, medium-, and high-
expansion foams from detergent foam concentrates. These concentrates consist
of synthetic hydrocarbon surfactants (syndets) (10). Some important characteristics of surfactant foams are shown below (3,10):

- Applicable to water-immiscible, non-polar hydrocarbons;

- Resistance to burnback and fire knockdown ability is not as good as
  that of other low-expansion foams;

- Capability of expansion ratios of up to 1000:1 for total flooding
  capability;

- Readily available surfactant foams are mostly anionic;

- Non-toxic and biodegradable;

- Unlimited shelf life if stored properly;

- Produce less odor when burned than do other foam types;

- Large variety of concentrates on market, with wide range of physical properties, which can make proper selection difficult; and

- Available in concentrates to yield from 1 to 6 percent foam solutions;

Aqueous Film Forming Foams (AFFF)--

Concentrations for AFFF foams are composed of fluorinated surfactants and hydrocarbon surfactants for fast knockdown of water-immiscible, non-polar hydrocarbon fuel fires. Additional characteristics of AFFF foam are listed below (2,3,6,10):

- Ability to be used in low- to high-expansion foam-making equipment;

- Ability to spread a water solution film across spill quickly to extinguish fire, exclude combustion air, and stop fuel from vaporizing or toxic vapors from being emitted;

- Water of composition lost more quickly after application than with other foams;

- Low viscosity;

- Available in concentrates to yield 3 and 6 percent foam solutions;

- Fresh or sea water can be used to make foam solution;

- Non-toxic and biodegradable after dilution; and

- Long shelf life.

Alcohol-Type Foam (ATF)--

Alcohol-type foam was developed to use on fires involving water-miscible, polar materials, where ordinary air foams exhibit rapid breakdown and loss of effectiveness. Examples of polar materials for which ATF foams have been used are alcohols, paint thinners, methyl ethyl ketone, acetone, isopropyl ether, acrylonitrile, ethyl and butyl acetate, and amines and anhydrides (10). ATF foams consist of a regular protein, fluoroprotein, surfactant, or AFFF concentrate with additives of a metal stearate or a polar material resistant polymer (2,3,6,10). ATF foams using polymers resistant to polar materials are applicable to both non-polar hydrocarbon and polar solvents. When these types of "universal" foams are applied to a spill, a gel is formed on the surface. This gel forms an insoluble, low-permeability polymeric layer between the foam and the spilled chemical that protects the film and foam from degradation, and mitigates the release of vapor from the spill. ATF foams are generally available in concentrations to yield 3 and 6 percent foam solutions. These foams are normally used at temperatures of from 35°F to 120°F.

Special Foams--

Special foams are those developed to suppress hazardous vapors from liquid spills that typical firefighting foams cannot control effectively. Examples[a] of these are Hazmat NF® Number 1 and Hazmat NF® Number 2 produced by National Foam, and Type V foam (VEE foam®) by MSA Research (11). Hazmat NF® Number 1 is designated to use on spills involving alkaline materials, while Hazmat NF® Number 2 is for acid material spills. MSAR Type V foam is for controlling hazardous vapors from water-reactive chemicals.

---

[a]These products are mentioned as examples only; such mention does not constitute an endorsement.

9.2.2   <u>Foam Quality</u>

Two important qualities of a foam will effect its ability to control vapors from hazardous volatile chemical spills: the expansion ratio and the quarter drainage time. The quarter drainage time is a measure of the stability of the foam. More stable foams can prevent vapor from being emitted from the spill for longer periods of time than can less stable foams.

As indicated in Subsection 9.1, expansion ratio classifications are somewhat subjective and there is no definite agreement on what constitutes a low-, medium-, or high-expansion ratio foam. Ratios can range from less than 20:1 to over 150:1. High-expansion foams have long quarter drainage times, sometimes greater than one hour, but can be blown away by the wind, while medium-expansion foams and low-expansion foams have quarter drainage times of between approximately 15 and 30 minutes, and between 3 and 12 minutes, respectively.

Medium-expansion foams may be the most effective for controlling vapors from spilled volatile materials. They can be applied more easily than high-expansion foams and will not be blown away as easily by wind. They also exhibit longer quarter drainage times than do low-expansion foams, and thus require less frequent reapplication.

9.2.3   <u>Foam Application Systems</u>

Foam systems consist of a stored quantity of foam concentrate, a water supply to mix with the concentrate and a nozzle system to introduce air into the foam solution to produce the expanded foam for application to the spill. Foam systems may be fixed, semifixed, or portable (6).

the foam solution to produce the expanded foam for application to the spill. Foam systems may be fixed, semifixed, or portable (6).

Regardless of the type of foam system selected, a proportioning system or proportioner is used to mix the foam concentrate with the make-up water to form the foam solution. The expanded foam is generated by mechanically mixing air with the foam solution within a nozzle. The specific types of equipment used to make the foam depends on the type of foam and the expansion ratio. Low- and medium-expansion foams can be generated using various foam types and nozzles. Examples of portable air aspirating foam-making equipment typically used to respond to hazardous chemical spills are described in References 3 and 10. Generally, foam application systems are sold as part of a total package from the foam vendors, and design of a custom system is not necessary.

9.3   DESIGN

The design of a foam system depends primarily on the type of hazard involved. In the case of a volatile chemical spill, an appropriate foam concentrate should be selected and on hand to control the release of hazardous vapors from that particular material. For handling emergency releases in chemical processing plants, the user needs to review the types of hazardous chemicals and the potential areas for spill releases in order to select the proper speciality foam and equipment. In the case of a transportation accident and subsequent chemical release, fire service departments may have a limited selection of foams to choose from, which may or may not be appropriate in controlling vapor releases from the spill. A table showing the qualitative performance abilities of various foams on several chemicals is shown in Subsection 9.5.

If the type of hazard and foam to control this hazard have been defined, the other major design criteria are:

- The water supply; and
- The magnitude of the hazard.

Without an adequate supply of water at the required rate and pressure, the foam application system may be inoperable or unable to produce the necessary quantity of foam to control the release (6).

The potential size of the chemical spill is also an important design criterion. If the size of a spill, for instance within a diked area, can be determined before a release, the amount of water at adequate pressure and flow rate can be planned for, along with the amount of foam concentrate that would fully contain the release.

For contained and uncontained chemical spills, the size of the hazard influences the selection of application equipment and methods, and suitable foams of the appropriate quality for controlling vapors from the spill. These are discussed below.

9.3.1 Application Equipment and Methods for Low-/Medium-Expansion Foams

Air-aspirating foam nozzles for applying low- and medium-expansion foams should be selected that minimize the drainage of liquid from the foam and allow maximum expansion. These include turret-mounted or hand-held nozzles that disperse the foam as a stream or spray; however, use of fog-foam nozzles can effectively reduce the expansion ability of the foam. Different nozzles may produce foam of varying quality; therefore, it is important to choose the nozzle that yields the desired quarter drainage time and expansion ratio with a given foam. Detailed discussions with foam and equipment manufacturers or tests may be required to evaluate equipment and the actual quality of the foam as applied.

Low-expansion foams should be applied in such a way that the foam is not sprayed directly into the spill. The foam stream or spray should be directed just in front of the spill and allowed to flow into the spill area or directed onto a wall or other surface just behind the spill (7,9). This allows the foam to blanket the surface with minimal disturbance of the spilled chemical.

The selection of application equipment and the foam application rate can
be determined from the desired foam depth, the area of the spill, the time
required to achieve complete blanketing of the spill, and the expansion ratio
of the foam, as given by the formula shown below:

$$R = \frac{7.48\ DA}{TE}$$

Where:

R = Discharge capacity of foam solution for nozzles (U.S. gal/min)

D = Depth of foam (ft)

A = Spill area (ft^2)

T = Time required to completely blanket spill (min)

7.48 = conversion factor (U.S. gal/ft^3)

E = Expansion ratio of foam

For a 5000-square-foot spill to be covered by a 0.5 foot thick foam blanket within five minutes, about 190 gallons per minute of foam solution with an
expansion ratio of 20:1 would be needed. This may require the use of several
smaller nozzles (i.e., 3-60 gallons per minute nozzles) or fewer larger flow
rate nozzles. For 60 minutes of operation using a foam concentrate that
yields a 6 percent foam solution, approximately 690 gallons (190 gal/min x 60
min x 0.06) of foam concentrate would be required to control the spill.

9.3.2   Application Equipment and Methods for High-Expansion Foams

High-expansion foams are applied to a chemical spill by means of a foam
chute. They are used indoors to totally flood an area to extinguish fires.
Use of these foams outdoors to control vapor releases from spills is limited
to situations where the wind speed is less than 10 miles per hour, unless
precautions are taken to restrain displacement of the foam by the wind.

The application rate for the generated foam should exceed 0.5 cubic feet per minute per square foot of spill area. Foams should be applied to flammable liquids to a thickness of at least 18 inches.

9.4 APPLICABILITY AND PERFORMANCE

Certain types of aqueous foams have been shown to be effective in controlling the release of flammable and/or toxic vapors from various chemical spills. A major criterion in selecting a foam is the compatibility of the foam with the spilled chemical. Foams have limited effectiveness in controlling vapor releases from flowing spills. To control vapor releases, the spill must first be physically contained.

Currently, it does not appear that foams are used to control airborne releases of hazardous chemicals; however, foams have been used for gas scrubbing (8). Some work has been done where a contaminated air stream has been sent into a foam generation machine, where the air is incorporated into the foam. The contaminant is then temporarily trapped in the foam, where its rate of release is slowed, or where it is absorbed by the water in the foam. Such a system could be used wherever the vapors from an accidental release could be contained and routed into the foam-generation machine. In such situations, however, an alternate treatment system such as a scrubber or flare could also be used.

The use of aqueous foams for controlling vapors from water-immiscible hydrocarbon fuels is well established because of their use in extinguishing fires. The use of typical commercial fire-fighting foams is limited to non-polar materials with dielectric constants of less than three; otherwise, rapid degradation of the foam will occur (3). For control of water-miscible, polar material spills, alcohol-type foams (ATF) have been developed that protect the foam blanket by creating a film barrier between the spilled material and the foam. Universal or multipurpose foam agents have also been produced that are effective against both polar and non-polar solvents, and

chemically neutral material spills (7). Specialty foams have been developed
to control hazardous vapor releases from spills of alkaline-, acid-, and
water-reactive chemicals.

Another major point in foam selection is the drainage of liquid from the
foam, or the quarter drainage time. The longer the quarter drainage time, or
the longer the foam retains its water, the better its fluidity, spreading
capability, collapse resistance, and heat resistance (in the case of fire)
(11).

Because of the variety of fire-fighting and specialty foams available and
the numerous chemicals in commercial use, it is important to evaluate which
foams are applicable for controlling vapor from spills of particular chemicals. Table 9-1 shows an overview of the types of foam which are useful for
controlling vapors from spills of specific chemicals. Shown in this table are
the chemical group, the particular chemical, the recommended type of foam,
satisfactory types of foam, and ineffective or dangerous types of foams. New
foams for vapor control will probably be developed for these and other chemicals. Therefore, foam manufacturers should be consulted before selecting foam
concentrates, particularly for chemicals not listed in Table 9-1. Table 9-2
shows typical application information about some universal and specialty foams
available from some manufacturers.

Depending on the type of foam used and the material spilled, foams can
reduce vapor concentrations, measured just above the foam layer, by over 90
percent (8). However, this reduction is temporary and eventually the vapor
concentration above the applied foam layer will reach its uncontrolled
concentration as the foam breaks down over time. The length of time that a
foam can effectively mitigate vapors without reapplication can range from 5 to
120 minutes (8). An illustration of the mitigation of vapor released from
liquid benzene is illustrated in Figure 9-1 (3).

TABLE 9-1.   FOAMS FOR CONTROL OF FLAMMABLE/TOXIC VAPORS

Group	Chemical	Recommended Foam Types (Minimum Thickness/ Reapplication Rate)(in/min)[a]	Satisfactory Foam Types (Minimum Thickness/ Reapplication Rate)(in/min)[a]	Ineffective/Dangerous Foam Types (Reason)
ALCOHOLS	Butanol, methanol, and propanol	Alcohol (ND)[b]	None	AFFF (collapse) Fluoroprotein (collapse) Protein (collapse) Surfactant H&L (collapse)
	Octanol	Alcohol (ND)	AFFF Fluoroprotein Protein Surfactant H&L	
ALDEHYDES AND KETONES	Acetone, methyl ethyl and methyl butyl ketones	Alcohol (10/60)	None	AFFF (collapse) Fluoroprotein (collapse) Protein (collapse) Surfactant H&L (collapse)
AMINES	Ethylamines, ethylene diamine, hydrazine, and methylamines	Alcohol (ND) Hazmat NF #1 (ND)[d] MSA Type V (ND)	None	AFFF (collapse) Fluoroprotein (early breakthrough) Protein (collapse) Surfactant H&L (collapse)

(Continued)

TABLE 9-1. (Continued)

Group	Chemical	Recommended Foam Types (Minimum Thickness/ Reapplication Rate)(in/min)[a]	Satisfactory Foam Types (Minimum Thickness/ Reapplication Rate)(in/min)[a]	Ineffective/Dangerous Foam Types (Reason)
ETHERS	Ethyl ether	Alcohol (5/25-120)	None	AFFF (collapse) Fluoroprotein (collapse) Protein (collapse) Surfactant (collapse)
ESTERS	n-Butyl acetate	AFFF (5/120) Alcohol (5/120) Fluoroprotein (5/120) Protein (5/120) Surfactant L (5/120)	None	Surfactant H (no information)
	Methyl acrylate	Alcohol (ND)	None	AFFF (collapse) Fluoroprotein (collapse) Protein (collapse) Surfactant H&L (untested)
	Vinyl acetate (monomer)	Alcohol (ND) Fluoroprotein (ND)	AFFF (ND)	Protein Surfactant H&L
HYDRO- CARBONS (ALI- PHATIC)	Ethane and ethylene	Surfactant H (ND)	Alcohol (ND) Fluoroprotein (ND) Protein (ND) Surfactant L (ND)	AFFF (accelerates boil- off)

(Continued)

TABLE 9-1.   (Continued)

Group	Chemical	Recommended Foam Types (Minimum Thickness/ Reapplication Rate) (in/min)[a]	Satisfactory Foam Types (Minimum Thickness/ Reapplication Rate) (in/min)[a]	Ineffective/Dangerous Foam Types (Reason)
HYDRO-CARBONS (ALI-PHATIC) (con't)	Heptane	Alcohol (ND) Fluoroprotein (ND) Protein (ND) Surfactant H&L (ND)	AFFF (ND)	None
	Hexane and octane	Alcohol	AFFF Fluoroprotein Protein Surfactant	None
HYDRO-CARBONS (AROMATIC)	Benzene	Alcohol (2.5/120)	Fluoroprotein (5/60) Protein (5/60) Surfactant H (25/50)	AFFF (early break-through) Surfactant L (early breakthrough)
	Ethylbenzene	AFFF (5/120) Alcohol (5/120) Fluoroprotein (5/120) Protein (5/120) Surfactant L (5/120)	Surfactant H (50/30)	None

(Continued)

TABLE 9-1.   (Continued)

Group	Chemical	Recommended Foam Types (Minimum Thickness/ Reapplication Rate) (in/min)[a]	Satisfactory Foam Types (Minimum Thickness/ Reapplication Rate) (in/min)[a]	Ineffective/Dangerous Foam Types (Reason)
HYDRO- CARBONS (AROMATICS) (Con't)	Toluene	Alcohol (ND) Fluoroprotein (ND) Protein (ND)	AFFF (5/40) Surfactant H (50/30) Surfactant L (5/30)	None
HYDRO- CARBONS (ALI- CYCLIC)	Cyclohexane	Alcohol (5/>120) Fluoroprotein (5/>120) Protein (5/>120)	AFFF (5/60) Surfactant H (ND) Surfactant L (5/30)	None
HYDRO- CARBONS (INDUS- TRIAL	Gasoline and kerosene	Alcohol (ND) Fluoroprotein (ND) Protein (ND) Surfactants (ND)	AFFF (ND)	None
	Naphtha	Alcohol (ND) Fluoroprotein (ND) Protein (ND)	AFFF (ND) Surfactants (ND)	None
LIQUEFIED ORGANIC GASES	Ethylene oxide	None	Alcohol (ND)	AFFF (collapse) Protein (collapse) Fluoroprotein Surfactant H (untested)

(Continued)

TABLE 9-1. (Continued)

Group	Chemical	Recommended Foam Types (Minimum Thickness/ Reapplication Rate) (in/min)[a]	Satisfactory Foam Types (Minimum Thickness/ Reapplication Rate) (in/min)[a]	Ineffective/Dangerous Foam Types (Reason)
LIQUIFIED ORGANIC GASES (con't)	Liquefied natural gas	Surfactant H (ND)	None	AFFF (early breakthrough) Alcohol (early break- through) Fluoroprotein (early breakthrough) Protein (early break- through) Surfactant L (early breakthrough)
INORGANICS	Carbon disulfide	Hazmat NF #2 (ND) MSA Type V (ND)	None	AFFF (untested) Alcohol (untested) Fluoroprotein (untested) Protein (early break- through) Surfactant H&L (untested)
	Hydrochloric acid and hydrogen chloride (anhydrous)	Hazmat NF #2 (ND) MSA Type (V (ND)	None	AFFF (untested) Alcohol (untested) Fluoroprotein (untested) Protein (untested) Surfactant H&L (untested)

(Continued)

TABLE 9-1.   (Continued)

Group	Chemical	Recommended Foam Types (Minimum Thickness/Reapplication Rate)(in/min)[a]	Satisfactory Foam Types (Minimum Thickness/Reapplication Rate)(in/min)[a]	Ineffective/Dangerous Foam Types (Reason)
INORGANICS (con't)	Nitric acid	Hazmat NF #2 (ND) MSA Type V (ND)	None	AFFF (untested) Alcohol (untested) Fluoroprotein (untested) Surfactant H&L (untested)
	Silicon tetrachloride	Surfactant H (ND)	None	AFFF (violent reaction) Alcohol (violent reaction) Fluoroprotein (violent reaction) Protein (violent reaction) Surfactant L (violent reaction)
	Sulfur trioxide	Hazmat NF #2 (ND) Surfactant H (ND) MSA Type V (ND)	None	AFFF (violent reaction) Alcohol (violent reaction) Fluoroprotein (violent reaction) Protein (violent reaction) Surfactant L (violent reaction)
	Titanium tetrachloride	Hazmat NF #2 (ND) MSA Type V (ND)	None	AFFF (untested) Alcohol (untested) Fluoroprotein (untested) Protein (untested) Surfactant H&L (untested)

(Continued)

TABLE 9-1.   (Continued)

Group	Chemical	Recommended Foam Types (Minimum Thickness/ Reapplication Rate)(in/min)[a]	Satisfactory Foam Types (Minimum Thickness/ Reapplication Rate)(in/min)[a]	Ineffective/Dangerous Foam Types (Reason)
INORGANICS (con't)	Ammonia	Surfactant H (30/10)	AFFF (8/5) Alcohol (8/5) Surfactant (8/5) Fluoroprotein (8/5) Protein (8/5)	None
INORGANIC	Bromine and chlorine	Hazmat NF #2 (ND) Surfactant H (ND) MSA Type V (ND)	None	Other foams are ineffective for use on chlorine, causing accelerated boil-off of heavier-than-air chlorine gas.

[a] Reapplication times are generally based on the amount of time before a 1-percent vapor concentration by volume is established above the spill surface.

[b] No data are available on foam thickness or reapplication rate.

[c] H – high-expansion surfactant foam; L = low-expansion or medium-expansion surfactant foam.

[d] Hazmat NF #1, Hazmat NF #2, and MSA Type V are specified by name because they were designed for controlling hazardous vapors. This is not an endorsement.

TABLE 9-2. PERFORMANCE SUMMARY OF THREE UNIVERSAL AND SPECIALTY FOAMS

Foam: National Universal Foam

General Applicability

Multi-purpose foam for extinguishing hydrocarbon, alcohol, and polar solvent fires; also for vapor control for hydrocarbons, alcohols, polar solvents, and chemically neutral materials.

Chemicals for Which this Foam is Ineffective:

Chlorosulfonic Acid
Fluorosulfonic Acid
Phosgene
Phosphorus Oxychloride
Phosphorus Trichloride
Sulfuryl Chloride

Foam: Hazmat NF● Foam Number 1

General Applicability

Foam for vapor control for alkaline materials spills.

Chemicals for Which this Foam is Ineffective:

(see Table 9-1)

Foam: Hazmat NF● Foam Number 2

General Applicability

Foam for vapor control for acid material spills.

Chemicals for Which this Foam is Ineffective:

(see Table 9-1)

Foam: MSA VEEFOAM™

General Applicability

Multi-purpose foam for extinguishing fires and control of vapors; provides effective vapor control for water-immiscible organic liquids and acid and alkaline inorganic materials; can be used in low- and high-expansion foam generating equipment.

Chemicals for Which this Foam is Ineffective:

Polar compounds (with dielectric constant greater than 15)
Some organic acids
Some inorganic acis
Uncontained spills of liquified gases

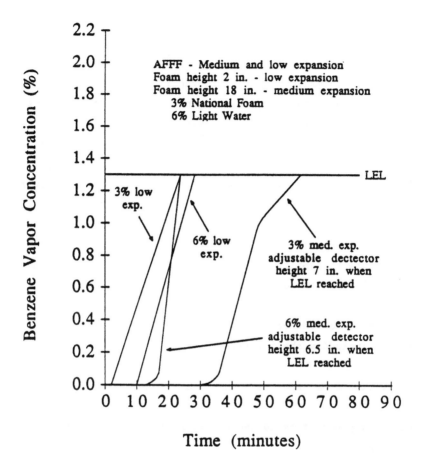

Figure 9-1. Benzene vapor concentration versus time.

Source: Adapted from Reference 3.

9.5 RELIABILITY

Information concerning the reliability of foam systems for use in mitigating the release of flammable and/or toxic vapors from chemical spills does not appear in the literature; however, in general, the reliability of any system depends on the reliability of the system's components, human factors, interactions, and controls, as discussed in the section on spray systems.

The reliability of foam systems in emergency situations, either for fire or spill control, depends on regular inspection, maintenance, and testing of foam concentrates and application equipment.

To evaluate the reliability of foam concentrates on hand for emergency responses the following should be considered (10):

- Storage in accordance with the manufacturer's instructions, including shelf life;

- Protection from exposure to extremes of heat and cold;

- Adequate precautions to avoid contamination with other materials; and

- Inspection of concentrate for formation of any precipitates that may affect the foam's performance or render the foam concentrate useless in an emergency.

For foam-generating and applicating equipment, adequate procedures should be included for periodic inspection, maintenance, and testing of various components of the system without actually producing the foam on a full-scale emergency basis. Valving, piping, and any electrical controls can be checked for proper operation and to ensure that corrosion has not rendered the system inoperative. If special test equipment and qualified personnel are available, some important equipment performance characteristics can be checked against the results of tests conducted when the equipment was first installed or accepted. These include (10):

- The foam equipment discharge pattern;

- The percent concentration of the foam concentrate in the foam solution;

- The expansion ratio of the foam;

- The quarter drainage rate of the foam; and

- The film-forming characteristic of the concentrate.

9.6 SECONDARY HAZARDS

The primary purpose of using foam for controlling vapors from hazardous chemical spills is to minimize hazards to plant personnel, emergency spill response teams, and the general public. However, if the selected foam is incompatible with the spilled material, more vapors may actually be released than controlled. For example, certain types of foams may lose their liquid quickly and form openings in the foam layer, causing rapid generation of flammable vapors from some chemicals (i.e., liquified natural gas spills). Selection of a slow-draining foam that maintains an effective blanket over the spill but that allows the gas to become buoyant enough for safe dispersal over a period of time can eliminate this major secondary hazard. Also, if the wrong foam is used to respond to a spill of an inorganic water-reactive chemical, more vapor could be released than if no foam had been used in the first place.

Foams that cause increased vaporization should not be used in totally enclosed areas by personnel without adequate safety precautions, including self-contained breathing apparatus and protective clothing. Alternatively, a foam should be selected that is specifically formulated to reduce this secondary vaporization hazard.

Protective clothing should be worn where mobile equipment is used. Appropriate breathing apparatus should also be used by response personnel. Personnel should be properly trained in the use of all safety equipment to minimize exposure to spilled liquids and hazardous vapors.

9.7   COSTS

Costs for foam and foam application systems vary widely and depend on the type of foam selected and the type and complexity of equipment used. Approximate costs, for illustrative purposes only, for various foams to industrial customers are presented in Table 9-3 (12).

Note that the prices in Table 9-3 are approximations; the actual prices may differ, depending on the volume of foam concentrate purchased by a customer and on the particular distribution system.

Costs for application equipment also vary, ranging from small manually operated portable systems to large automatically operated fixed systems and self-contained mobile foam trucks. Examples of foam application equipment costs are shown in Table 9-4 (12,13).

TABLE 9-3. APPROXIMATE COSTS FOR VARIOUS FOAMS

Type of Foam	Cost ($/Gallon)
Regular protein foam (3 percent formulation)	$8/gallon
Surfactant foam (1.5 percent formulation)	$11/gallon
Aqueous film forming foam (3 and 6 percent formulations)	$9 - $12/gallon (depending on formulation)
Alcohol type foam (3 and 6 percent formulations)	$14 - $15/gallon (depending on formulation)
Specialty foams (6 percent formulations)	$15/gallon

TABLE 9-4. EXAMPLES OF FOAM APPLICATION EQUIPMENT COSTS

Equipment	Cost 1987 ($)
Smallest manually operated portable systems	$ 125
Larger manually operated portable system with nozzle and eductor	$ 600
Modification of city fire department pumper for 1 to 2 foam discharge lines	$1,200 to $1,500
Skid-mounted foam system for semi-fixed or permanent installation	$15,000 to $100,000, depending on whether full automatic control is required
Trailer-mounted foam system	$6,000 to $30,000, depending on size
Self-contained mobile foam trucks with storage for foam concentrate and water	$120,000 to $400,000, depending on additional safety and spill control equipment needed
Fully automatic high-expansion foam generators for total flooding	$8,000 to $25,000
Small portable high-expansion foam generators	$1,000 to $1,500

9.8   REFERENCES

1.   Vervalin, C.H.   Role of Foam in Fighting Flares.   Fire Protection Manual for Hydrocarbon Processing Plants, Volume 1, Third Edition.   Gulf Publishing Company, Houston, TX, 1985.

2.. Basically Speaking, Foam Systems.   The Sentinel, Second Quarter, 1986.

3.   Gross, S.S., and R.H. Hiltz.   Evaluation of Foams for Mitigating Air Pollution from Hazardous Spills.   EPA-600/2-82-029 (NTIS PB82-227117), U.S. Environmental Protection Agency, March 1982.

4.   Lees, F.P.   Loss Prevention in the Process Industries, Butterworth's, London, England, 1983.

5.   Chandnani, M.F.   Design Fundamentals.   Fire Protection Manual for Hydrocarbon Processing Plants, Volume 1, 3rd Edition.   Gulf Publishing Company, Houston, TX, 1985.

6.   Gillespie, P.J. and L.R. DiMaio.   How Foam Can Protect HPI Plants.   Fire Protection Manual for Hydrocarbon Processing Plants, Volume 2.   Gulf Publishing Company, Houston, TX, 1981.

7.   National Foam Systems, Inc.   Controlling Hazardous Vapors, advertising literature, Section XIV, 1986.

8.   Evans, M.L., and H.A. Carroll.   Handbook for Using Foams to Control Vapors from Hazardous Spills.   EPA-600/8-86/019 (NTIS PB87-145660), U. S. Environmental Protection Agency, July 1986.

9.   Hiltz, R.H.   The Potential of Aqueous Foams to Mitigate the Vapor Hazard from Released Volatile Chemicals.   International Symposium on Preventing Major Chemical Accidents.   American Institute of Chemical Engineers, Washington, D.C., February 1987.

10. McKinnon, G.P., (ed.), The National Fire Protection Association, National Fire Prevention Handbook, 15th Edition.   Quincy, MA, 1981.

11. MSA Research Corporation.   Advertising Literature, Data Sheet 18-01-01.

12. Personal communication with E.C. Norman, National Foam Systems, Inc., Lionville, PA, April 1987.

13. Personal communication with J. Headrick, Global Fire Equipment Company, Ft. Worth, TX, April 1987.

Glossary

This glossary defines selected terms used in the text of this manual which might be unfamiliar to some users or which might be used differently by different authors.

Accidental release: The unintentional spilling, leaking, pumping, purging, emitting, emptying, discharging, escaping, dumping, or disposing of a toxic material into the environment in a manner that is not in compliance with a plant's federal, state, or local environmental permits and results in toxic concentrations in the air that are a potential health threat to the surrounding community.

Alkane: A chemical compound consisting only of carbon and hydrogen in which the carbon atoms are joined to each other by single bonds.

Assessment: The process whereby the hazards which have been identified are evaluated in order to provide an estimate for the level of risk.

Autocatalytic: A chemical reaction which is catalyzed by one of the products of the reaction.

Carcinogen: A cancer causing substance.

Containment/Control: A system to which toxic emissions from safety relief discharges are routed to be controlled. A caustic scrubber and/or flare can be containment/control devices. These systems may serve the dual function of destructing continuous process exhaust gas emissions.

Contingency Plan: A plan which describes the actions that facility personnel will take to minimize the hazards to human health or the environment from fires, explosions or accidental releases of hazardous materials.

Control System: A system designed to automatically maintain all controlled process variables within a prescribed range.

Creative Checklist: A list of major hazards and nuisances designed so that when an individual item from the list is associated with a particular material or a significant part of a unit, an image of a specific hazard or nuisance is generated as a stimulus to the imagination of members of a multidisciplinary team.

Creative Checklist Hazard and Operability Study: A Hazard and Operability Study which uses a Creative Checklist to stimulate a systematic, yet creative search for hazards.

Emergency Response Plan: A plan of action to be followed by source operators after a toxic substance has been accidentally released to the atmosphere. The plan includes notification of authorities and impacted population zones, minimizing the quantity of the discharge, etc.

Event Tree: A logic diagram which depicts all pathways (success and failure) originating from an initiating event.

Exothermic: A term used to characterize the evolution of heat. Specifically refers to chemical reactions from which heat is evolved.

Facility: A location at which a process or set of processes are used to produce, refine or repackage chemicals, or a location where a large enough inventory of chemicals are stored so that a significant accidental release of a toxic chemical is possible.

Fault Tree: A logic diagram which depicts the interrelationships of various primary events and subevents to an undesired top event.

Fire Monitor: A mechanical device holding a rotating nozzle, which emits a stream of water for use in firefighting. Fire monitors may be fixed in place or may be portable. A fire monitor allows one person to direct water on a fire whereas a hose of the same flowrate would require more than one person.

Guide Word Hazard and Operability Study: A Hazard and Operability Study which uses Guide Words to stimulate a systematic yet creative search for hazards.

Hazard: A source of danger. The potential for death, injury or other forms of damage to life and property.

Hazard and Operability Study: The application of a formal systematic critical examination to the process and engineering intentions of the new facilities to assess the hazard potential of maloperation of individual items of equipment and the consequential effects on the facility as a whole.

Hygroscopic: Readily taking up and retaining moisutre (water).

Identification: The recognition of a situation, its causes and consequences relating to a defined potential, e.g. Hazard Identification.

Lachrymator: A substance which increases the flow of tears.

Mitigation: Any measure taken to reduce the severity of the adverse effects associated with the accidental release of a hazardous chemical.

Mutagen: An agent that causes biological mutation.

Plant: A location at which a process or set of processes are used to produce, refine, or repackage, chemicals.

Prevention: Design and operating measures applied to a process to ensure that primary containment of toxic chemicals is maintained. Primary containment means confinement of toxic chemicals within the equipment intended for normal operating conditions.

Primary Containment: The containment provided by the piping, vessels and machinery used in a facility for handling chemicals under normal operating conditions.

Probability/potential: A measure, either qualitative or quantitative, that an event will occur within some unit of time.

Process: The sequence of physical and chemical operations for the production, refining, repackaging or storage of chemicals.

Process machinery: Process equipment, such as pumps, compressors, heaters, or agitators, that would not be categorized as piping and vessels.

Protection: Measures taken to capture or destroy a toxic chemical that has breached primary containment, but before an uncontrolled release to the environment has occurred.

Pyrophoric: A substance that spontaneously ignites in air at or below room temperature without supply of heat, friction, or shock.

Qualitative Evaluation: Assessing the risk of an accidental release at a facility in relative terms; the end result of the assessment being a verbal description of the risk.

Quantitative Evaluation: Assessing the risk of an accidental release at a facility in numerical terms; the end result of the assessment being some type of number reflects risk, such as faults per year or mean time between failure.

Reactivity: The ability of one chemical to undergo a chemical reaction with another chemical. Reactivity of one chemical is always measured in reference to the potential for reaction with itself or with another chemical. A chemical is sometimes said to be "reactive", or have high "reactivity", without reference to another chemical. Usually this means that the chemical has the ability to react with common materials such as water, or common materials of construction such as carbon steel.

Redundancy: For control systems, redundancy is the presence of a second piece of control equipment where only one would be required. The second piece of equipment is installed to act as a backup in the event that the primary piece of equipment fails. Redundant equipment can be installed to backup all or selected portions of a control system.

Risk: The probability that a hazard may be realized at any specified level in a given span of time.

Secondary Containment: Process equipment specifically designed to contain material that has breached primary containment before the material is released to the environment and becomes an accidental release. A vent duct and scrubber that are attached to the outlet of a pressure relief device are examples of secondary containment.

Teratogenic: Causing anomalies of formation or development.

Toxicity: A measure of the adverse health effects of exposure to a chemical.

Metric (SI) Conversion Factors

Quantity	To Convert From	To	Multiply By
Length:	in	cm	2.54
	ft	m	0.3048
Area:	in^2	cm^2	6.4516
	ft^2	m^2	0.0929
Volume:	in^3	cm^3	16.39
	ft^3	m^3	0.0283
	gal	m^3	0.0038
Mass (weight):	lb	kg	0.4536
	short ton (ton)	Mg	0.9072
	short ton (ton)	metric ton (t)	0.9072
Pressure:	atm	kPa	101.3
	mm Hg	kPa	0.133
	psia	kPa	6.895
	psig	kPa*	(psig)+14.696)x(6.895)
Temperature:	°F	°C*	(5/9)x(°F-32)
	°C	K*	°C+273.15
Caloric Value;	Btu/lb	kJ/kg	2.326
Enthalpy:	Btu/lbmol	kJ/kgmol	2.326
	kcal/gmol	kJ/kgmol	4.184
Specific-Heat Capacity:	Btu/lb-°F	kJ/kg-°C	4.1868
Density:	lb/ft^3	kg/m^3	16.02
	lb/gal	kg/m^3	119.8
Concentration:	oz/gal	kg/m^3	
	quarts/gal	cm^3/m^3	25,000
Flowrate:	gal/min	m^3/min	0.0038
	gal/day	m^3/day	0.0038
	ft^3/min	m^3/min	0.0283
Velocity:	ft/min	m/min	0.3048
	ft/sec	m/sec	0.3048
Viscosity:	centipoise (CP)	Pa-s (kg/m-s)	0.001

*Calculate as indicated

ACIDIC EMISSIONS CONTROL TECHNOLOGY AND COSTS

by

T.E. Emmel, J.T. Waddell, R.C. Adams

Radian Corporation

Pollution Technology Review No. 168

This book describes acidic emissions control technology and costs. The objectives are: 1) to identify and characterize stationary combustion and industrial sources of directly emitted acidic materials in the United States; 2) to evaluate the feasibility of control technologies for these sources; and 3) to estimate the costs of applying these control technologies.

The book gives results of estimates, using a model plant approach, of costs for retrofitting selected acidic emission control systems to utility and industrial boilers, Claus sulfur recovery plants, catalytic cracking units, primary copper smelters, coke oven plants, primary aluminum smelters, and municipal solid waste incinerators.

Sources of directly emitted acidic materials were identified via a literature search. For most source categories, emissions were estimated using emission factors and combustion and process capacities found in the literature. To focus on source categories with the greatest emissions, model units were developed for those sources which emit 4,500 Mg (5,000 tons) or more of acidic material per year. These model units were then used as bases to establish control techniques and determine control costs.

Utility and industrial boilers are the largest U.S. sources. emitting approximately 760,000 Mg (830,000 tons) and 180,000-250,000 Mg (200,000-275,000 tons) of acidic material per year, respectively. Total direct emissions of acidic materials represent an estimated two percent of annual acid precipitation precursor (SO_2, NO_x, and VOC) emissions from stationary sources.

Results of this study can be used to evaluate the merits of controlling directly emitted acidic materials as part of a policy evaluation of overall acid deposition control strategies. For example, if it were determined that, for a region, local emissions of directly emitted acid materials were more significant than long range precursor emissions, the information in this book could be used to evaluate the cost effectiveness of controlling local sources of directly emitted acidic materials versus sources of long range precursor emissions.

CONTENTS

1. INTRODUCTION

2. RESULTS AND RECOMMENDATIONS

3. SUMMARY OF ACIDIC EMISSIONS ESTIMATES

4. DEVELOPMENT OF MODEL UNITS
Industrial and Utility Boilers
  Model Boilers
  Control of Acidic Emissions from Boilers
  Model Control Systems
Claus Plants
Fluid Catalytic Cracking
Primary Copper
Coke Ovens
Primary Aluminum
Municipal Solid Waste Incinerators

5. COST ANALYSIS
Industrial and Utility Boilers
Claus Plants
Fluid Catalytic Cracking
Primary Copper
Coke Ovens
Primary Aluminum
Municipal Solid Waste Incinerators

6. RESEARCH AND DEVELOPMENT
Research and Development in Particulate
  Control
    Electrostatic Precipitators
    Fabric Filtration
    Impact of Particulate R&D on Acidic
    Materials
  Electron-Beam Irradiation
  Granular Bed Filter

APPENDICES

ISBN 0-8155-1208-2 (1989)

155 pages

Other Noyes Publications

HAZARDOUS WASTE MANAGEMENT FACILITIES DIRECTORY
Treatment, Storage, Disposal and Recycling

by

U.S. Environmental Protection Agency

Versar, Inc.

Camp Dresser & McKee, Inc.

This book provides a listing of 1045 commercial hazardous waste management facilities, along with information on the types of commercial services offered (e.g., treatment, storage, disposal or recycling), and types of wastes managed. It is a compilation of recent data from EPA data bases. Facility name, address, contact person and phone number are listed for each site as available.

The purpose of the book is to assist in locating pertinent waste management facilities in specific geographical areas.

The directory has been prepared in a form which will assist the reader in locating facilities that commercially process specific groups of wastes. It begins with an alphabetic listing of all facility names and their directory identification numbers. Next, the reader will find a listing of all facilities with commercial processes available, organized by states. Following this section are several appendices which group facilities with like waste groups or management practices.

Also included in the book are five Addenda containing the names of Additional Commercial Facilities, Limited Commercial Facilities, Company Captive Facilities, Safety-Kleen Operations, and Vendors of Mobile Waste Treatment Technologies.

An important feature of this directory is that, if you have an immediate toxic waste disposal problem, you can easily find a waste disposal facility in your geographical area.

CONTENTS

ISBN 0-941459-02-0 (1989)

pages

CONTROL OF EMISSIONS FROM MUNICIPAL SOLID WASTE INCINERATORS

by

F. Thomas DePaul and Jerry W. Crowder

DePaul and Associates, Inc.

Pollution Technology Review No. 169

This study was prepared to identify those pollutant emissions from municipal solid waste (MSW) incinerators which pose the greatest potential human health risks, and to document which existing control scenarios have the best demonstrated performance in reducing these pollutant emissions. The study was done for the State of Illinois, but it would apply as well to similar installations throughout the country. A brief review of other state air-toxics programs is included in the book.

For example, to minimize air quality impacts from future incineration sources, an Illinois Public Act requires best available control technology (BACT) on new MSW incinerators burning 25 tons or more per day. A survey of state air toxic regulators was undertaken to determine the impact of health based regulations on MSW incinerator emissions and to help identify the pollutant emissions of greatest concern. The pollution control scenarios which have been permitted by state environmental control agencies in order to comply with these regulations are presented and discussed. In addition, information on uncontrolled and controlled emissions has been collected for 61 incineration facilities, encompassing mass burn, starved air, and RDF-fired combustion technology. The uncontrolled emissions database has been analyzed to determine expected emission levels for both criteria and non-criteria pollutants, and the controlled emissions database has been analyzed to determine expected control levels. Data on the cost of control equipment for each type of combustor are included. This information, together with the documented levels of control, is presented in an effort to assist regulatory agency personnel responsible for making BACT determinations. Lastly, the impacts of several alternative control scenarios on the reduction of those pollutants with the highest potential risk to human health are discussed.

CONTENTS

ISBN 0-8155-1209-0 (1989)

275 pages

ODOR AND CORROSION CONTROL IN SANITARY SEWERAGE SYSTEMS AND TREATMENT PLANTS

by

Robert P.G. Bowker and John M. Smith
J.M. Smith & Associates

Neil A. Webster
FW Consultants

Pollution Technology Review No. 165

This comprehensive manual brings together available information on odor and corrosion control procedures for sanitary sewerage systems and treatment plants. The manual will satisfy the needs of those designing new systems or applying odor and corrosion control measures to existing systems.

Wastewater is known to the public for its potential to create odor nuisance. Sometimes it is the odors escaping from sewer manholes that cause complaints; more commonly, the odor source is a wastewater treatment facility. Yet there are wastewater treatment facilities that are free from this stigma, and techniques to prevent odor nuisances are available to those committed to construct odor-free treatment works.

Traditional sanitary sewer design practice has not fully acknowledged the importance of corrosion and odor control, as evidenced by the widespread occurrence of sulfide and odor control problems throughout the United States for sanitary sewers serving both small and large tributary areas.

Substantial information on odors and corrosion in municipal sewerage systems has been reported in recent years. In addition, significant developments have evolved for controlling odors and corrosion in wastewater treatment plants. In particular, the use of chemicals for control has increased substantially.

The listing below gives **chapter titles and selected subtitles.**

1. INTRODUCTION

2. THEORY, PREDICTION, AND MEASUREMENT OF ODOR AND CORROSION
Compounds Causing Odor and Corrosion
Mechanisms for the Generation of Hydrogen Sulfides
Mechanisms of Corrosion

Predicting Sulfide Buildup and Corrosion in Sewers
Approach to Investigating Odor and Corrosion
Measurement and Monitoring of Corrosion and Odor
Toxicity and Safety Practices

3. ODOR AND CORROSION CONTROL IN EXISTING WASTEWATER COLLECTION SYSTEMS
Improving the Oxygen Balance
Chemical Addition
Case Histories

4. ODOR AND CORROSION CONTROL IN EXISTING WASTEWATER TREATMENT PLANTS
Sources of Odors in Wastewater Treatment Plants
Corrosion in Wastewater Treatment Plants
Corrosion Control Techniques at Existing Wastewater Treatment Plants
Case Histories

5. DESIGNING TO AVOID ODOR AND CORROSION IN NEW WASTEWATER COLLECTION SYSTEMS
Hydraulic Design
Ventilation of Sewers
Selection of Materials

6. DESIGNING TO AVOID ODOR AND CORROSION IN NEW WASTEWATER TREATMENT FACILITIES
Common Sites of Odor Generation
General Design Considerations for Avoiding Odor Generation and Release
Design Procedures for Specific Odor-Producing Unit Processes
General Design Considerations for Avoiding Corrosion
Paint and Coatings
Selection of Materials

ISBN 0-8155-1192-2 (1989)

7"x10"   130 pages

EXTREMELY HAZARDOUS SUBSTANCES
Superfund Chemical Profiles

U.S. Environmental Protection Agency

This comprehensive reference guide contains chemical profiles for **each of the 366 chemicals listed as "extremely hazardous substances"** by the USEPA in 1988. The EPA developed this set of documents for use in dealing with Section 302 of Title III of the Superfund Amendments and Reauthorization Act (SARA). Each profile contains a summary of documented information, which has been reviewed for accuracy and completeness.

The profile for each hazardous substance includes chemical identity information and nine sections detailing:

   I. REGULATORY INFORMATION
   Toxicity data from NIOSH/RTECS TPQS, RQs, Section 313 status
   II. PHYSICAL/CHEMICAL CHARACTERISTICS
   III. HEALTH HAZARD DATA
   OSHA Permissible Exposure Limits (PELs)
   ACGIH Threshold Limit Values (TLVs)
   NIOSH Immediately Dangerous to Life and Health (IDLH) levels
   IV. FIRE AND EXPLOSION HAZARD DATA
   NFPA ratings
   V. REACTIVITY DATA

   VI. USE INFORMATION
   VII. PRECAUTIONS FOR SAFE HANDLING AND USE
   VIII. PROTECTIVE EQUIPMENT FOR EMERGENCY SITUATIONS
   IX. EMERGENCY TREATMENT INFORMATION

In addition, **many of the profiles include an Emergency First Aid Treatment Guide (EFATG)**, containing expanded and more detailed guidance for emergency treatment based on information obtained from the proprietary database of the Rocky Mountain Poison Center. The emergency treatment information is geared to use by first responders; thus references to signs and symptoms of exposure, as well as procedural guidance, avoid the use of highly technical medical language.

A discussion of **Personal Protective Equipment**, a **List of Abbreviations**, and a **Glossary of Medical Terminology** are included in the introductory material to further assist the reader in gaining rapid access to the information in the book.

The 366 extremely hazardous substances are presented alphabetically. The first 57 substances are listed below:

Acetone cyanohydrin
Acetone thiosemicarbazide
Acrolein
Acrylamide
Acrylonitrile
Acrylyl chloride
Adiponitrile
Aldicarb
Aldrin
Allyl alcohol
Allylamine
Aluminum phosphide
Aminopterin
Amiton
Amiton oxalate
Ammonia
Amphetamine
Aniline
Aniline, 2,4,6-trimethyl-
Antimony pentaflouride
Antimycin A
ANTU
Arsenic pentoxide

Arsenous oxide
Arsenous trichloride
Arsine
Azinphos-ethyl
Azinphos-methyl
Benzal chloride
Benzenamine, 3-(trifluoro-
   methyl)-
Benzenearsonic acid
Benzene, 1-(chloromethyl)-
   4-nitro-
Benzimidazole, 4,5-dichloro-
   2-(trifluoromethyl)-
Benzotrichloride
Benzyl chloride
Benzyl cyanide
Bicyclo[2.2.1]heptane-2-carbo-
   nitrile, 5-chloro-6-(((methyl-
   amino)carbonyl)oxy)-imino)-,
   (1S-(1 alpha, 2 beta, 4 alpha,
   5 alpha, 6E))-
Boron trichloride
Boron trifluoride

Boron trifluoride compound
   with methyl ether (1:1)
Bromadiolone
Bromine
Cadmium oxide
Cadmium stearate
Calcium arsenate
Camphechlor
Cantharidin
Carbachol chloride
Carbamic acid,methyl-, O-
   (((2,4-dimethyl-1-3-dithiolan-
   2-yl)methylene)amino)-
Carbofuran
Carbon disulfide
Carbophenothion
Chlordane
Chlorfenvinfos
Chlorine
Chlormephos
Chlormequat chloride
   *plus 309 other extremely
   hazardous substances*

ISBN 0-8155-1166-3 (1988)   6"x9"   2 volumes   **1807 pages**

INDUSTRIAL HYGIENE ENGINEERING
Recognition, Measurement, Evaluation and Control
Second Edition

Edited by

John T. Talty
National Institute for Occupational Safety and Health

This book provides an advanced level of study of industrial hygiene engineering situations with emphasis on the *control* of exposure to occupational health hazards. Primary attention is given to industrial ventilation, noise and vibration control, heat stress, and industrial illumination. Other topics covered include industrial water quality, solid waste control, handling and storage of hazardous materials, personal protective equipment, and costs of industrial hygiene control.

The book will serve as a single reference source on the fundamentals of industrial hygiene engineering as related to the design of controls for exposure to health hazards in the workplace. The control of occupational health hazard exposures requires a broad knowledge of a number of subject areas. To provide a text that includes the necessary theoretical foundation as well as the practical application of the theory is a significant undertaking. This text will be a valuable and needed addition to the literature of industrial hygiene, providing the reader with a systematic approach to problem solving in the field of industrial hygiene.

The eight sections of the book are self-contained. Each covers a particular subject area, thus allowing for reference to a single topic area without the need to consult other sections of the book.

CONTENTS

ISBN 0-8155-1175-2 (1988)

831 pages

Printed and bound by CPI Group (UK) Ltd, Croydon, CR0 4YY

03/10/2024

01040335-0010